THE LIBRARY
GUILDFORD COLLEGE
of Further and Higher Education

KT-547-203

Basic Engineering

W. Bolton

Basic Engineering

W. Bolton

WITHDRAWN

BH NEWNES

620·1 BOL
83285

Newnes
An imprint of Butterworth-Heinemann
Linacre House, Jordan Hill, Oxford OX2 8DP
A division of Reed Educational and Professional Publishing Ltd

A member of the Reed Elsevier plc group

OXFORD BOSTON JOHANNESBURG
MELBOURNE NEW DELHI SINGAPORE

First published 1995
Reprinted 1996

© W. Bolton 1995

All rights reserved. No part of this publication may be reproduced in any material form (including photocopying or storing in any medium by electronic means and whether or not transiently or incidentally to some other use of this publication) without the written permission of the copyright holder except in accordance with the provisions of the Copyright, Designs and Patents Act 1988 or under the terms of a licence issued by the Copyright Licensing Agency Ltd, 90 Tottenham Court Road, London, England W1P 9HE. Applications for the copyright holder's written permission to reproduce any part of this publication should be addressed to the publishers

British Library Cataloguing in Publication Data
Bolton, W.
1. Basic Engineering
I. Title

Library of Congress Cataloguing in Publication Data
A catalogue record for this book is available from the Library of Congress

ISBN 0 7506 2584 8

Printed and bound in Great Britain by
Martins the Printers Ltd, Berwick upon Tweed

Contents

Preface

This book has been written to provide a comprehensive coverage of all the mandatory units for the Intermediate GNVQ in Engineering; namely Engineering Materials and Processes; Graphical Communication in Engineering; Science and Mathematics in Engineering; and Engineering in Society and the Environment. The aim of this book is to enable the reader to:

- Comprehend the reasons for the selection, and select, materials and components for engineered products.
- Comprehend the reasons for the selection, and select, processes to make engineered products.
- Use graphical methods for communicating and interpreting engineering information.
- Use science and mathematics relevant to engineering systems.
- Gain an understanding of the influence of engineering in society and the environment.

The book, as such, is also seen as being of relevance to other courses where a knowledge of basic engineering is required.

The aims of the chapters and their relationship to the elements of the GNVQ unit are:

	Aim	*GNVQ element*
Chapter 1	Identify the purposes for which materials are needed and the conditions under which they are used. Select materials for particular applications	1.1
Chapter 2	Identify mechanical fixing and electrical components and select components	1.1
Chapter 3	Identify the processes used to make engineered products and select processes for particular applications	1.2, 1.3
Chapter 4	Describe safety procedures and equipment	1.2, 1.3
Chapter 5	Select graphical methods for communicating information	2.1
Chapter 6	Produce scale and schematic drawings for engineering applications	2.2
Chapter 7	Interpret information presented in engineering drawings	2.3

Chapter 8	Use mathematical techniques for the manipulation of data in physical relationships	3.3
Chapter 9	Describe engineering systems in terms of scientific laws	3.1
Chapter 10	Identify measurement techniques for physical quantities used in engineering	3.2
Chapter 11	Describe the application of engineering technology in society	4.1
Chapter 12	Describe the effects of engineering activities on the environment	4.1
Chapter 13	Describe the main sectors of the engineering industry and carreer options and pathways in engineering	4.2

There are worked examples in the text. In addition, at the end of each chapter there are multiple-choice questions and problems. Answers are given for all the multiple-choice questions and guidance given as to the answers for all the problems.

W. Bolton

Acknowledgements

Extracts from BS 308: Part 1: 1993, BS 308: Part 2: 1985 (1992) and BS 499: Part 2: 1980 (1989) are reproduced with the permission of BSI. Complete copies can be obtained by post from BSI Customer Services, 389 Chiswick High Road, London W4 4AL.

Figure 4.1 has been adapted from *Essentials of Health and Safety at Work,* 4th revised edition, Health and Safety Executive, 1994, and is reproduced with the permission of the Controller of Her Majesty's Stationery Office.

1 Engineering materials

1.1 Engineers and manufacturing

Manufacturing involves the combining together of materials, people and equipment in order to obtain economical production of products. Engineers are involved in the designing of the products that are to be manufactured. They are thus involved with the selection of materials and manufacturing methods. The design engineer has to consider what functions are required of the product and hence such factors as what loads the product will encounter when in use and the environments the product will be subject to. In order to meet these requirements, appropriate materials must be selected. Linked with the choice of materials will be the processing methods needed to obtain the product in the required form. The selection of materials and processing methods go hand-in-hand since the selection of the material may dictate the processing method to be used, or the selection of the processing method may dictate the materials to be used. Engineers will also be concerned with the selecting and coordinating of machines needed for the manufacturing processes, supervising and managing their use, checking that the materials used are to specification, checking that the machines deliver the products to the required design specifications. Some engineers will design the machines and tools used in the manufacturing, some will be concerned with the development of new and better materials.

This chapter is about the selection of materials. Chapter 2 links the selection of materials with the selection of components in engineering design and Chapter 3 considers the selection of manufacturing processes.

1.2 Properties of materials

The following are some of the basic properties of materials which influence their selection for engineered products.

1 *Physical properties*
These properties include:

Density, i.e. the mass per unit volume of material. A high density material has a high mass for a given volume. For example, aluminium has a lower density than steel, thus for the same size product, the one made with aluminium will be much lighter than one made with steel.

2 *Mechanical properties*
These properties include:

Strength, i.e. the ability to support loads without breaking. A strong material requires a large load to break it. For example, steels tend to have higher strengths than aluminium alloys. Thus for the same size, a strip of steel will require more force to break it than a strip of aluminium alloy.

Stiffness, i.e. the ability to resist bending. The stiffer a material is the greater the forces required to bend it. To illustrate this, consider a wooden ruler and a plastic ruler. The plastic ruler bends more easily than the wooden one. Thus the wood is said to be stiffer than the plastic.

Hardness, i.e. the ability to withstand scratching, abrasion or indentation by other materials. The harder a material, the more difficult it is to scratch it or make indentations in it. Steel tends to be harder than plastics. A plastic surface is much more easily scratched than a steel surface. A harder material will scratch a softer material but a softer material will not scratch a harder one. Thus a steel point will scratch a plastic surface but a plastic point will not scratch a steel surface.

Toughness, i.e. the ability to resist the propagation of cracks and so break. It is also used to describe the ability to withstand impacts without breaking. Mild steel is reasonably tough. Thus a car body made with mild steel can suffer reasonable impact loads without breaking, though it may dent. A car body made of glass would break very easily if subjected to impact loads, glass not being a tough material.

Elasticity, i.e. the ability to resume the original shape after being stretched or compressed. Rubber is an obvious example of an elastic material, as it is capable of being stretched to considerable lengths and then springing back to its original shape when the load is removed.

Ductility, i.e. the ability to change shape under load. A ductile material can be easily bent into a shape without breaking. The term *brittle material* is used for one that cannot be bent into a shape without breaking. Mild steel is an example of a ductile material, sheets of it can be bent over formers into shapes to form the body of a car. Glass at room temperature is an example of a brittle material. Thus when a glass rod is bent at room temperature it breaks.

3 *Electrical properties*
These properties include:

Electrical resistivity, i.e. the ability of a material to resist the flow of electricity. The higher the electrical resistivity of a material the smaller the current when a voltage is applied across a particular size piece of the material. Plastics tend to have very high resistivities, so high that they are termed insulators. Thus electrical circuits can be mounted on plastic boards with virtually no current leaking through the plastic.

Electrical conductivity, i.e. the ability of a material to conduct electricity. The higher the conductivity of a material the higher the current when a voltage is applied across a particular size piece of the material. Copper has a high conductivity and is thus widely used for electric cables.

4 *Thermal properties*
These properties include:

Expansivity, i.e. the extent to which a length of a material will expand when its temperature increases. A high expansivity is for a material that expands a lot when the temperature increases. Plastics tend to have high expansivities and thus products made from plastics will markedly increase in size when the temperature increases and thus an item that perhaps fitted at room temperature no longer fits at a higher temperature.

Heat capacity, i.e. the extent to which the temperature of an object increases for a given input of heat. The higher the heat capacity the smaller the rise in temperature for a given heat input.

Thermal conductivity, i.e. the extent to which the material conducts heat. A high thermal conductivity material is one which conducts heat well so that when one end of a strip is heated the other end rapidly rises in temperature. Metals tend to have high thermal conductivities. Thus a kitchen pan with a metal handle can present problems when the pan contains a hot liquid because the handle also becomes hot. Wood and plastics have lower themal conductivities and thus pan handles made with these materials are more comfortable to hold.

5 *Chemical properties*
These properties include:

Resistance to chemical attack, i.e. the ability of a material to resist attack by specified chemicals. The term *corrosion* is often used for chemical attack on metals as a result of exposure to the environment. The rusting of steel in air is an obvious example of a material which is not resisting chemical attack. Glass is a material which has good resistance to chemical attack, hence the wide use of glass for containers.

6 *Optical properties*
These properties include:

Transparency, i.e. the extent to which a material transmits light. Glass is an example of a material with a high transparency, while metals are opaque with no transparency.

Example

A material is said to be tough with high thermal conductivity and low electrical resistivity. How would you expect the material to behave in use?

A tough material is one that resists breaking when subjected to an impact load or when it contains cracks. Thus you would expect such a material to be useful in situations where it is likely to be subjected to knocks or abrasions which result in surface scratches, i.e. fine cracks. A high thermal conductivity means that the material is a good conductor of heat. Thus if one end of a strip of the material is heated, the other end very rapidly becomes hot. A low electrical resistivity means that the material is a good conductor of electricity. Thus if a length of the material is connected across a voltage supply then a high current might be expected.

1.3 The range of materials

Materials can be classified under the headings of metals, polymers (or plastics) and elastomers, ceramics and glasses, and composites. The following are the basic properties typical of such materials.

1 *Metals*
In general, metals have high electrical conductivities, high thermal conductivities, can be ductile and thus permit products to be made by being bent into shape, and have a relatively high stiffness and strength. Engineering metals are generally alloys. The term *alloy* is used for metallic materials formed by mixing two or more elements. For example, mild steel is an alloy of iron and carbon, stainless steel is an alloy of iron, chromium, carbon, manganese and possibly other elements. The reason for adding elements to the iron is to improve the properties. Pure metals are very weak materials. The carbon improves the strength of the iron. The presence of the chromium in the stainless steel improves the corrosion resistance. Engineering metals are also subdivided into ferrous metals and non-ferrous metals. *Ferrous metals* are those that are primarily iron, e.g. steel. *Non-ferrous metals* are those that contain little if any iron, e.g. aluminium and copper and their alloys.

2 *Polymers and elastomers*
Polymers can be classified as either *thermoplastics* or *thermosets*. Thermoplastics soften when heated and become hard again when the heat is removed. The term implies that the material becomes 'plastic' when heat is applied. Thermosets do not soften when heated, but char and decompose. Thus thermoplastic materials can be heated and bent to form required shapes, thermosets cannot. Thermoplastic materials are generally flexible and relatively soft. Polyethylene is an example of a thermoplastic, being widely used in the forms of films or sheet for such items as bags, 'squeeze' bottles, and wire and cable insulation. Thermosets are rigid and hard. Phenol formaldehyde, known as Bakelite, is a thermoset. It is widely used for electrical plug casings, door knobs and handles. The term *elastomers* is used for polymers which by their structure allow considerable extensions which are reversible. The material used to make rubber bands is an obvious example of such a material.

All thermoplastics, thermosets and elastomers have low electrical conductivity and low thermal conductivity, hence their use for electrical and thermal insulation. Compared with metals, they have lower densities, expand more when there is a change in temperature, are generally more corrosion resistant, have a lower stiffness, stretch more, and are not as hard. When loaded they tend to creep, i.e. the extension gradually changes with time. Their properties depend very much on the temperature so that a polymer which may be tough and flexible at room temperature may be brittle at 0°C and show considerable creep at 100°C.

3 *Ceramics and glasses*
Ceramics and glasses tend to be brittle, relatively stiff, stronger in compression than in tension, are hard, chemically inert, and bad conductors of electricity and heat. Glass is just a particular form of ceramic, with ceramics being crystalline and glasses non-crystalline. Examples of ceramics and glasses abound in the home in the form of cups, plates, and glasses. Alumina, silicon carbide, cement and concrete are examples of ceramics. Because of their hardness and abrasion resistance, ceramics are widely used as the cutting edges of tools.

4 *Composites*
Composites are materials composed of two different materials bonded together. For example, there are composites involving glass fibres or particles in polymers, ceramic particles in metals (referred to as cermets), and steel rods in concrete (Figure 1.1(a)) (referred to as reinforced concrete). Concrete itself is a composite, consisting of gravel in a matrix of cement (Figure 1.1(b)). Cement is weak. Cracks easily propagate through it. The presence of the gravel in the concrete does, however, hinder crack propagation and so gives a stronger material. Wood is a natural composite consisting of tubes of cellulose in a natural polymer called lignin (Figure 1.1(c)). Plywood is a composite material made by gluing together thin sheets of wood with their grain directions at right angles to each other (Figure 1.1(d)). The grain directions are the directions of the cellulose fibres in the wood and thus the resulting structure has fibres in mutually perpendicular directions. Thus, whereas

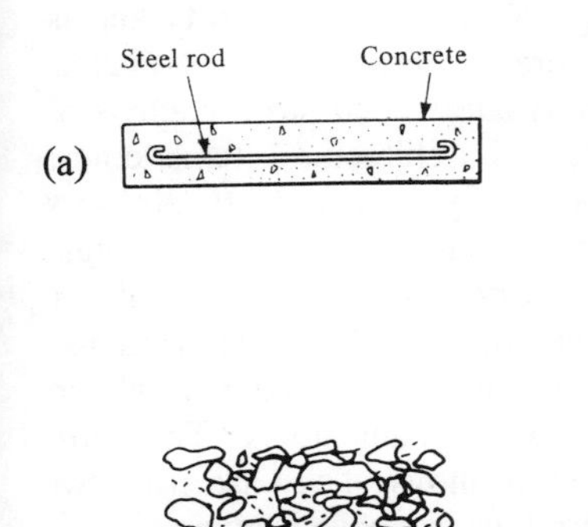

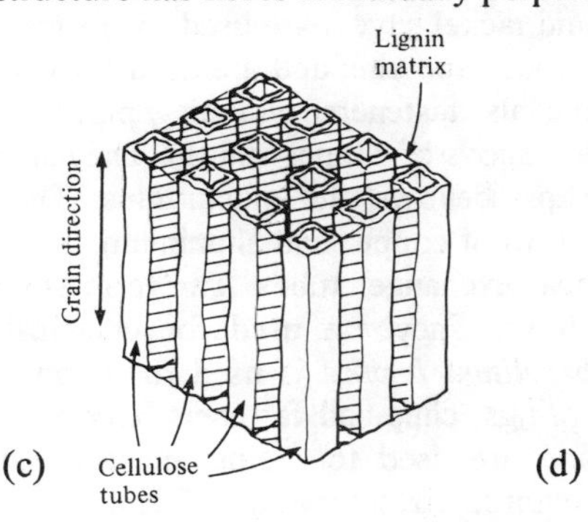

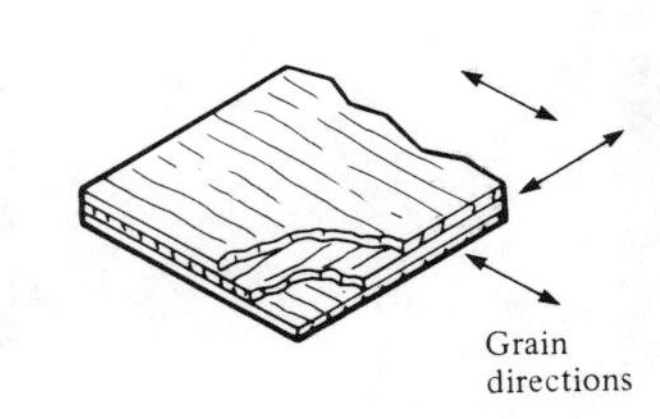

Figure 1.1 *(a) Reinforced concrete, (b) concrete, (c) wood, (d) plywood*

the thin sheet has properties that were directional, the resulting composite has no such directionality. Composites made with fibres embedded all aligned in the same direction in some matrix will have properties in that direction markedly different from properties in other directions. Composites can be designed to combine the good properties of different types of materials while avoiding some of their drawbacks.

Example

A material with high electrical conductivity is required. What group of materials is likely to be considered?

Metals have high electrical conductivities compared with polymers, ceramics or composites and thus a metal is likely to be used.

1.3.1 Commonly used engineering materials

A wide range of materials are used for engineered products. The following are examples of commonly used ones and their typical properties and uses.

Metals

Aluminium
This is used in commercially pure form and alloyed. Aluminium and its alloys have low density, high electrical and thermal conductivity and excellent corrosion resistance. They are used for such applications as engine parts, car trims, aircraft structures, fan blades, typewriter frames, cooking utensils, storage tanks, pressure vessels and chemical equipment.

Copper
Copper is widely used in the commercially pure form and alloyed. Copper and its alloys have good corrosion resistance, high electrical and thermal conductivity, and can be joined by soldering, brazing and welding. Copper in its commercially pure form has very high electrical conductivity and is widely used for electrical cables. The terms brasses, bronzes, cupro-nickels and nickel silvers are used for forms of copper alloys. *Brasses* are alloys of copper and zinc and are used for decorative and architectural items, coins, medals, fasteners, plumbing pipes, locks, hinges, pins and rivets. *Bronzes* are alloys of copper and tin. They are used for screws, bolts, rivets, springs, clips, bellows and diaphragms. The term *aluminium bronze* is used for alloys of copper and aluminium. They are used for nuts, bolts, bearings and heat exchanger tubes. The term *silicon bronze* is used for copper–silicon alloys. They are used for chemical and marine plant items. The term *beryllium bronze* is used for copper–beryllium alloys. They are used for springs, clips and fasteners. Copper–nickel alloys are called *cupro-nickels* and are used for coins, medals and where high corrosion resistance is required to sea water. If zinc is added to the copper–nickel alloy, the

resulting alloy is called a *nickel-silver* and is used for clock and watch components, rivets, screw, clips and decorative items.

Iron

The term *steel* is used for alloys of iron with between 0.05% and 2% carbon and *cast irons* for 2% to 4.3% carbon. The term *carbon steel* is used for those steels in which essentially just iron and carbon are present. Such steels with between 0.10% and 0.25% carbon are termed *mild steels*, between 0.20% and 0.50% carbon are *medium-carbon steels* and 0.50% to 2% carbon are *high-carbon steels*. Mild steels are general-purpose steels with reasonable ductility and are used for such applications as joists in buildings, bodywork for cars and ships, screws, nails and wire. Medium-carbon steels are more brittle but stronger than mild steels and are used for shafts and parts in car transmissions, suspensions and steering. High-carbon steels are even more brittle and stronger and are used for machine tools, saws, axes, hammers, cold chisels, punches and drills. The term *alloy steel* is used for steels when there are significant amounts of other elements present. Stainless steels are alloy steels with more than 12% chromium and have excellent corrosion resistance. They are used in chemical and food processing equipment.

Plastics

Polyamides or nylons

Nylons are translucent materials with, for a polymer, high tensile strength and medium stiffness. They are used as fibres in clothing. Additives such as glass fibres are added to nylon to increase the strength. Such materials are used for housings for power tools, electrical plugs and sockets. Nylons have low coefficients of friction, which can be further reduced by suitable additives. Such a material is widely used for gears, rollers, bearings and bushes.

Polyesters

The thermoset form of polyesters are widely used with glass fibres to form composite materials which are used for boat hulls, building panels and stackable chairs. The composites can be formed by spreading glass fibres, in the form of glass cloth, over a former. Then the polyester is used to coat it. A further layer of glass cloth is then added and yet more polyester. By repeating this process the required thickness of material can be built up. When the polyester has set then the composite has much greater strength and stiffness than would be obtained by the plastic alone.

Polyethylene

This is a thermoplastic polymer with fairly low strength and stiffness. It has good impermeability to gases and very low absorption rates for water. It is used for bags, 'squeeze' bottles, ball-point pen tubing, wire and cable insulation, piping, toys and household ware.

Polystyrene
This is a thermoplastic with moderate strength and reasonable stiffness. It is, however, fairly brittle and exposure to sunlight results in yellowing. It is attacked by many solvents. A toughened grade is produced by blending it with rubber. Such material is used for vending machine cups, casings for cameras, radios and TV sets. Polystyrene containing large numbers of air bubbles is termed expanded polystyrene and is widely used for heat insulation and as a packaging material. Its use for heat insulation is because of its very low thermal conductivity. Its use for packaging is because it is able to withstand impact loads and not transmit them to the packaged items.

Polyvinyl chloride (PVC)
This is a thermoplastic which has, for a plastic, high strength and stiffness. It is a fairly rigid material. It is frequently combined with other materials, termed plasticizers, to give lower strength, less rigid materials. Without plasticizer, it is used as piping for waste and soil drainage systems, rain water pipes, lighting fittings and curtain rails. With plasticizer, it is used for plastic rain coats, bottles, shoe soles, garden hose pipes and inflatable toys.

Ceramics and composites

Cement
When water is mixed with cement a chemical reaction occurs which results in a ceramic structure being formed. The resulting material has, however, very low strength in tension. The strength can be improved by using a composite. Concrete is a composite involving a mixture of cement, gravel and sand. This still has relatively low strength in tension. The strength can be improved with reinforced concrete. This is a composite involving concrete reinforced with steel rods.

Clays
Many ceramic products are made primarily from clay to which other materials have been added. The materials are mixed with water and the required shape formed for the product. They are then dried and fired to form the ceramic. Drain pipes, china and porcelain are examples of such products.

Glasses
The basic ingredient of most glasses is silica. Glasses tend to be brittle, have a strength which is markedly affected by small defects and surface scratches, low thermal expansivity and thermal conductivity, good resistance to chemicals and high electrical resistivity. Glass fibres are frequently used in composites with polymers (see the details given above for polyesters).

1.3.2 Forms of supply

The ferrous and non-ferrous metals used in workshops are purchased from suppliers in a range of basic shapes and sizes (Figure 1.2). For example, the forms might be sheets and plates of various thicknesses, bars with rectangular cross-sections (termed flats), square cross-sections and round, and special cross-sections such as the I-section used for joists, tubing lengths and lengths with rectangular hollow sections, L-shaped sections and a wide variety of other shaped sections.

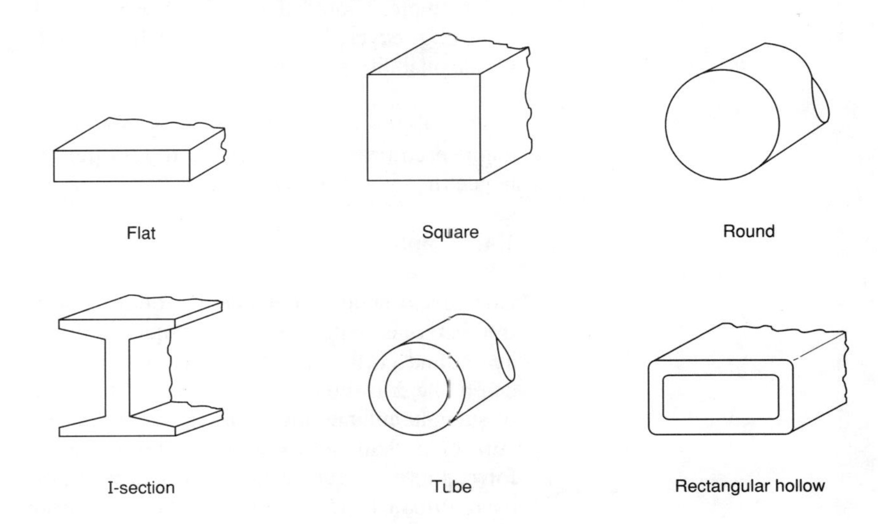

Figure 1.2 *Forms of supply*

1.4 Materials and engineered products

The term *engineered product* is used for products that are manufactured to fulfil some function or functions. The term *engineering materials* is used for the materials designed, made and used for engineered products and hence for a practical purpose such as carrying loads, providing thermal insulation, carrying electrical currents, resisting chemical attack, etc. In considering materials and their uses we need to consider the function or functions required of the product being made and hence the conditions under which the materials will be used and the properties required of them. For example: for a particular product, do we want a material to have the property of being strong and not breaking under high loads? Do we want a material to have the property of being a bad conductor of heat? Do we want a material to have the property of being a good conductor of electricity?

The selection of a material from which a product can be manufactured depends on a number of factors. These are often grouped under four main headings, namely:

1. The requirements imposed by the conditions under which the product is used, i.e. the service requirements. Thus if a product is to be subject to forces then it might need strength, if subject to a corrosive environment then it might require corrosive resistance.
2. The requirements imposed by the methods proposed for manufacturing the product. For example, if a material has to be bent as part of its processing, the material must be ductile enough to be bent without breaking. A brittle material could not be used.
3. The requirements imposed by environmental considerations. For example, should the material used be selected because it is capable of being recycled or does not result in pollution?
4. Availability and cost

The following examples illustrate how the functions required of engineered products determine the properties required of the materials and hence the choice of materials.

1.4.1 Cups

Consider a simple engineered product – a cup. The main requirements of the materials that are used for cups are that they will hold hot liquids, such as tea and coffee, and can be formed into a suitable shape so that the cup can be held in the hand and used for drinking. There are a number of forms of such an engineering product. We might thus have a pottery cup, a plastic cup, or perhaps an expanded polystyrene beaker. The materials can all be formed into suitable shapes and be leak proof. The cup has to be picked up by a human hand without burning or causing discomfort. Thus we might choose a material that is a bad conductor of heat. The polystyrene beaker is an example of where a material with the property of being a very bad conductor of heat is used. Alternatively, we might solve this problem by adding a handle with the result that the cup is held at a point remote from the hot liquid. This would mean that we could use a material which is a poor but not excessively bad conductor of heat, e.g. as with the pottery cup. A variety of materials are used and are able to give a product which fulfils the requirements.

It is necessary to distinguish between the functions required of the product and those required of the material. We might be able to meet a particular requirement by product design methods, as for example by designing a handle for a cup for hot liquids. Alternatively we might adopt a materials solution to the problem, as by the use of polystyrene as the material for a beaker.

1.4.2 Domestic kitchen pan

Consider the problem of the material for a domestic kitchen pan. The functions of a domestic kitchen pan may be deemed to be: to hold liquid and allow it to be heated to temperatures of the order of 100°C. The material used for the pan must therefore:

1 Not deform when heated to these temperatures.
2 Be a good conductor of heat.
3 Not allow liquid to leak through it.
4 Not ignite when in contact with a flame or hot electrical element.

In addition there may be other requirements which are not so essential, but certainly desirable. For the pan we might thus require an attractive finish for the material.

From these requirements we can now consider possible materials. The requirement that the material be a good conductor of heat would seem to reduce the consideration to metals, particularly when taken together with the requirement that the material can be put in contact with a flame and contain hot liquids. This would effectively rule out polymers. But what properties are required of the metal? For the body of the pan to be shaped as a single entity, a manufacturing process in which a sheet of the metal is pressed out into the required shape is suggested, a so-called deep drawing process. A ductile material is required in sheet form for such a process.

The above represents one line of argument regarding the design of pans. It is instructive to examine a range of pans and consider the materials used and what reasons might be advanced for them being chosen. Why for example are some pans made of glass, of a ceramic, of a steel coated on the outside with an enamel and on the inside with a non-stick polymer polytetrafluoroethylene (PTFE)?

The above is only the consideration of the container part of the pan. There is still the handle to consider. The function required is that it can be used to lift the pan and contents, even when they are hot. The properties required are thus poor thermal conductivity, ability to withstand the temperatures of the hot pan, stiffness and adequate strength. The handle can be considered to be essentially a cantilever with a load, the pan and contents, at its free end. Before going too far with considering the design and materials for the handle, British Standards can be consulted. BS 6743 gives a standard specification for the performance of handles and handle assemblies attached to cookware. This sets the levels of safe performance against identified tests simulating hazards experienced in normal service. The need for the handle to have low thermal conductivity indicates that metal would not be a good choice. A polymer is thus a possibility. It needs however to be able to withstand a temperature of the order of 100°C at the pan end, have a reasonable stiffness, and reasonable strength. These requirements suggest a thermoset is more likely to be feasible than a thermoplastic. Another possibility is wood.

In the above considerations of the pan and the handle the item that has so far not been discussed is the life of the items. The purchaser of the pan wants it to last, without problems, for a reasonable period of time. This is likely to be years. The handle should not break during this time or discolour or deteriorate when used and washed a large number of times. The pan should not wear thin or change its mechanical properties with frequent heating, exposure to hot liquids and washing-up liquids.

1.4.3 Car bodywork

The functions required of car bodywork are that it protects the engine and car occupants from the weather and provides a pleasing appearance. The requirements for the material are thus that:

1. It can be formed to the shapes required.
2. It has a smooth and shiny surface.
3. Corrosion is not too significant.
4. In service it is sufficiently tough to withstand small knocks.
5. It is stiff.
6. It is cheap and can be used in the mass production of car bodies.

The shapes required, together with the need for mass production, would suggest forming from sheet as the manufacturing process. This process involves pressing sheets over formers so that plastic deformation gives the required shape. Thus a ductile metal could be used. The material generally used for car bodywork is a low carbon steel. It is relatively cheap, is available in sheets with a smooth finish, is reasonably tough, is reasonably stiff and can be protected against corrosion by coating it with paint.

Polymer materials could be used for the car body work. The problem with such materials is obtaining enough stiffness – polymeric materials are much less stiff than metals. One way of overcoming this is to form a composite material with glass fibre or cloth in a matrix of a thermoset. This could be done by laying a layer of glass fibre cloth over a former, coating it with the polymer, putting another layer of cloth over the former and more polymer and so building up the required thickness of material. Unfortunately, such a process of building up bodywork is a manual rather than a mechancial process and so very slow and labour intensive. While it can be used for one-off bodies, it is not suitable for mass production.

1.4.4 Electrical resistors

Consider the requirements for resistors for general use in electrical and electronic circuits. They are required to obey Ohm's law, i.e. the current through the resistor is proportional to the voltage across it, and have resistances which do not markedly change when the temperature changes. Metals have such properties. Unfortunately, metals have low resistivities. This means that to obtain an electrical resistor of, say, 100 Ω, we need a long length of very fine wire, the greater the diameter of the wire the longer the length required. One possible method by which we can use metals is to deposit very thin layers on an insulating substrate. This can then give a very thin thickness of metal which we can then make into a reasonable length by etching a suitable pattern in the metal. Thus a spiral groove might be cut through the metal deposited on a cylindrical substrate. The resistance value can be adjusted to some required value by stopping the groove cutting when that value is obtained. Nickel-chromium alloys (nichrome) are widely used for resistors manufactured in this way.

Another alternative is to mix a conductive powder with an insulator and organic solvent, the resulting mixture then being spread over an insulating substrate as a thin film. The mixture is fired so that organic solvents evaporate and the insulator and conductive material are left bonded to the substrate. The dispersed conductive particles are considered to form convoluted chains through the insulator. Thus we effectively end up with a number of exceedingly small cross-section conductors. The resistance value for the resistor is then determined by the concentration of the conductive powder in the insulator.

There is an alternative to using a metal and that is to use carbon in the form of graphite. Resistors using graphite all tend to be about the same size, but can have a wide range of values. The process used for making carbon resistors involves powdered graphite being mixed with naphthalene and an insulating filler such as china clay. The mix is then pressed into little cylinders and then fired. This gets rid of most of the naphthalene and leaves a porous graphite structure. The amount of naphthalene, and hence the degree of porosity, determine the resistance of the cylinder.

1.4.5 Coca-cola containers

Consider another example. What materials could be used for the containers for Coca-Cola? Well you can buy Coca-Cola in aluminium cans, in glass bottles and in plastic bottles. What makes these materials suitable and others not? The primary function of the container is to be able to hold a liquid. We might then consider the need for the container material to be:

1 Rigid, so that the container does not become stretched unduly, i.e. become floppy, under the weight of the Coca-Cola.
2 Strong, so that the container can stand the weight of the Coca-Cola without breaking.
3 Resistant to chemical attack by the Coca-Cola.
4 Low density so that the container is not too heavy.
5 Able to keep the 'fizz' in the Coca-Cola, i.e. not to allow the gas to escape through the walls of the container.
6 Cheap.
7 Easy and cheap to process to produce the required shape.

You can no doubt think of more requirements, e.g. that the container should be capable of being recycled and so reduce the demand on the earth's resources. The selection of a material involves balancing a number of different requirements and making a choice of the material which fulfils as many requirements as possible, as well as possible.

1.4.6 Electrical cables

Consider the requirements for the materials used for the conductors in an electric cable. These are likely to include:

1 A very good conductor of electricity.

2 Flexible so that cables can easily be bent round corners.
3 Can be produced with a circular cross-section in long lengths with small diameters.
4 Cheap.

The most commonly used material is copper, this material having one of the highest electrical conductivities.

Example

Metal cups are often used by campers who have to carry their camping gear in a rucksack. What are the functions required of the material which makes a metal to be a reasonable choice for such a cup?

One of the main requirements for such a cup is that it will not break when subjected to being carried around in a rucksack. For this reason a metal is a good choice, since it does not break so easily as pottery or plastics. There is, however, the problem that metals are good conductors of heat and so there can be problems in holding such a cup in the hand when it is full of a hot liquid. However, in this case this disadvantage is outweighed by the other advantages.

Example

What functions might be required of a material for the rainwater gutters and drainpipes used with houses.

The service requirements might be considered to be for a rigid, durable material which is capable of withstanding an outdoor environment without deteriorating. It should be capable of being processed in reasonable long lengths and be cheap. A material that is used is the polymer unplasticized PVC.

Problems

Questions 1 to 13 have four answer options: A, B, C and D. You should choose the correct answer from the answer options.

Questions 1 to 3 relate to the following purposes for which materials might be needed:

A Protection from environmental damage
B Act as a supporting member
C Conduct electricity
D Conduct heat

Select the most likely purpose from the above list of answer options for which the material to be used in each of the following instances is selected:

1 The connecting wires to a resistor.
2 The leg of a chair.

3 The glass envelope of an electric light bulb.

Questions 4 to 6 relate to the following properties that materials might have:

A Strength
B Density
C Brittleness
D Hardness

Select the property that would indicate the suitability of a material for:

4 Steel used for load bearing structures.
5 Steel used for the point of a drill bit.
6 Plastic used for a washing-up bowl.

Questions 7 to 9 relate to the following properties that materials might have:

A Conductivity of heat
B Conductivity of electricity
C Transparency
D Hardness

Select the property that would indicate the suitability of a material for:

7 A hacksaw blade.
8 The casing for an electrical plug.
9 The lens material for a vehicle rear light.

10 Decide whether each of the following statements is TRUE (T) or FALSE (F).

In general, metals are:
(i) good conductors of heat.
(ii) good conductors of electricity.

A (i) T (ii) T
B (i) T (ii) F
C (i) F (ii) T
D (i) F (ii) F

11 Decide whether each of the following statements is TRUE (T) or FALSE (F).

In general, ceramics are:
(i) Good conductors of heat.
(ii) Strong in compression and weak in tension.

A (i) T (ii) T
B (i) T (ii) F
C (i) F (ii) T
D (i) F (ii) F

Questions 12 and 13 relate to the following key properties which might determine the selection of a material for a particular use are. A key property is one that must be taken account of, without it being right a material cannot be used.

A Ductility
B Toughness
C Electrical conductivity
D Thermal conductivity

12 Select the key property of relevance in determining the selection of a material for use as the casing for a portable telephone.

13 Select the key property of relevance in determining the selection of a material for use as the casing for a mains electric plug.

14 Investigate the materials used with the following products and give reasons why they might have been chosen in preference to others:

(a) The casing for mains electric plugs
(b) Spades
(c) Domestic cold and hot water pipes
(d) The casing for the body of a vacuum cleaner
(e) Joists to support floors in a small house

2 Engineering components

2.1 Engineering components

In designing and making products, engineers use standard components such as nuts, bolts, screws, electrical resistors, electrical wiring, etc. The components have to be chosen to fit the requirements of the product. Consider the problem you might face as a student in fastening sheets of paper together. You might use a paper clip, or perhaps rings through punched holes, or staples, or perhaps glue. If you had only a few sheets of paper and wanted a temporary fastening, the paper clip would be suitable. However, if you had a large number of sheets and wanted a fastening which was easily demountable then rings through punched holes would be more suitable. If you wanted to permanently fix sheets of paper together then staples or glue would be possible solutions. Similar situations occur in engineering. Thus, for example, it is no use using rivets to fix two items together if a temporary fastening system is required. A nut and bolt would be a better proposition.

This chapter is a brief consideration of commonly used mechanical fastening components and electrical circuits components.

2.2 Fastening systems

The purpose of a fastener is to provide a clamping force between two pieces of material. There are a wide range of fastening systems that can be employed when two items have to be fixed together. The choice depends on a number of factors:

1 *Service requirements* Is the fastener to be permanent or demountable? Is there to be frequent assembly and disassembly?
2 *Nature of the external loading on the fastener* Are the fastened items to be subject to tensile loads trying to pull the two apart, or perhaps compressive loads which are pushing the two materials together?
3 *The method of assembly of the fastener* How easy is it to make the fastening?
4 *Environmental conditions* Under what environmental conditions is the fastener to be used, e.g. temperature and corrosive conditions?
5 *The quantity of fasteners required and their cost*

Fasteners can be classified into three types – threaded, non-threaded and special purpose. The following sections give common examples of such fasteners.

2.2.1 Threaded fasteners

Examples of *threaded fasteners* used to fix two pieces of material together are bolts mated with nuts and bolts or screws mated with threaded holes. With a bolt and a nut (Figure 2.1(a)), the clamping force holding materials

THE LIBRARY
GUILDFORD COLLEGE
of Further and Higher Education

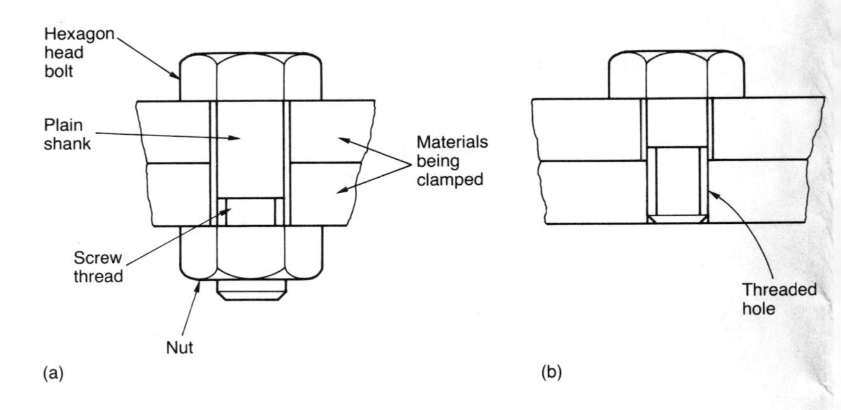

Figure 2.1 *(a) Bolt and nut, (b) bolt and threaded hole*

together is provided by the bolt being stretched when the nut is tightened. With a bolt mated with a threaded hole (Figure 2.1(b)), the clamping force is produced by rotation of the bolt causing the bolt to become stretched. A stretched bolt in trying to contract to its original length clamps the pieces of materials together. Such fasteners are widely used for temporary joints which have to be frequently assembled and dismantled.

The performance of a threaded fastener is affected by many factors, notably:

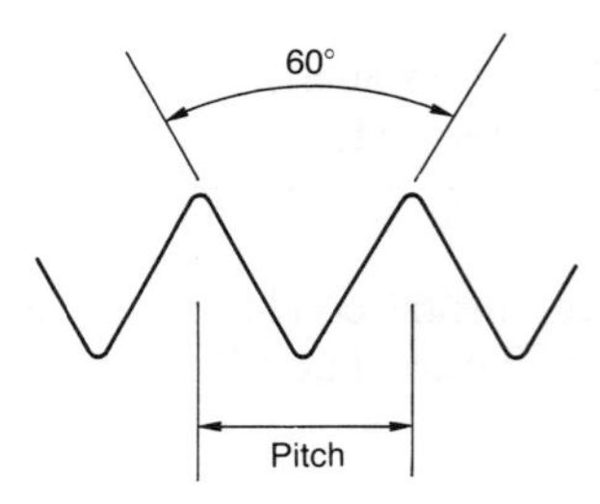

Figure 2.2 *Standard metric thread*

1 *The thread form* There are a number of screw thread systems. The *international standard metric thread system* (*ISO metric*) is based on a V-thread form, the Vs having an angle of 60° (figure 2.2). Bolts, nuts and screws marked with a letter M are specified in the ISO metric system. A thread specified as M6 × 1 is ISO metric with a diameter of 6 mm (this is the major diameter, see Figure 2.3) and thread of 1 mm pitch. Other thread systems that might still be found are: British Standard Whitworth (BSW), an obsolete V-thread form with an angle of 55°; British Standard Fine (BSF), a Whitworth form with a finer pitch; British Association (BA), a V-thread form with an angle of 47.5° which was originally introduced for miniature screwed fastening for scientific instruments; British Standard Pipe (BSP), the original Whitworth form with metric dimensions; Unified National Coarse (UNC) and Unified National Fine (UNF), these being systems widely used in the United States and having the same proportions as the ISO metric system but with dimensions expressed in inches.

2 *The pitch of the thread* The pitch is the amount by which one rotation of the bolt causes it to advance into the nut or a screw into the screwed

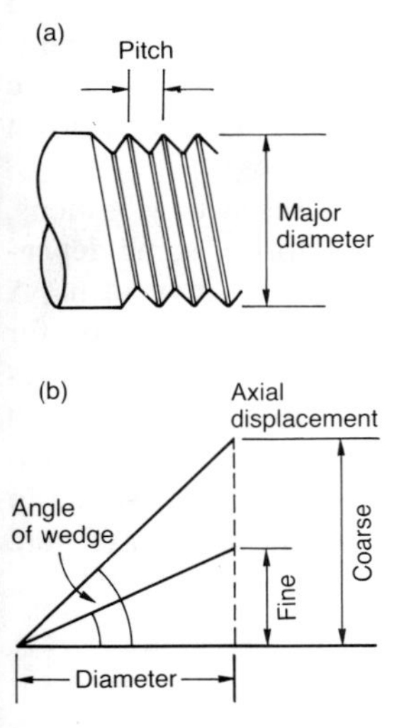

Figure 2.3 *Pitch*

material (Figure 2.3(a)). The action of a screw thread is like that of a wedge (Figure 2.3(b)) with the wedge angle being given by the axial displacement or pitch relative to the diameter of the bolt or screw. A fine pitch means that the axial movement is quite small for one rotation. Such a pitch has a better locking action than a coarse pitch, in that more rotations are needed to unlock the clamping action. A coarse pitch means that the axial movement is bigger for one rotation. Such a pitch gives a greater clamping force than a fine pitch in that there is greater axial movement per bolt rotation. Coarse threads are more difficult to strip off than fine threads, there being less material with the threads on a fine thread than a coarse thread. Thus for material which has low strength, coarse threads should be used rather than fine threads. Aluminium and plastics are examples where coarse threads should be used.

3 *The loading* Threaded fasteners can withstand tensile loading and shear loading without losing their clamping force but compressive loads result in the clamping force being reduced or lost. Vibrations can result in fastening slackening unless locking devices are used.

4 *The fastener material* See the note above on the pitch of the threads. Performance can also be affected by coatings on the fastener material. For example, steel bolts might be coated with zinc or cadmium to make them more resistant to corrosion.

There are a range of heads for screwed fastenings. Figure 2.4 shows some common examples. The hexagon head (Figure 2.4(a)) is probably the main one used in general engineering applications. Slotted cap heads (Figure 2.4(b), (c), (d), (e)) are another form of head. The form of the slot determines whether a conventional screwdriver, a posidrive screwdriver or perhaps an Allen key (for the socket head) is required for tightening. Such heads can be recessed to provide a flush surface.

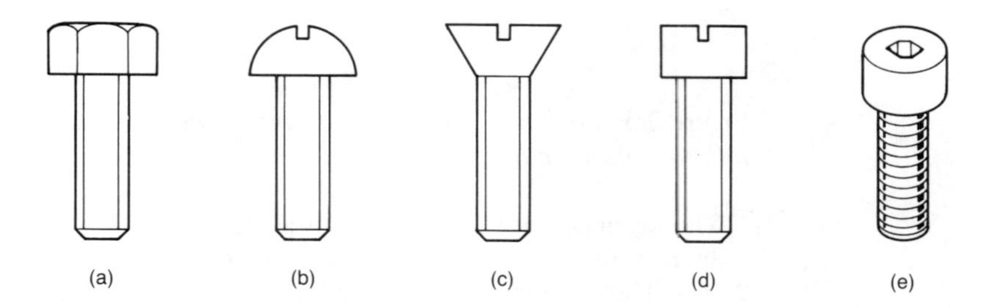

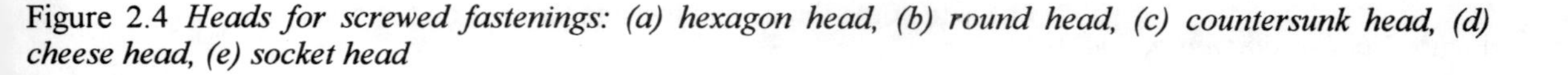

Figure 2.4 *Heads for screwed fastenings: (a) hexagon head, (b) round head, (c) countersunk head, (d) cheese head, (e) socket head*

When a bolt is tightened up to provide the clamping force, there is a danger that the bolt might be bent or distorted and scour the surface against which it is being forced. To reduce such possibilities, soft steel *washers* are inserted between the nut and the surface. These allow nuts to bed down uniformly and protect the surface from scouring.

To prevent screwed fastenings from slackening as a result of vibrations, *locking devices* are used. These can take a number of forms, some depending on friction and some on a positive action to prevent movement. A common method is a *spring washer* (Figure 2.5(a)). This uses friction for its locking action. Another form is the *tab washer* (Figure 2.5(b)). This uses a positive action to prevent rotation. *Slotted nuts* (Figure 2.5(c)) have a split pin through a hole in the screw and fit in a slot in the nut, so preventing rotation of the screw relative to the nut. This uses a positive action to prevent rotation. Another method that can be used is a *locking nut* (Figure 2.5(d)).

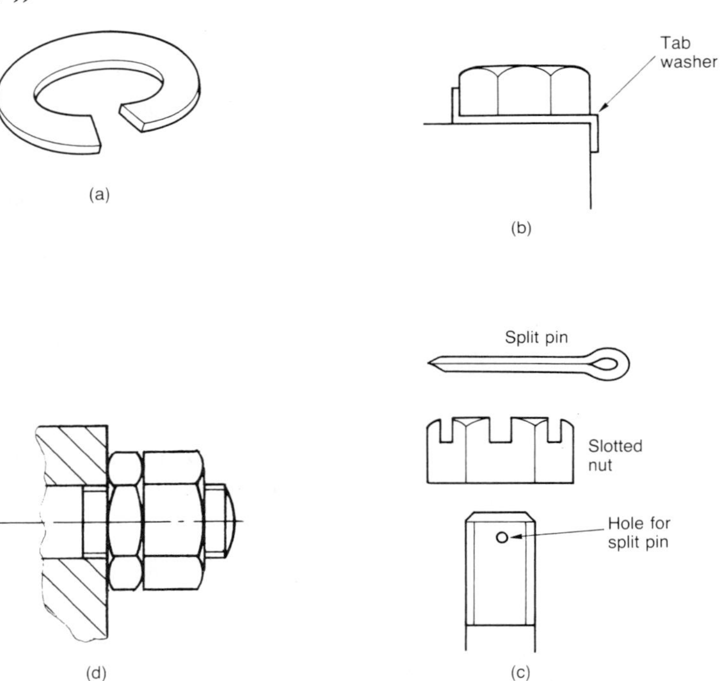

Figure 2.5 *Locking devices: (a) spring washer, (b) tab washer, (c) slotted nut, (d) locking nut*

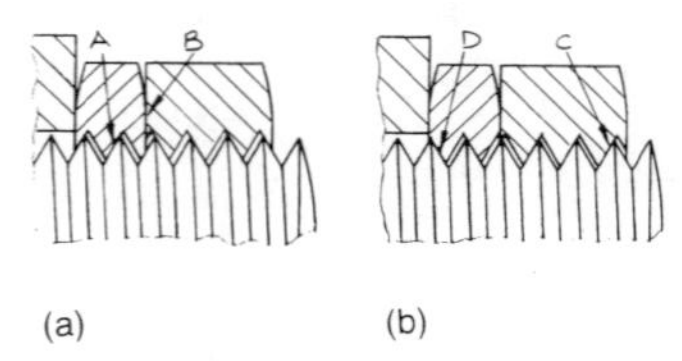

Figure 2.6 *Locking nut action*

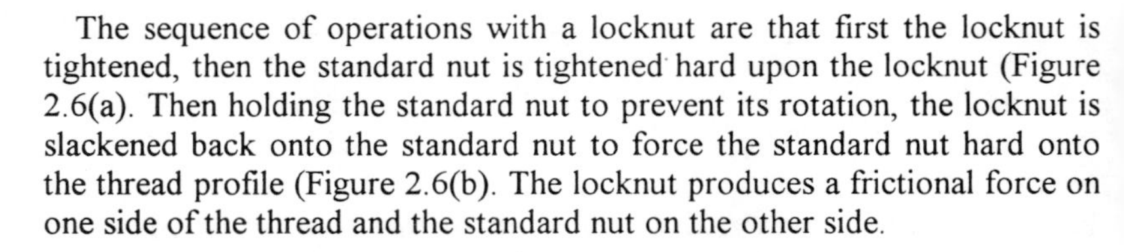

The sequence of operations with a locknut are that first the locknut is tightened, then the standard nut is tightened hard upon the locknut (Figure 2.6(a). Then holding the standard nut to prevent its rotation, the locknut is slackened back onto the standard nut to force the standard nut hard onto the thread profile (Figure 2.6(b). The locknut produces a frictional force on one side of the thread and the standard nut on the other side.

A positive method of locking a group of nuts is to use a *locking wire*. This is a soft iron wire which passes through the slots in slotted nuts and holes in the screws (the arrangement of each being as shown in Figure 2.5(c)) and is then tightened up. Figure 2.7(a) shows the arrangement when

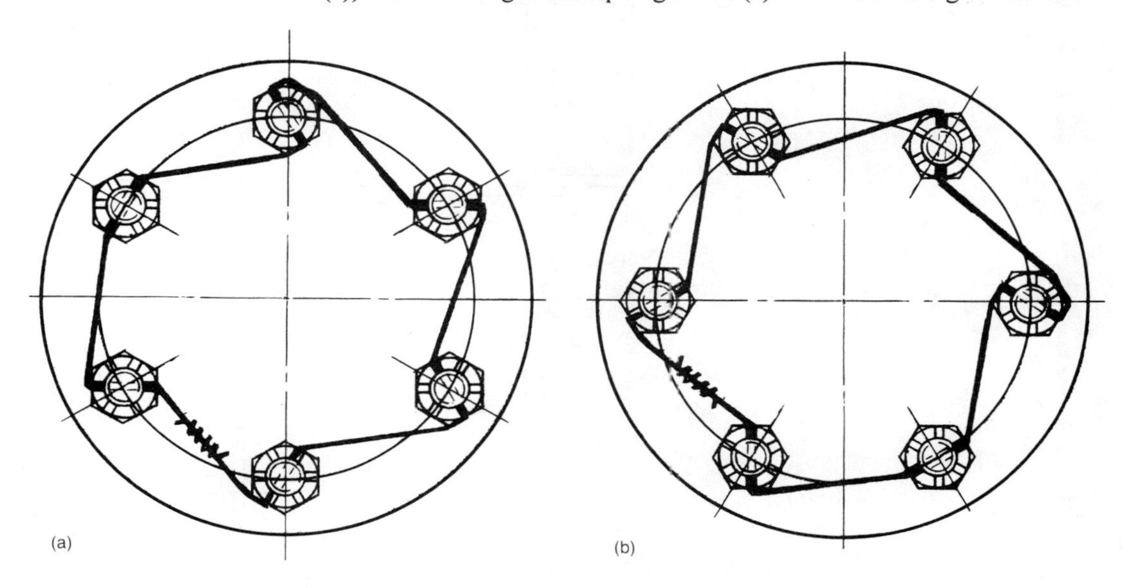

Figure 2.7 *Wired nuts: (a) correct, (b) incorrect*

the wire is in such a position as to prevent the loosening of any one nut. If the nuts had been wired in the way shown in Figure 2.7(b), then if nut A loosened, the wire connecting A to B would become slack and so nut B would no longer be locked and can loosen, hence the wire between B and C would become slack.

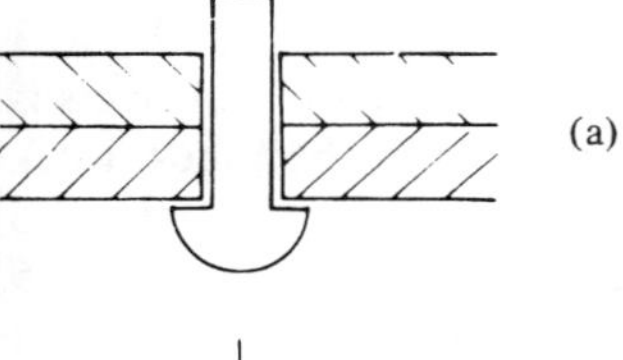

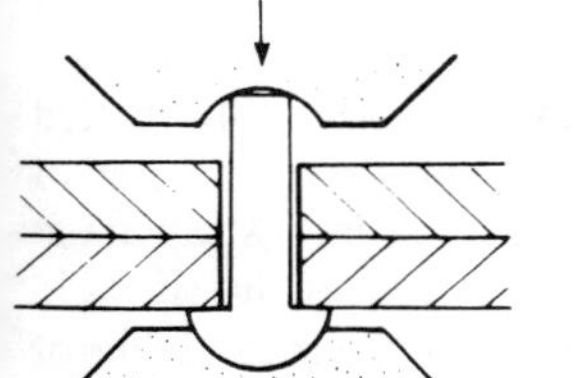

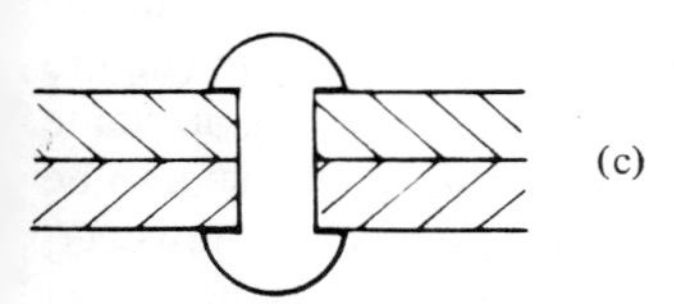

Figure 2.8 *Typical stages in riveting*

2.2.2 Non-threaded fasteners

Examples of non-threaded fasteners are rivets, pins and spring-retaining clips.

Riveting is used to make permanent joints. Such joints are designed to take shear forces and not tensile forces. Figure 2.8 shows the basic stages of a riveting process. When compressive forces are applied to a rivet, as the shank of the rivet decreases in length the shank increases in diameter. A ductile material is used for a rivet so that when the force applied to the rivet is sufficiently high, the rivet permanently deforms until it fills the hole. Then the unsupported part of the shank outside the hole continues to deform until a head is formed. Ductile materials that are used for rivets are mild steel, copper, brass, aluminium and aluminium alloys. For some materials, in order to achieve sufficiently ductility, the rivets are used in the cold state, in other cases, while hot.

Figure 2.9 shows some of the forms of riveted joints. Joints can be made by lapping over the edges of the plates and fastening with one or more rows of rivets (Figure 2.9(a) and (c)). Where butt joints between materials are required, butt straps are used (Figure 2.9((b) and (d)). These are thin sheets of material placed between the rivet heads and the materials being joined.

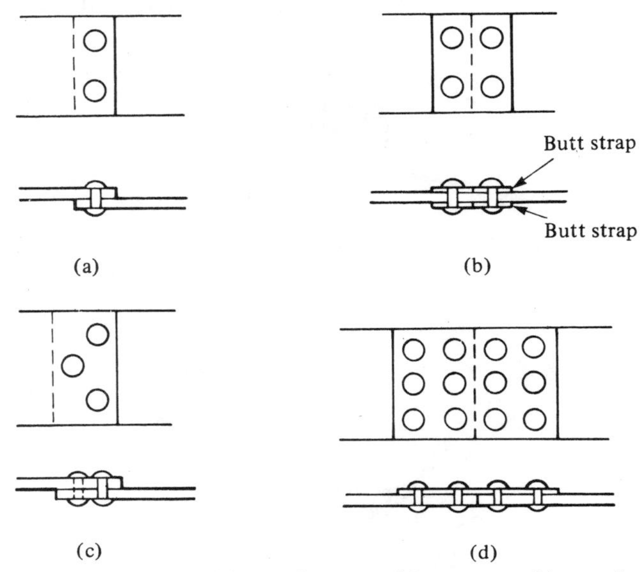

Figure 2.9 *Riveted joints: (a) single-riveted lap joint, (b) single-riveted butt joint, (c) double-riveted lap joint, (d) double-riveted joint with a single butt strap*

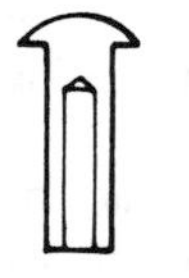

Figure 2.10 *A tubular rivet*

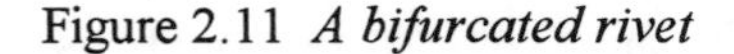

Figure 2.11 *A bifurcated rivet*

With regard to the rivets, the shank diameter of a rivet should be not less than the thickness of the thickest sheet being joined, nor more than 2.5 to 3 times the thickness of the outside sheet in the joint. The spacing of rivets is important. If they are too close together the material becomes weak because of the holes in it. If the rivets are too far apart they are unable to withstand the shear forces involved. Rivets should be spaced generally 1.5 times their diameter from the edges of the plates and 3 times their diameter apart. Thus, with a lap joint and a single row of rivets, the material must overlap by 2×1.5 times the diameter of the rivets. With a double-riveted lap joint, the material must overlap by $(2 \times 1.5 + 3)$ times the diameter of the rivets. With a single-rivetted strapped butt joint, the butt strap must have a width of $(1.5 + 1.5 + 1.5 + 1.5)$ times the diameter of the rivets in order to ensure that each rivet is 1.5 times the diameter from the edge of the sheets of materials being joined and the edges of the strap.

In applications where the riveting force might be large enough to damage or distort the materials being joined, tubular or semi-tubular rivets (Figure 2.10) might be used instead of solid rivets. Tubular rivets can, in some circumstances, be self-piercing. This means that a hole does not have to be drilled prior to inserting the rivet. Self-piercing tubular rivets are, however, limited to materials such as leather and fibre. Bifurcated rivets have two prongs (Figure 2.11) and are used as self-piercing in a wider range of materials, e.g. plywood, plastics and fibres.

The term *pop rivet* is used for those rivets that are installed from just one side of the workpiece, requiring no operator or any holding tool on the other side of the piece. The pop rivet is a hollow rivet with a steel mandrel (Figure 2.12). When the rivet has been inserted into the hole in the workpiece, a special tool is used to pull the mandrel and cause the hollow end of the rivet to expand on the blind side and so clamp the materials together.

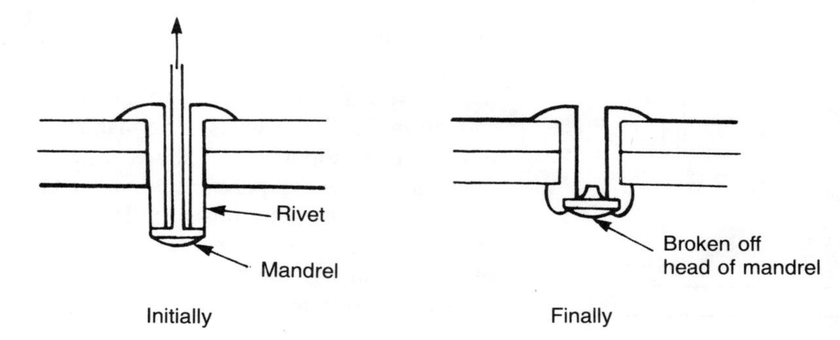

Figure 2.12 *A pop rivet*

Figure 2.13 *Taper pin*

Figure 2.14 *Split pin*

Pins, either in the solid or tubular forms, are widely used for fastening. Thus *taper pins* (Figure 2.13) are used to join components such as collars onto shafts. When the collar is correctly positioned a hole is drilled through it and the shaft. A taper pin is then driven through holes in the two parts until it is fully home and giving a tight fit. The term *cotter pin* is used for a taper pin secured by a nut. An example of this is the pedal crank of a bicycle which is fastened to its shaft by a cotter pin. Another form of pin is the *split pin* (Figure 2.14). This is used where freedom of movement in the joint is required, or as a locking device for slotted nuts (Figure 2.5(c)).

There is a variety of forms of *spring-retaining clips*. A simple form is a C-clip. This can be used to lock and retain components on shafts, the clip generally fitting into a groove on the shaft (Figure 2.15).

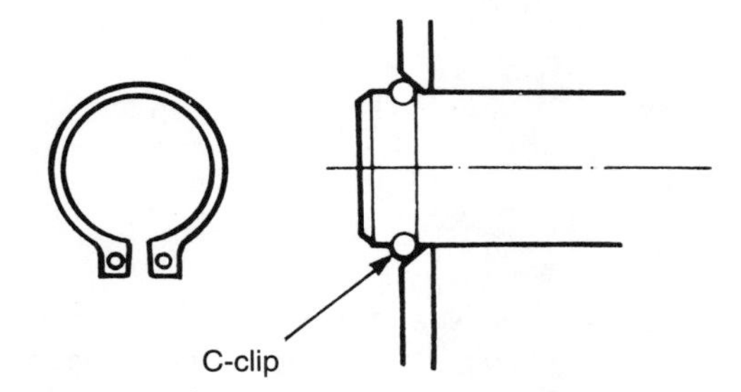

Figure 2.15 *C-clip*

2.2.3 Snap- and press-fits

One of the advantages of plastics is that they can be subjected to quite severe elastic distortion and still return to their original shape when the load is removed. Press- and snap-fits rely on this characteristic. Such joints may be designed for permanent or recoverable use and are widely used with plastics.

A common form of snap-fit is the hook joint, Figure 2.16 showing an example. When the component is pushed into the hole, the end is deformed so that it can slide through the hole until it emerges from the other end. then it expands and locks the component in position. The type of hook end shown in the figure gives a permanent joint in that it is not possible to disengage it. Figure 2.17 shows how the hook end varies when it is designed for use as a permanent joint and a recoverable joint.

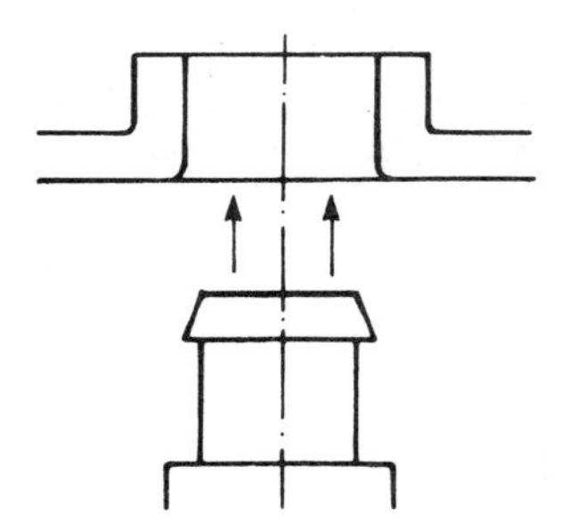

Figure 2.16 *A snap-fit*

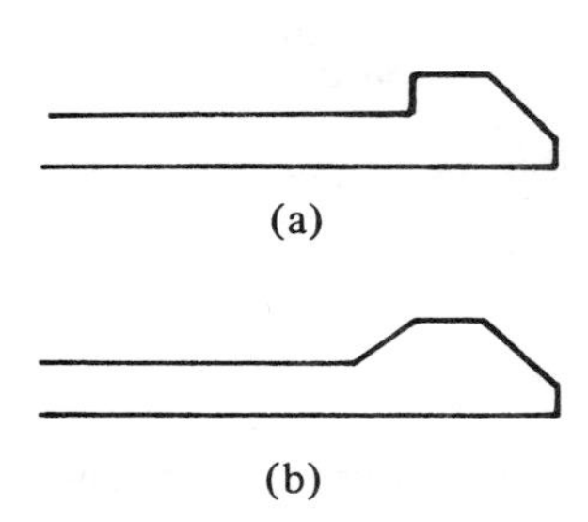

Figure 2.17 *(a) Permanent fixing, (b) recoverable use*

Figure 2.18 shows a press-fit. The component is slightly larger diameter than the hole. Thus when the component is pushed into the hole it is squashed and the resulting forces exerted by the component on the hole walls are responsible for its locking action. The component is a tight fit in the hole.

Snap-fits are stronger and more dependable than press fits, relying on a mechanical interlocking of two components, as well as friction, whereas press fits rely only on friction.

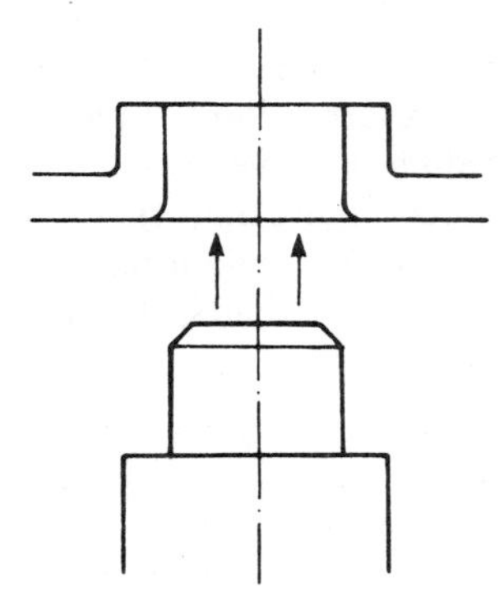

Figure 2.18 *A press-fit*

Example

A catch is required for a cupboard door. The catch is to hold the door closed until such time as the handle is pulled to open the cupboard. Suggest a catch that could be used.

A possibility is a press-fit or a snap-fit with a hook designed for recoverable use. Another alternative which is widely used is a small magnet fixed on the door and a piece of soft iron on the door frame. There are other forms of catch which will do the same job. Examine the catches on cupboard doors.

2.3 Electrical components

This section is a discussion of commonly used electrical components: cables, circuit boards, resistors, capacitors, inductors and integrated circuits. In quoting the values of resistors, capacitors, etc. prefixes are often used to save writing out a lot of zeros in the number for the value. Table 2.1 lists some of the commonly used prefixes.

Table 2.1 Prefixes

Multiplication factor	Prefix	
$1\,000\,000 = 10^6$	mega	M
$1\,000 = 10^3$	kilo	k
$100 = 10^2$	hecto	h
$10 = 10$	deca	da
$0.1 = 10^{-1}$	deci	d
$0.01 = 10^{-2}$	centi	c
$0.001 = 10^{-3}$	milli	m
$0.000\,001 = 10^{-6}$	micro	μ
$0.000\,000\,001 = 10^{-9}$	nano	n
$0.000\,000\,000\,001 = 10^{-12}$	pico	p

Thus, for example, 1000 Ω can be written as 1 kΩ, 1 000 000 Ω as 1 MΩ, and 0.000 001 F (farad, unit of capacitance) as 1 μF.

Example

Express the resistance value 5000 Ω in kilohms.

5000 Ω is 5 kΩ.

Example

Express the resistance value 2.2 MΩ in full.

2.2 MΩ is 2 200 000 Ω.

2.3.1 Cables

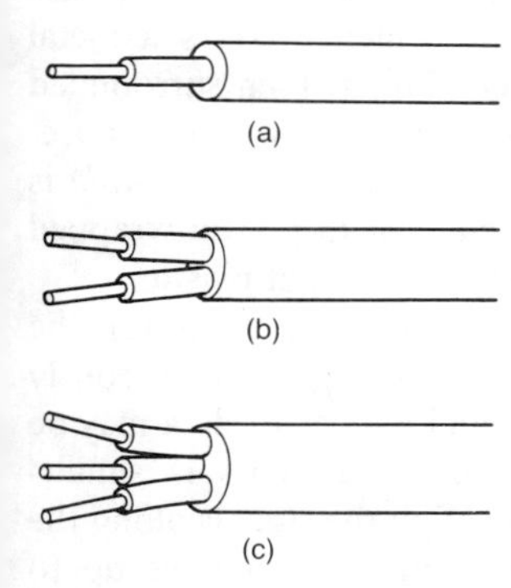

Figure 2.19 *(a) Single core, (b) twin core, (c) three core*

Cables are used for current paths in electrical circuits, e.g. the domestic electrical wiring system, and for signal and data transmission, e.g. TV signals and data transmission between computers. The form of cable used depends on the current that has to be carried and whether it may cause interference, or there might be interference, with signals in neighbouring cables or other installations.

Cables used in electrical installations, e.g. house wiring, tend to be *PVC insulated, PVC sheathed cables*. Such cables consists of single, twin or three PVC insulated wires sheathed in PVC (Figure 2.19). The conducting wires may be single wires or consist of a number of fine wires. The criterion determining the cross-sectional size of the conducting wire is the current that it is designed to carry. A large current requires a large cross-section wire if the wire is not to overheat. In cabling used for installations in houses, twin conductor cables are used with a third conductor as an earth wire. For cabling used to connect appliances to the supply via a plug and socket, three-core cable is used, the live cable being sheathed in brown plastic, the neutral in blue plastic and the earth in green/yellow plastic.

Cable used in industrial and commercial premises where more robust cables are required, can be mineral-insulated metal-sheathed cables or armourplated PVC-insulated, PVC sheathed cables. *Mineral-insulated metal sheathed cables* consist of insulated copper or aluminium conductors contained in a compressed mineral powder, magnesium oxide, in a metal sheath (Figure 2.20(a)). *Armoured cables* consist of PVC insulated conductors, in armour protection. This consists of galvanised steel wire secured between PVC bedding and a tough PVC outer sheath (Figure 2.20(b)).

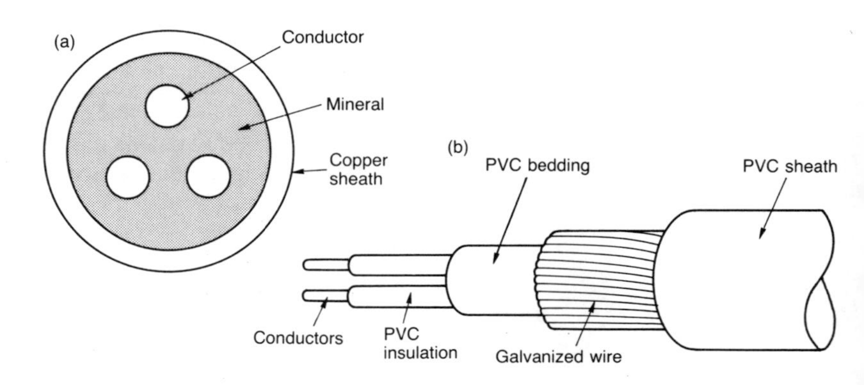

Figure 2.20 *(a) Mineral-insulated cable, (b) armoured cable*

Cables designed to carry currents at frequencies higher than that of the mains supply must be able to prevent the currents interfering with those in other nearby cables or appliances and also exclude interference from currents in other cables or appliances. For low-power signal transmission, up to audio frequencies, *screened cable* can be used (Figure 2.21). This has one or more plastic sheathed conductors around which there is a metal screen of helical wound wire or woven wire braid. This is then surrounded by outer insulation. The aim of the metal screen is to prevent interference. Where signal transmission at radio frequencies is involved *coaxial cable* is used (Figure 2.22). This consists of an inner conductor in plastic wrapped in a metal braid outer conductor and then an outer covering of plastic.

With microprocessor and computer systems there is a need for the transmission of data with a number of signals being simultaneously transmitted. Computers work with binary signals and a piece of data may be transmitted as, for example, 10010110. Each of these digits is simultaneously transmitted by a cable, so that the result of all the signals along the cables constitutes the piece of data. When operating at distances up to about 5 to 10 m *ribbon cables* are used. These consist of a number of parallel insulated cables in the form of a ribbon (Figure 2.23).

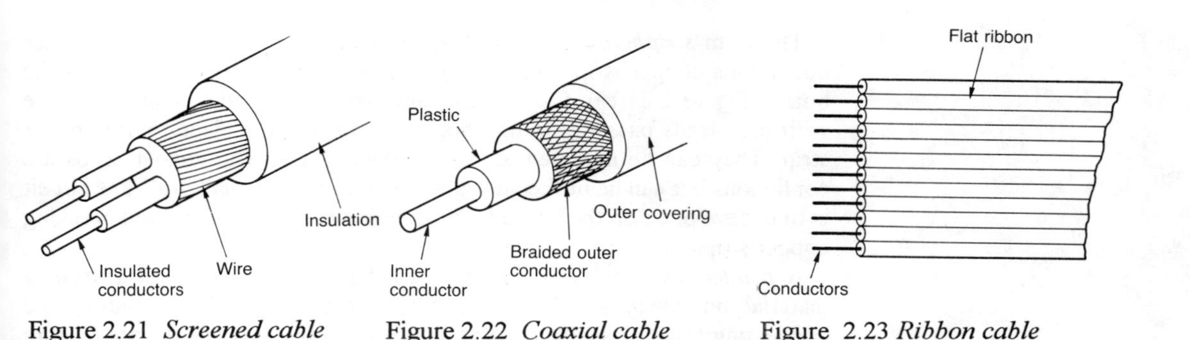

Figure 2.21 *Screened cable*

Figure 2.22 *Coaxial cable*

Figure 2.23 *Ribbon cable*

Example

Which type of cable would be needed for connecting (a) the mains power to a TV set, (b) the aerial to a TV set?

(a) Three-core cable with one of the cores being the earth.
(b) Coaxial cable because radio frequency signals are involved.

2.3.2 Circuit boards

Permanent circuits are generally assembled on some type of insulated board. There are three types of board: matrix, strip and printed circuit board.

Matrix board consists of an insulated board which has a matrix of holes in it (Figure 2.24(a)). The holes are 1 mm diameter with 2.54 mm (0.1 inches) between centres. Boards are available in various sizes, e.g. 160 × 100 mm with 39 × 62 holes, 100 × 75 mm with 39 × 39 holes, 300 × 100 with 39 × 117 holes. Matrix pins press into the holes and provide a terminal post to which components and connecting wires can be soldered.

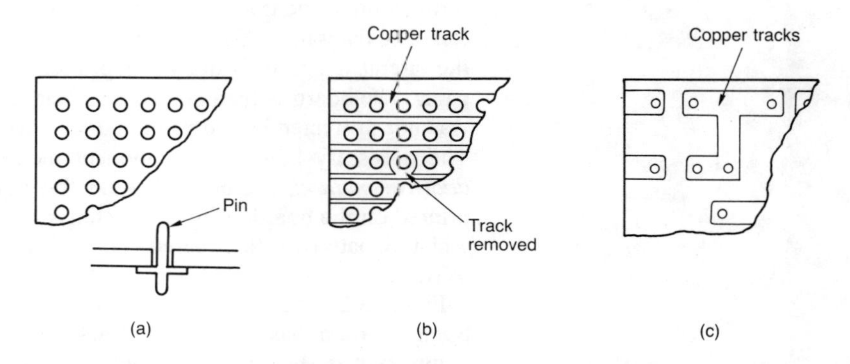

Figure 2.24 *(a) Matrix board, (b) strip board, (c) printed circuit board*

The term *stripboard* or *Veroboard* is used for matrix board which has continuous copper strips linking together rows of holes on one side of the board (Figure 2.24(b)). Components are assembled on the plain board side with their leads passing through holes and soldered in place to the copper strip. They can be mounted across or along strips. The copper strips are continuous but can be broken using a strip cutter or a small drill. The circuit is thus developed on the board by positioning the components and breaking copper strips.

A *printed circuit board* consists of a base, a thin board of insulating material on which all the circuit components are to be mounted, and conducting tracks, usually copper, on one or both sides of the base (Figure 2.24(c)). The conducting tracks are laid down in the form required by the circuit so that they make the interconnections between components. Components are connected to the tracks at the appropriate points to make the required circuit.

The following are some of the properties required of the material used for circuit boards:

1 It should be an electrical insulator.
2 It must give boards which are mechanically stiff and strong, even though quite thin.
3 The board material should be hard enough to be machined cleanly and have tiny holes drilled in it.
4 Conducting tracks should adhere to it.
5 It must not char or distort when components are fastened to the tracks by soldering.
6 It should not absorb water and so spoil its insulation.
7 It should be able to withstand voltages between conductors which are comparatively close together without the insulation breaking down.
8 It must be cheap.

Circuit boards are generally composites made of polymers impregnating woven glass fibre, cotton fabric or paper. For rigid base boards it is usually an epoxy resin reinforced with glass fibre cloth.

With printed circuit boards, two basic processes can be used for the production of the tracks. With the *subtractive method*, a base is coated with foil. The diagram indicating the position of the tracks, i.e. the conductors in the circuit diagram, is drawn or printed on the foil using an acid-resisting paint. The board is then dipped into acid which dissolves the copper which was not protected by the paint and so leaves the required pattern of tracks. The necessary holes for component mounting are then drilled. With the *additive method*, the board has no foil covering. The circuit is drawn or printed on the board using conducting paint. Copper is then deposited from a plating bath over the treated conducting areas to give the required pattern of tracks.

Figure 2.25 shows the basic form of a *single-sided printed circuit board*. Such a board has conductive tracks on only one side of the board. A component is shown as mounted by the component leads passing through holes in the board, the leads then being soldered to make contact with the tracks which are on the opposite side of the board to the component. This is referred to as *through-hole technology*. An alternative, newer, form of mounting is to attach component leads to the tracks of the board without passing them through holes. This is referred to as *surface mount technology*. This avoids the problem of having holes accurately drilled and enables components to be mounted closer together. Single-sided boards are mainly used for low-cost, low-component-density circuits used in such applications as portable radios.

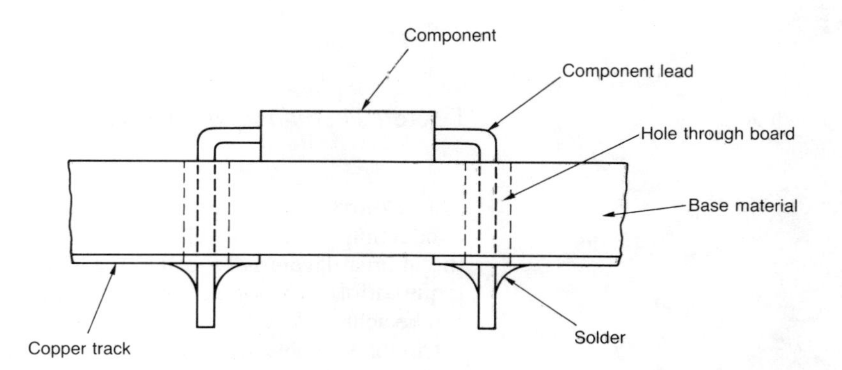

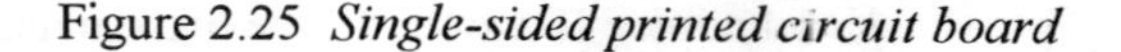
Figure 2.25 *Single-sided printed circuit board*

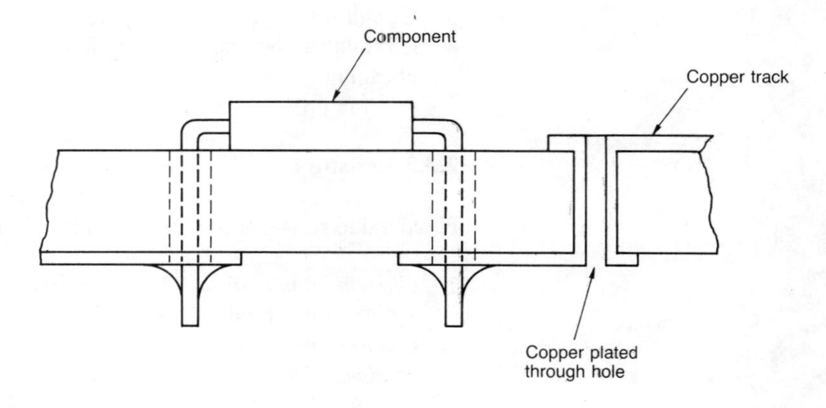

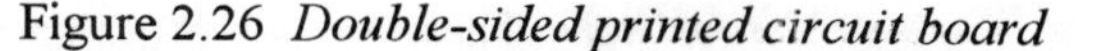
Figure 2.26 *Double-sided printed circuit board*

Figure 2.26 shows the basic form of a *double-sided printed circuit board*. Such a board has conductive tracks on both sides of the board. The boards generally have plated-through holes, i.e. holes which have copper deposited on the inside walls, to enable interconnections to be made between the two sides of the board. The fact that components can be mounted on both sides of the board enables a higher component-density to be achieved.

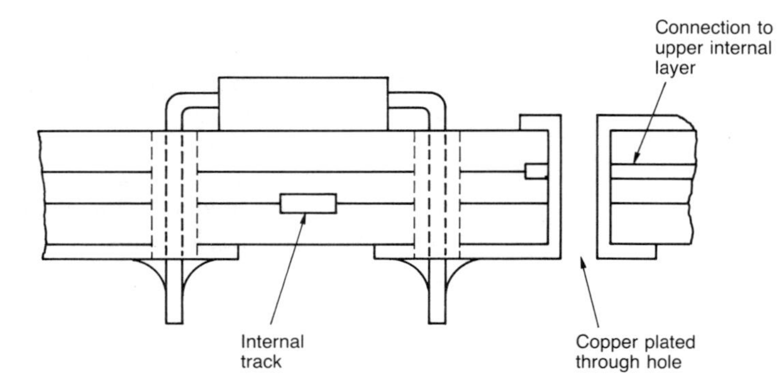

Figure 2.27 *Multi-layer printed circuit board*

Multi-layer printed circuit boards (Figure 2.27) have internal layers of conducting tracks, as well the tracks on the outer surfaces. Connection to the internal layers is by plated-through holes. The manufacturing methods required for such boards are more complex but higher component densities can be achieved.

Circuit assembly using printed circuit boards enables:

1 Exact copies of a circuit to be made many times.
2 Insertion of components in circuits and soldering to be automated.
3 Circuit testing can be automated.
4 Circuits can be manufactured in large quantities with high reliability and cheaply.

2.3.3 Resistors

Fixed value resistors can take a number of forms:

1 *Carbon composition* These consist of rods of carbon (see section 1.4.4). This type of resistor was very widely used but is now less common because of its inferior stability and the fact that its resistance changes quite markedly when the temperature changes.

2 *Film resistors* These consist of a thin film of metal, carbon, a mixture of metals and a glass (so termed metal glaze) or a metal and an insulating oxide (so termed metal oxide) deposited on a ceramic rod (see section 1.4.4). For many applications, this type of resistor is preferred. It offers a resistance range from about 10 Ω to 1 MΩ, has resistances which change only a little when the temperature changes and good stability.

3 *Wire-wound* These consist of resistance wire wound on an insulating former. The other types of resistors tend to be not able to cope with

electrical power greater than about 250 mW. This is the power developed when a current of 0.016 A flows through a 1 Ω resistor or 0.005 A through a 10 Ω resistor (power = I^2R). For dissipating more heat, wire-wound resistors can be used. These can dissipate powers from about 1 W to 25 W. 1 W is the power dissipated when a current of 1 A flows through a 1 Ω resistor. Because long lengths of very thin wire are required to achieve quite small resistances, resistance values are limited to about 100 kΩ. Wire-wound resistors can have high stability and resistances which change little with changes in temperature.

Resistors have values specified within *tolerance bands*. For example, a resistor may be specified as having a tolerance band of ±10%. For a 100 Ω resistor, this means that resistance lies within plus or minus 10% of 100 Ω, the specified value, i.e. since 10% of 100 is 10 then the resistance is between (100 – 10) = 90 Ω and (100 + 10) = 110 Ω. For a 100 Ω resistor with a ±20% tolerance, since 20% of 100 is 20, the resistance lies within (100 – 20) = 80 Ω and (100 + 20) = 120 Ω.

Fixed resistors can be made to have any resistance value. In most cases, however, it is possible to use preferred values. These are standard values that have been adopted in order to provide coverage of a wide range of resistance values by the use of a limited number of values. *Preferred values* are chosen so that, taking into account specified tolerances, the upper value of one preferred resistor is either equal to or just overlaps the lower value of the next higher preferred resistor. For example, a 10 Ω resistor with 10% tolerance will have a value between 9.0 Ω and 11.0 Ω, while the next preferred value of 12 Ω has a value between 10.8 Ω and 13.2 Ω. Table 2.2 indicates some preferred values.

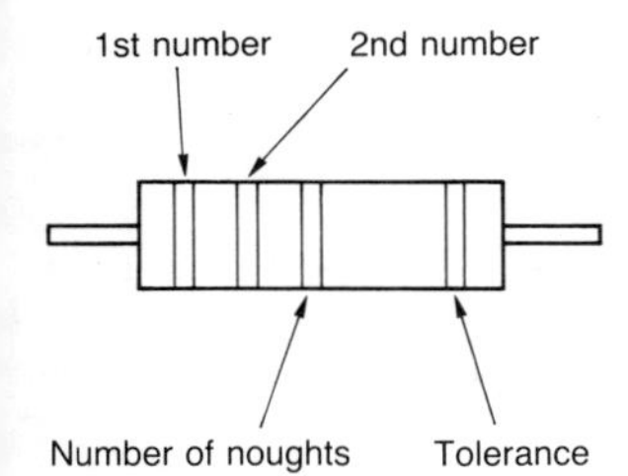

Figure 2.28 *Resistors*

Resistors have been traditionally coded by means of a colour code (Figure 2.28). The first band of colour indicates the first figure of the resistance, the second band the second figure and the third band the number of zeros following the second figure. A tolerance colour band is used to indicate the tolerance, though the absence of a band indicates that the tolerance is ±20%. Table 2.3 shows the code colours (these colours are also used for coding capacitor values). Thus, for example, the colour band sequence of yellow, violet, red, silver indicates a resistance of 4700 Ω with a tolerance of ±10%.

A mnemonic that can be used to remember the sequence of the colours in the code is:

Bye **B**ye **R**osie **O**ff **Y**ou **G**o, **B**ristol **V**ia **G**reat **W**estern

The first letter of each word gives the first letter of the colour associated with it.

Table 2.2 Preferred resistance values

Resistance values with tolerances:		
5%	10%	20%
10	10	10
11		
12	12	
13		
15	15	15
16		
18	18	
20		
22	22	22
24		
27	27	
30		
33	33	33
36		
39	39	
43		
47	47	47

Table 2.3 Code colours

Figure	Colour	Figure	Colour
0	black	*Used in 3rd band as multipliers*	
1	brown		
2	red	0.01	silver
3	orange	0.1	gold
4	yellow	*Tolerance*	
5	green	5%	gold
6	blue	10%	silver
7	violet		
8	grey		
9	white		
0.1	gold		

Resistance values are now often coded (British Standard 1852) by printed codes on resistors. In this scheme no decimal points are used. The value in ohms is indicated by the letter R, kilohms by K, megohms by M. The letters R, K and M are inserted in the number in the place that would be occupied by the decimal point. For example, 10R is 10 Ω, 1K5 is 1.5 kΩ, 2M2 is 2.2 MΩ. If the value is less than 1 Ω, a zero is used for the first digit, e.g. 0R5 is 0.5 Ω. After the value code a letter is added to indicate the tolerance: F is ±1%, G is ±2%, J is ±5%, K is ±10% and M is ±20%. Thus 4K7J is 4.7 kΩ with a tolerance of ±5%.

The criteria for the selection of resistors are basically the resistance value, and its acceptable tolerance, and the power dissipation that a resistor can withstand. When a current passes through a resistance, heat is produced and so power dissipated. Where circuits involving large currents are involved, and high power dissipation, wire-wound resistors tend to be used. Where low currents are involved, the choice tends to be mainly films on ceramics.

Example

A resistor has the colour code: black, red, black and silver. What does this mean?

Black indicates 0, red 2. Thus the resistance value is 020 Ω. The tolerance is ±10%.

Example

A resistor has the marking 33RM. What does this mean?

The R indicates that the number is the resistance value with no multiples and that the decimal point is after the 33. Thus the resistance is 33 Ω. The M indicates that the tolerance is ±20%.

2.3.4 Capacitors

A *capacitor* is essentially just two parallel conducting plates separated by an insulator called a *dielectric*. The unit of capacitance is the farad (F). However, capacitors tend to have values considerably smaller than 1 F. In electronics the capacitances are often microfarads (μF), nanofarads (nF) or picofarads (pF). The following are common types of capacitors:

1 *Paper* This type of capacitor has a paper dielectric and tends to have a rather bulky tubular form. They have capacitance values in the range 10 nF to 10 μF with tolerances of 10%. They can be used with d.c. voltages up to 600 V and a.c. voltages of 250 V. They have reasonable stability.

2 *Ceramic* These have a ceramic dielectric and are made in disc, tubular and rectangular plate forms. There a number of forms and together they have capacitances values in the range 5 pF to 47 μF, with tolerances ranging from 10 to 20% Some are used with d.c. voltages up to kilovolts and others with a.c. voltages of 250 V. They have generally good stability.

3 *Plastic film* These have a film of a plastic such polystyrene, polyester or propylene as the dielectric. They have capacitance values in the range of about 50 pF to 100 μF, the range depending on the plastic used. Tolerances tend to be 1 or 5%. Typically, they can be used with d.c. voltages of up to 500 V and a.c. voltages of about 400 or 500 V. Polystyrene capacitors have excellent stability, the other plastics having only reasonable stability.

4 *Silver mica* These consist of a mica dielectric with silver coatings on each side of the sheet. They have capacitance values in the range 5 pF to 10 nF, tolerances of ±0.5% and can be used with d.c. voltages up to about 600 V. They have excellent stability.

5 *Electrolytic* These have a dielectric formed by electrolytic action and have very large capacitance values, typically in the range 1 μF to 22 000 μF. There are two basic types, aluminium electrolytic and tantalum electrolytic. Their main disadvantage is that they are polarised, i.e. they can only be connected so that one side is always connected to the positive side and the other to the negative side in a circuit. They can thus only be used with d.c. voltages, and then only the correct way round, and are limited as to the voltages they can be used with, the voltages being values less than 100 V. They have excellent stability.

The criterion for the selection of capacitors are essentially the capacitor value and the working voltage The working voltage is, with some safety margin, the maximum voltage the capacitor can withstand without breakdown of the dielectric occurring. With some capacitors working voltages may be as low as 6 V, others as high as 20 kV. The working voltage must not be exceeded, the margin of safety is generally fairly small.

Some capacitors are marked with their values, in pF, in the form of three digits and a letter. The first digit gives the first number, the second digit the second number and the third digit the number of zeros. The letter F indicates a tolerance of ±1%, G a tolerance of ±2%, H a tolerance of ±2.5%, J a tolerance of ±5%, K a tolerance of ±10% and M a tolerance of ±20%. Thus a capacitor marked with 102K has a capacitance of 1000 pF with a tolerance of ±10%.

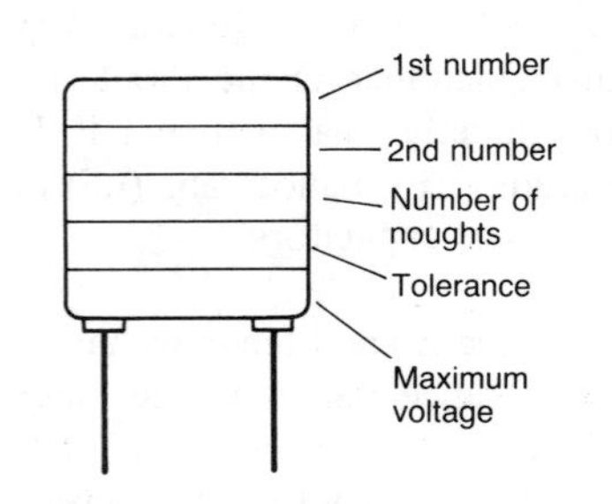

Figure 2.29 *Plastic film capacitors*

Plastic film capacitors often use colour coding to indicate their values, tolerances and working voltages in pF (Figure 2.29). The colour furthest from the leads gives the first digit, the second band the second digit and the third band the number of zeros (the same colour code is used as for resistors, see table 2.3). The fourth band indicates the tolerance with black indicating ±20%, white ±10% and yellow ±5%. The fifth band, i.e. the one nearest to the leads, gives the maximum working voltage, brown being 100 V, red 250 V and yellow 400 V.

Tantalum electrolytic capacitors sometimes use colour coding to indicate values and working voltage. The colour coding is the same as that used for resistors, see table 2.3. The colour code gives the values in μF. The capacitors look like a raindrop with two leads protruding from the bottom (Figure 2.30). The first colour at the end opposite the leads gives the first digit, the second colour the second digit and the colour of a spot the number of zeros. The band nearest to the leads gives the working voltage with black being 10 V, yellow 6.3 V, green 16 V, blue 20 V, grey 25 V, white 3 V and pink 35 V.

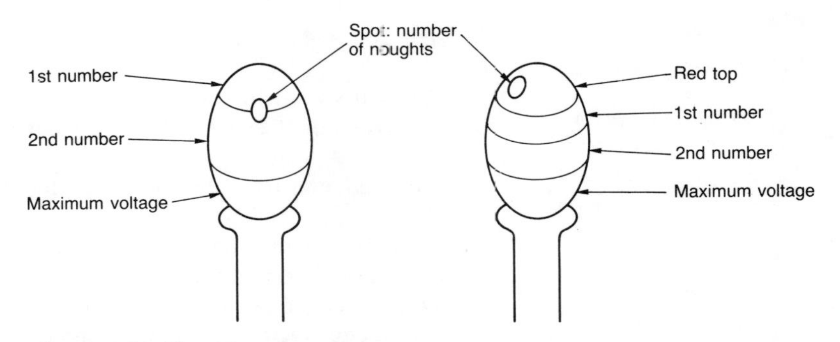

Figure 2.30 *Tantalum capacitors*

Example

A capacitor is required which will have a capacitance of 1000 μF. Suggest a possibility and state any restrictions on its use.

An electrolytic capacitor can be used to give such a large capacitance. However, such a capacitor often has a low working voltage and cannot be used with a.c. voltages.

Example

A polyester capacitor is marked with the colour bands, starting from the end furthest from the leads, of yellow, violet, yellow, green, yellow. What information does this give?

The first two digits give the numbers 47. The third digit gives the number of zeros as 4 and so the value is 470 000 pF or 0.47 μF. The fourth band indicates that the tolerance is ±5% and the fifth band that the working voltage is 400 V.

2.3.5 Inductors

Inductors are basically just coils of wire, generally wound on a core of some magnetic material. The unit of inductance is the henry (H). A range of pot cores can be bought on which the user can wind wire to give the required inductance, the inductance increasing the larger the number of turns of wire. To calculate the inductance L the formula $L = n^2 A_L$ is used, n being the number of turns and A_L the inductance of the core when just one turn of wire is used. For example, a core is available with an inductance of 400 nH, i.e. 400×10^{-9} H. When wound with 500 turns of wire the inductance is

$$L = 500^2 \times 400 \times 10^{-9} = 0.1 \text{ H}$$

Inductors used for radio frequencies may be colour coded in the same way as resistors (see table 2.3), the value of the inductance being in μH.

2.3.6 Semiconductor devices and integrated circuits

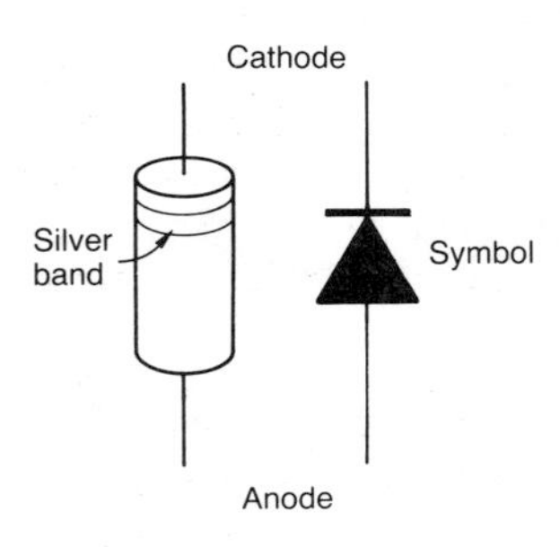

Figure 2.31 *Diode*

Diodes are circuit components which basically have a high resistance to current flow in one direction through them and a low resistance to current flow in reverse direction. With many diodes, a silver band is round the end of the diode referred to as the cathode (Figure 2.31), the other end being termed the anode. If the anode is made positive with respect to the cathode, the diode has very little resistance and a current will flow. This is referred to as *forward bias*. If the anode is made negative with respect to the cathode, virtually no current will flow. This is referred to as *reverse bias*. When connected in the reverse direction, if too high a voltage is used then breakdown occurs. Generally this is disastrous for the diode, but in the case of the Zener diode the design is such that it can be used in the breakdown region without damage. The main types of diodes used in modern electronic circuits are:

1 *Signal diodes* This term is used for all diodes which have been designed for use in circuits where large currents and/or voltages do not occur.

2 *Power diodes* These are generally used for the conversion of alternating current to direct current. This is termed *rectification* and the diodes are called *rectifiers*.

3 *Zener diodes* A Zener diode is a special form of diode which has a pre-determined reverse breakdown voltage. When this voltage is reached, the diode which had been previously of high resistance suddenly becomes low resistance. Zener diodes are available with a range of breakdown voltages, e.g. 2.7 V, 4.7 V, 5.1 V, 6.2 V, 6.8 V, 9.1 V, etc. up to 200 V.

4 *Varactor diodes* A varactor diode is used for its capacitance. It is designed to be operated with voltage applied in the reverse direction. The capacitance of the diode is then inversely proportional to the square root of the voltage.

The *transistor* is a semiconductor device that can be used to amplify an electrical signal or act as an electronic switch. Transistors are generally classified in the following groups:

1 Small-signal, low-frequency
2 Low-power and medium-power, low-frequency
3 High-power, low-frequency
4 Small-signal, high-frequency
5 Medium- and high-power, high-frequency
6 Switching

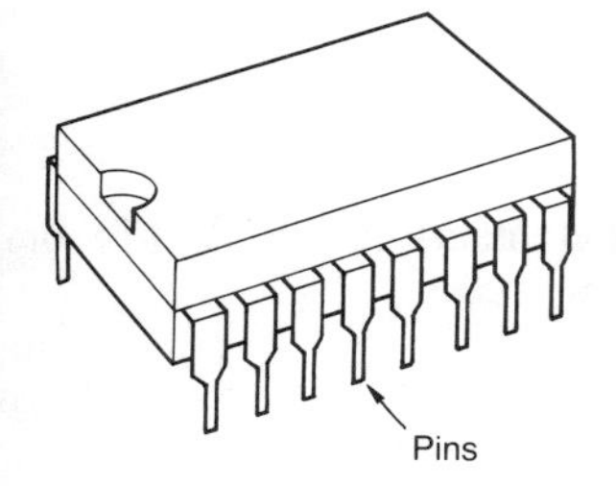

Figure 2.32 *Integrated circuit*

The term *integrated circuit* is used for a, very often complex, electronic circuit involving resistors, capacitors, diodes, transistors, etc. all on a single chip of semiconductor material. A typical integrated circuit may contain very large numbers of such components, perhaps hundreds or more of transistors, in a chip only a few millimetres across. Integrated circuits have had a profound effect on the design of electronic products. They permit complex electronic circuitry to be contained in a very small space, are reliable and are cheap. The chip is mounted in a package with connections made to the various parts of its circuit via external pins, Figure 2.32 showing a common form. The form shown is referred to as a DIL packaged integrated circuit since it has dual in line pins.

Semiconductor devices and integrated circuits are fairly easily damaged by physical or thermal shock and by electrostatic discharge. They are easily damaged by overheating during soldering and precautions have to be taken to avoid the heat from the hot solder reaching the component. Heat-absorbing pliers can be used to hold the leads during soldering. When you rub a piece of plastic against, say, nylon it becomes charged and can attract dust and small pieces of paper. Such static electric charges must be avoided when handling semiconductor devices since they can damage them. A common method of reducing such possibilities is for the person handling the components to wear an earthing strap which is connected to earth. Static charge thus has a path from the person to earth and so does not build up.

2.3.7 Semiconductor materials

Metals are good conductors of electricity because they have large numbers of free electrons which, under the action of an applied voltage, can move through it. Pure semiconductors have, at room temperature, only a relatively few free electrons. The electrons have been released from atoms and leave behind 'holes' into which electrons can move. With a pure semiconductor there are as many holes as free electrons. However, by adding small amounts of certain materials, the process is termed *doping*, we can change this situation. If we end up with more free electrons than holes we have n-type material, if we have more holes than free electrons p-type material. The semiconductor diode consists of a single crystal of a semiconductor in which there is a transition from a p-type material to n-type material, hence the term *pn junction*. Such a junction can be made by taking a small chip of an n-type semiconductor and dissolving a material such as indium in part of the surface and so converting part of it to p-type.

Problems

Questions 1 to 20 have four answer options A, B, C and D. You should choose the correct answer from the answer options.

1 A permanent, not readily demountable, form of joint between two materials is provided by:

A A riveted joint
B A bolted joint
C A cotter pin joint
D A screwed joint

2 Decide whether each of the following statements is TRUE (T) or FALSE (F).

A bolted joint can be locked by:
(i) A plain washer.
(ii) A spring washer.

A (i) T (ii) T
B (i) T (ii) F
C (i) F (ii) T
D (i) F (ii) F

3 Decide whether each of the following statements is TRUE (T) or FALSE (F).

Screwed fastenings with fine, rather than coarse, threads are used where:
(i) Better locking action is required.
(ii) Maximum thread strength is required.

A (i) T (ii) T
B (i) T (ii) F
C (i) F (ii) T
D (i) F (ii) F

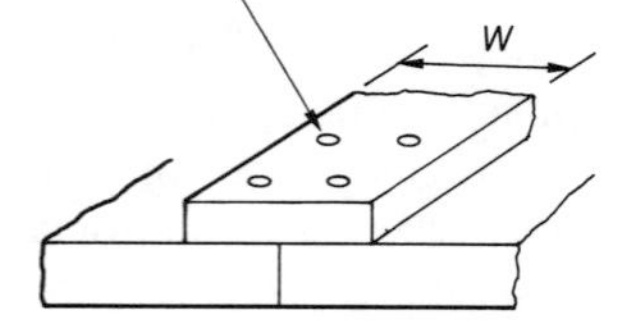

Figure 2.33 *Problems 4/5*

4 Decide whether each of the following statements is TRUE (T) or FALSE (F).

The riveted joint shown in figure 2.33 is:
(i) A butt joint.
(ii) Has a butt plate.

A (i) T (ii) T
B (i) T (ii) F
C (i) F (ii) T
D (i) F (ii) F

5 The width *W* for the rivetted joint shown in figure 2.33 is ideally:

A 1.5*D*
B 3.0*D*
C 4.5*D*
D 6.0*D*

6 Screwed joints hold two components in close contact because they bolt is made of a material with:

A High ductility
B High compressive strength
C High elasticity
D High hardness

Questions 7 to 10 relate to the following forms of joints:

A Slotted nuts so they are locked on a shaft.
B Collars to shafts.
C Two sheets of material in a butt joint.
D A demountable joint between two overlapping sheets of material.

Select the joint which frequently would be made with:

7 A cotter pin
8 Rivets
9 A bolt and nut
10 A split pin

11 Snap-fit joints made with plastics rely on the plastic having:

A High elasticity
B High ductility
C High hardness
D High strength

12 Decide whether each of the following statements is TRUE (T) or FALSE (F).

A resistor has printed on it the code 22KJ. This indicates that the resistor has:
(i) A resistance of 22 kΩ.
(ii) A tolerance of ±5%.

A (i) T (ii) T
B (i) T (ii) F
C (i) F (ii) T
D (i) F (ii) F

13 Decide whether each of the following statements is TRUE (T) or FALSE (F).

A resistor is marked with the colour bands: green brown black silver. These indicate that the resistance has:
(i) A resistance of 510 Ω.
(ii) A tolerance of 10%.

A (i) T (ii) T
B (i) T (ii) F
C (i) F (ii) T
D (i) F (ii) F

14 Decide whether each of the following statements is TRUE (T) or FALSE (F).

Electrolytic capacitors have the characteristics of:
(i) Only being able to be used with d.c. voltages.
(ii) Large capacitance values.

A (i) T (ii) T
B (i) T (ii) F
C (i) F (ii) T
D (i) F (ii) F

15 Decide whether each of the following statements is TRUE (T) or FALSE (F).

The criteria which have to be considered in selecting a material for a printed circuit board base are:
(i) Its electrical resistivity.
(ii) Its mechanical stiffness.

A (i) T (ii) T
B (i) T (ii) F
C (i) F (ii) T
D (i) F (ii) F

16 A resistor is required for a circuit where large currents, and hence large power dissipation, are involved. The optimum form of resistor is:

A Metal film
B Carbon composition
C Carbon film
D Wire-wound

Questions 17 to 20 relate to an electronic component which is to be manufactured in large quantities and involve a circuit containing capacitors, resistors, diodes and transistors mounted in a box.

17 The optimum choice of assembly method for the circuit is:

A With components connected by suitable cables.
B With components mounted on a matrix board.
C With components mounted on Veroboard.
D With components mounted on a printed circuit board.

18 Components/circuit boards are to be mounted in a box so that they can be replaced if, at some stage, they become defective. A suitable fixing device is:

A Rivets
B Cotter pins
C Screwed joint
D Adhesive

19 The material used for the box casing is required to be a good conductor of electricity. This suggests that:

A A plastic should be used.
B A metal should be used.
C A ceramic should be used.
D A composite of glass fibres in a plastic matrix should be used.

20 One of the capacitors to be used must be capable of being used with alternating current. A type of capacitor that should *not* be used in such a situation is:

A Mica capacitor
B Plastic film capacitor
C Electrolytic capacitor
D Ceramic capacitor

21 Examine three of the following products and identify the materials and fixing components used and any electrical components used for circuit mounting and connecting cables. Write a report detailing the components and materials used and reasons for their selection.
(a) A table lamp
(b) A mains electric plug
(c) An electric power drill
(d) The bonnet of a car
(e) A computer

3 Manufacturing processes

3.1 Manufacturing processes

In the manufacturing of engineered products, the main methods used to shape metals are:

1. *Casting* A product is formed by pouring liquid metal into a mould.
2. *Manipulative processes* A product shape is produced by deformation processes. This includes such methods as rolling, drawing, extrusion and forging.
3. *Powder techniques* A product shape is produced by compacting a powder.
4. *Cutting and grinding* A product shape is produced by metal removal.

The above shaping processes are one way of producing a product. Another way is metal joining, of which the main processes are:

1. Adhesives
2. Soldering and brazing
3. Welding
4. Mechanical fastening systems, e.g. rivets, bolts and nuts.

Polymers may be supplied in a powder, granule or sheet form, the supplier having mixed the polymer with suitable additives and even other polymers in order that, after processing, the finished material should have the required properties. The main processes used to shape polymers are:

1. *Casting* This can involve the mixing of the constituent parts of the plastic in a mould and then allowing the resulting chemical reaction to produce the polymer. This method can be used with thermosets and thermoplastics.
2. *Moulding* With thermoplastics the polymer might be melted and forced into a mould, the process being called injection moulding. With thermosets, the powdered polymer may be compressed between the two parts of the mould and then heated under pressure. This process is known as compression moulding. With transfer moulding the powdered thermoset is heated in a chamber before being transferred by a plunger into the mould.
3. *Forming* This process involves sheet polymers, thermoplastics, which are heated and pressed into or around a mould.
4. *Extrusion* With this process a thermoplastic polymer is forced through a die.

In addition, products may be formed by polymer joining. The main processes are:

1 Adhesives
2 Welding
3 Various forms of fastening systems, e.g. riveting, press and snap fits, screws.

The shaping or assembly method used for a particular product will depend on:

1 The material to be used. For example, a ductile metal can be shaped in ways that a brittle material cannot.
2 Its form of supply. The commercial forms of supply of materials might be as bar, sheet, ingot, or pellet (see section 1.3.2).
3 The tolerances required. The tolerance is the difference between the maximum limit of size and the minimum limit of size, being the amount by which deviation occurs from the desired dimension.
4 The surface finish required. Some processes might leave a rough or semi-rough surface while others leave a smooth surface.
5 The quantity of the product required. A process which might be economic for very large numbers of components might be very uneconomic for small numbers.
6 The cost of the product.

This chapter is an overview of the characteristics of the processes outlined above for metals and polymers.

3.2 Metal shaping processes

The following are brief outlines of the metal shaping processes: casting, manipulative processes, powder techniques and cutting. There is also a brief consideration of chemical machining.

3.2.1 Casting

Casting, in the case of metals, is the shaping of an object by pouring the liquid metal into a mould and then allowing it to solidify and form a product with the internal shape of the mould. The products formed in this way fall into two main groups, those for which the solid shape produced is just a convenient form for further processing and those for which the shape produced is that of the final product. Where the shape produced is for further processing, simple geometrical shapes are used and the products are known as ingots. Where the casting is used to produce the shape of the final product, with perhaps only some surface finishing to give the required product, a mould is used which has the appropriate internal shape. The mould has to be designed in such a way that the liquid metal can easily and quickly flow to all parts. This has implications for the finished casting in that sharp corners and abrupt changes in sections have to be avoided, the metal having difficulty in flowing into such parts of the mould (Figure 3.1).

A number of casting methods are available, the choice of method depending on such factors as:

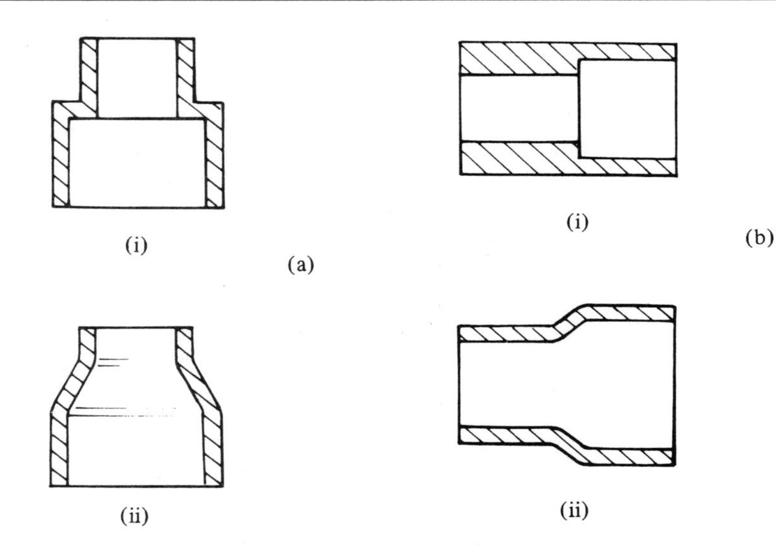

Figure 3.1 *(a) The need for no sharp corners means that (ii) is preferred to (i), (b) the need for no abrupt changes in section means that (ii) is preferred to (i).*

1 The size of casting required.
2 The number of castings required.
3 The complexity of the casting.
4 The mechanical properties required of the casting.
5 The surface finish required.
6 The dimensional accuracy required.
7 The metal to be used.
8 The cost per casting.

Two commonly used methods are sand casting and die casting.

Sand casting involves using a mould made of a mixture of sand and clay. The mixture is packed around a pattern made of the required casting to give the mould, which is usually made in two main parts so that the pattern can be extracted (Figure 3.2). The mould must be designed so that when the liquid metal is introduced into the mould, all air or gases can escape and the mould can be completely filled. After the casting has solidified, the mould is broken open. The mould is used for just one casting. Some machining is generally necessary with this form of casting, such as the trimming off of the metal in the feeder and riser. This method can be used for a wide range of casting sizes and for simple or complex shapes with holes and inserts. However, the mechanical properties, surface finish and dimensional accuracy of the casting are limited. The cost of the mould is relatively cheap but, since the mould can only be used for one casting, it is a cost which has to be borne by every item made. For small number production, sand casting is the cheapest casting process, but with large production runs this mould cost for each item produced adds up and makes the process uneconomic.

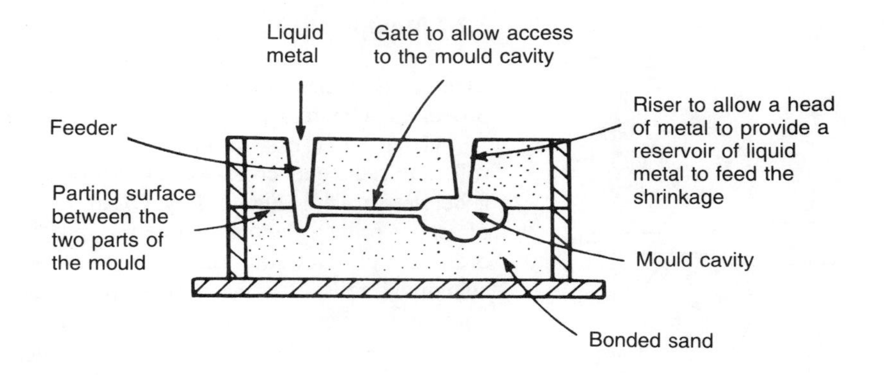

Figure 3.2 *Sand casting*

Die casting involves the use of a metal mould. With *gravity die casting* the liquid metal is poured into the mould, gravity being responsible for causing the metal to flow into all parts of the mould. With *pressure die casting* the liquid metal is injected into the mould under pressure. This enables the metal to be forced into all parts of the mould and very complex shapes with high dimensional accuracy can be produced. There are limitations as to the size of casting that can be produced by pressure die casting, the mass of the casting generally being less than about 5 kg. Gravity die casting can, however, be used for larger castings. Unlike the mould in sand castings, the metal mould can be used for many castings and so its cost can be defrayed over a large number of castings. The metal mould is more expensive than the sand casting mould, thus it is not economic for one-off castings or small runs but becomes economic when large production runs are involved. The castings produced by this method have very good mechanical properties, dimensional accuracy and finish.

With regard to the choice of casting metal, think of the problems of pouring treacle into a mould compared with pouring water. With the water it is much easier to fill all parts of the mould. Thus alloys used for casting where gravity is used to get the liquid metal to flow, as in sand casting and gravity die casting, have their alloying constituents chosen to give good flowing properties, i.e. low viscosity. Where pressure is used to force the liquid metal into the mould, as in pressure die casting, then a more viscous alloy can be used.

Casting is likely to be the optimum method for manufacturing a product when:

1 It has a large internal cavity. There would be a considerable amount of metal to be removed if machining were used. Casting removes this need.
2 It has a complex internal cavity. Such an item might be difficult or perhaps impossible to machine.
3 It has a complex shape. Casting may be then more economical than machining or assembling from a number of individual parts.
4 It is made of a material which is difficult to machine.
5 The metal used is expensive and so there is to be little waste.

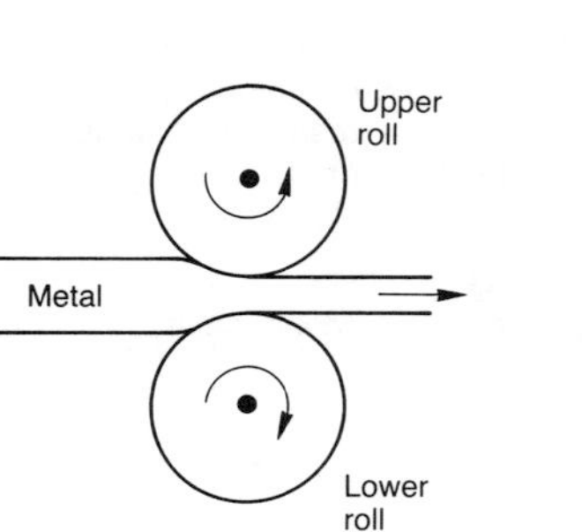

Figure 3.3 *Rolling*

3.2.2 Manipulative processes

Manipulative methods involve the shaping of a material by deformation processes. The main processes are rolling, drawing, forging and extrusion.

Metal is first cast into ingots. The hot ingots are then passed between rollers to reduce its cross-section. *Rolling* is the shaping of metal by passing it between rollers (Figure 3.3). Often the metal passes through a sequence of sets of rollers, the material getting progressively thinner at each stage, in order to produce the required shape and size product. With parallel cylindrical rolls, flat sheet or strip can be produced in long lengths. If contoured rollers are used, channel sections, rails, etc. can be produced. Figure 3.4 shows the shapes of rolls that might be used at the various stages in the rolling of a H-shaped section and some examples of the shapes of rolled sections.

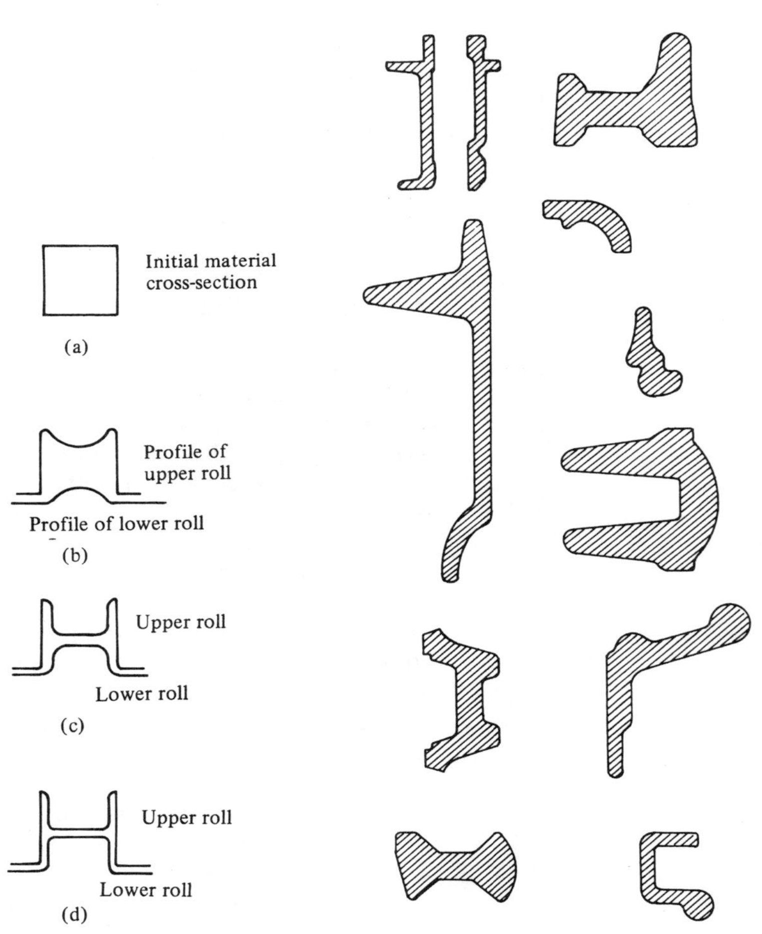

Figure 3.4 *Rolls sequence for H-shaped (a), (b), (c), (d) and examples of rolled sections*

Steel products produced by rolling hot material have a reddish blue, oxide, mill scale on the surfaces. However, after hot rolling the surface scale may be removed by pickling, i.e. immersing the material in an acid bath, and then a cold rolling process used to finish off the product. This improves the surface finish, gives better dimensional accuracy and improves the mechanical properties.

Drawing involves the pulling of metal through a die, Figure 3.5 illustrating this process for wire drawing. *Deep drawing* (Figure 3.6) involves sheet metal being pushed through an aperture by a punch. The more ductile metals such as aluminium, brass and mild steel can be used to shape products by this method.

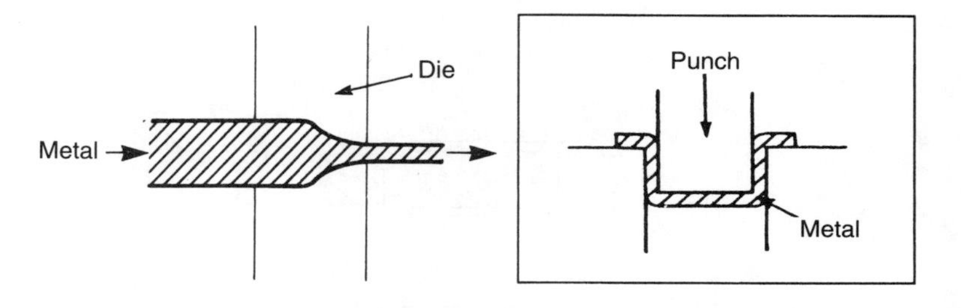

Figure 3.5 *Drawing*

Figure 3.6 *Deep drawing*

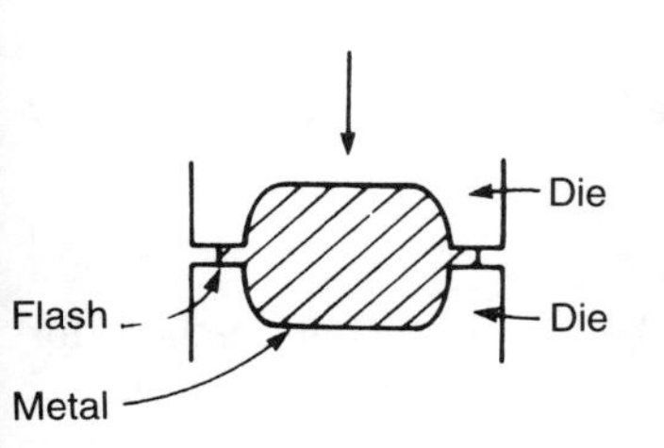

Figure 3.7 *Closed die forging*

Forging involves squeezing, either by pressing or hammering, a ductile metal between a pair of dies so that it ends up assuming the internal shape of the dies. Figure 3.7 illustrates this process in relation to what is termed closed-die forging. In order to fill the cavity completely, a small excess of metal is allowed to squeeze outwards to form a flash which is later trimmed away. Generally, for the material to be ductile enough during the operation the forging has to take place at a high temperature. Because of the cost of dies, the cost of forging is high for small number production. Since the dies can be used for many items, large number production is necessary to spread the cost and make production economic.

Extrusion is rather similar to the squeezing of toothpaste out of the toothpaste tube. The shape of the extruded toothpaste is determined by the shape of the nozzle through which it is ejected. With extrusion, hot metal is forced under pressure through a die, i.e. a shaped orifice. Two basic methods are used, direct extrusion and indirect extrusion. Figure 3.8 shows the basic principles and examples of some of the sections that can be produced. Considerably more complex sections are possible. The process is used extensively with aluminium, copper and magnesium alloys to produce very complex shaped sections and with steel for less complex sections. High production rates are possible and the products have a good surface finish. Extrusion dies are expensive and thus large quantity production is necessary so that the cost can be spread.

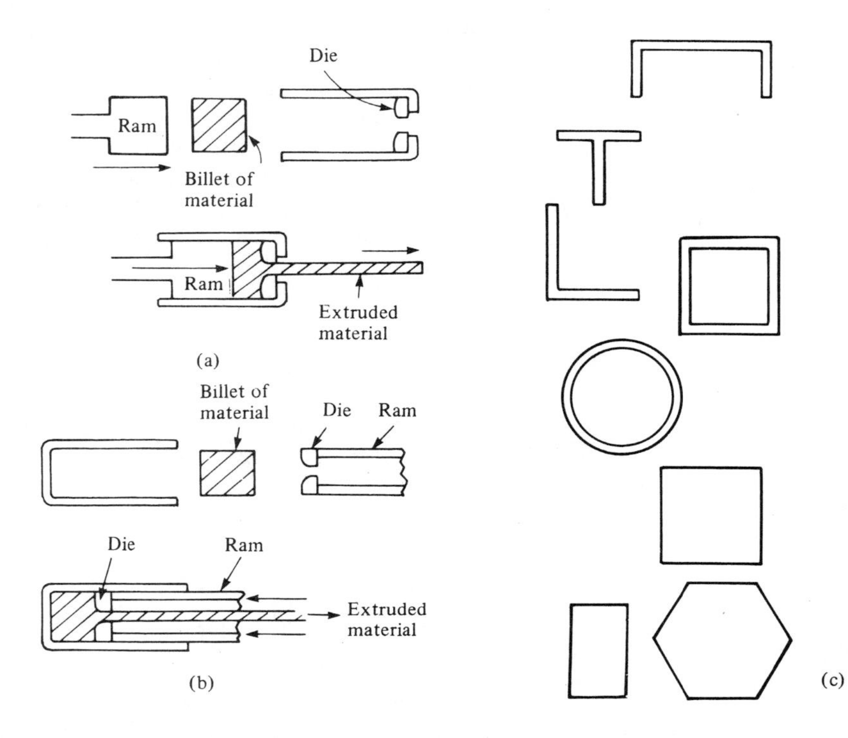

Figure 3.8 *(a) Direct extrusion, (b) indirect extrusion, (c) examples of sections*

Manipulative processes are likely to be the optimum method for manufacturing a product when:

1 The part is to be formed from sheet metal. Drawing or bending may be appropriate.
2 Long lengths of constant cross-section are required. Extrusion or rolling can be used.
3 The part has no internal cavities. Forging can be used.
4 Seamless cup-shaped objects or cans have to be produced. Deep drawing can be used.

Example

Suggest a process for the production of long lengths of railway tracks.

The processes that could be used for long lengths are rolling and extrusion. However, for such large cross-sections and with steel, rolling is used.

Example

What types of properties are required for metals used with the deep drawing process?

For deep drawing, high ductility is required. Thus, for example, in the case of carbon steels only those with low amounts of carbon can be used as only they have sufficient ductility. Thus mild steel can be used but not a high-carbon steel, it being too brittle.

3.2.3 Powder techniques

Shaped metal components can be produced from a metal powder. The process, called *sintering*, involves compacting the powder in a die and then heating it to a sufficiently high temperature for the particles to knit together. Sintering is a useful method for the production of components from brittle materials for which manipulative processes would be difficult and high melting point materials for which melting for casting becomes too expensive.

3.2.4 Cutting

All cutting tools, whether for hand or machine use, conform to the same basic principles. Thus, for example, we might refer to a cold chisel or a drill bit or a milling cutter. In all cases the tool is defined by certain angles (Figure 3.9):

1 The *wedge angle* is the angle to which the tool is ground.
2 The *clearance angle* is the angle between the work piece and the base of the tool. The reason for having such an angle is to prevent the tool rubbing on the cut surface of the work piece. The angle is typically of the order of 5° to 7°.

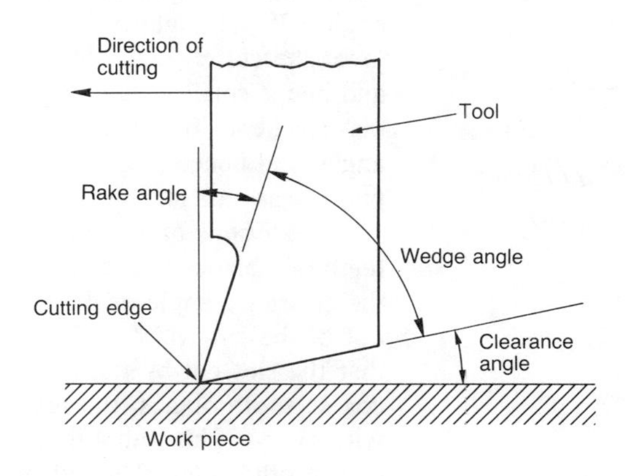

Figure 3.9 *Tool angles*

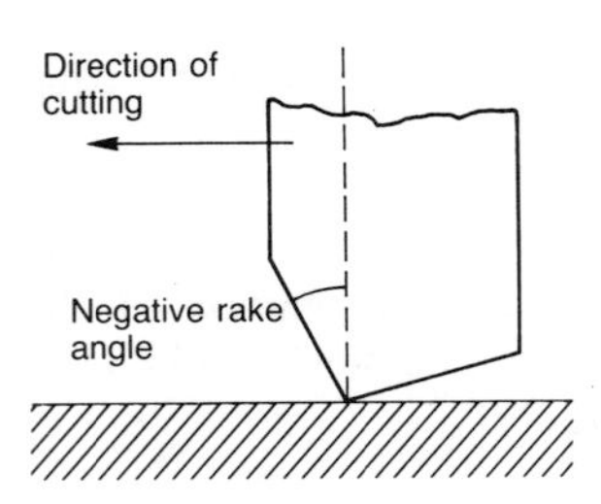

Figure 3.10 *Negative rake angles*

3 The *rake angle* is the angle between a line at right angles to the tool and the upper surface of the tool. This angle controls the cutting action of the tool. The smaller the rake angle the smaller the wedge angle and so the weaker the tool. The rake angle shown in Figure 3.9 is said to be a positive rake angle. Figure 3.10 shows a negative rake angle. With positive rake angles, the sum of the rake angle, the wedge angle and the clearance angle is 90°.

For a ductile material such as an aluminium alloy, the rake angle may be about 30°; for mild steel, also ductile, about 25°. For a less ductile steel, e.g. medium carbon steel, the angle may be about 15°. For a brittle material such as cast iron, the angle may be 0°. When a tool with zero rake angle is used to cut a brittle metal, it will cut quite easily and give a good, smooth surface finish. However, if it is used on a more ductile metal, a rough finish is obtained and tearing of the metal occurs. The optimum rake angle thus depends on the material being cut. Negative rake angles (Figure 3.10) are used with very hard but brittle cutting tool materials. This is necessary if the tool is to have sufficient strength.

The metal removed from the work piece by the tool is called the *chip*. This can take the form of small granules of metal in the case of relatively brittle materials, e.g. cast iron, to almost continuous ribbons of metal for highly ductile materials. The chips removed by the tool are also referred to as *swarf*. The large rake angles used with ductile materials allows the continuous chip to flow over the rake surface of the tool, a smaller angle would hinder this and make it more difficult to use the tool. The zero rake angle with cast iron, a brittle material, is because granular chips are produced and there is no need for them to flow over the face of the tool, they just break away. Cutting tool materials need to be harder than the work piece, otherwise the tool is very readily blunted when used. However, a very hard tool is likely to be brittle and this can present problems with tool breakage.

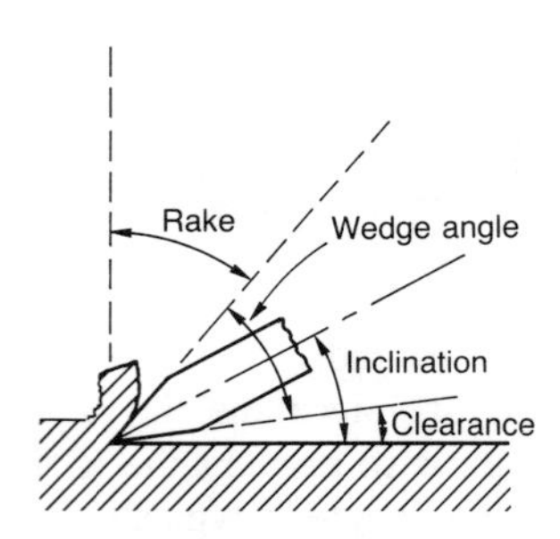

Figure 3.11 *Cold chisel*

Consider the use of a *cold chisel*. Figure 3.11 shows the cutting action. The angle at which the tool is inclined to the work piece surface is important. Changing the angle of inclination changes the rake and clearance angles. If the angle of inclination is too small the chisel point rises, if too great the chisel point digs in. For cutting soft materials a large rake angle, and hence small wedge angle, can be used and the angle of inclination can perhaps be as low as 25°. A hard material, however, requires a smaller rake angle, and hence larger wedge angle, and so a smaller angle of inclination, perhaps as high as 35°.

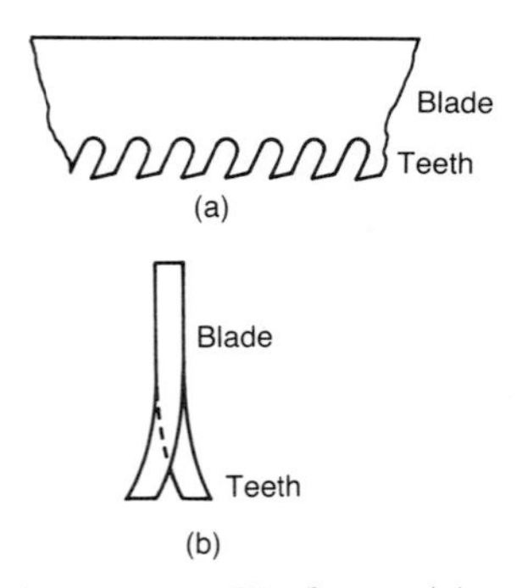

Figure 3.12 *Hacksaw, (a) form of teeth, (b) set of teeth*

A *hacksaw blade* is another example of a tool. Figure 3.12 shows the teeth of a blade, each tooth being a cutting tool. With a hand-held hacksaw, the clearance angle of the teeth is made larger so that chips may be carried out of the cut without clogging the saw. Saw teeth are also given a *set* so that the slot cut by the blade is slightly wider than the blade thickness. In coarse tooth blades, the teeth are bent alternately to the left and the right with every fifth tooth left straight in order to clear the slot being cut. With finer tooth blades, the teeth are wave set.

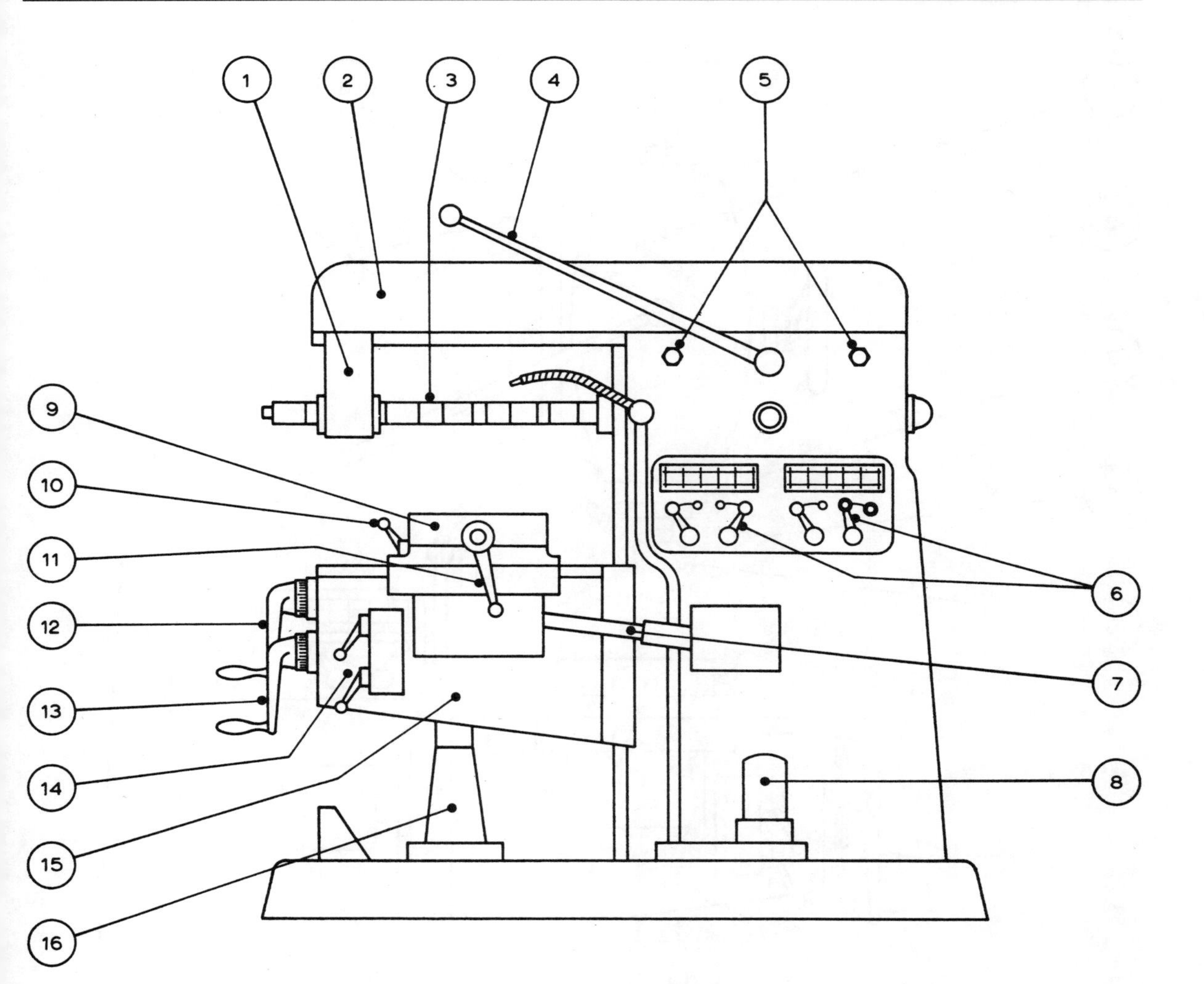

1 Arbor support or yoke (which slides on overarm)

2 Overarm

3 Arbor (on which the cutter is mounted)

4 Clutch lever

5 Overarm clamping screws

6 Speed and feed selector switches

7 Telescopic feedshaft

8 Coolant pump

9 Table which contains tee slots for clamping work or a vice

10 Automatic table feed (longitudinal)

11 Longitudinal feed (hand)

12 Vertical feed (hand) which raises or lowers knee assembly

13 Cross or transverse handle (hand)

14 Vertical and horizontal automatic-feed levers

15 Knee which supports table and moves up and down on dovetail slides

16 Knee elevating screw

Figure 3.13 *Horizontal milling machine*

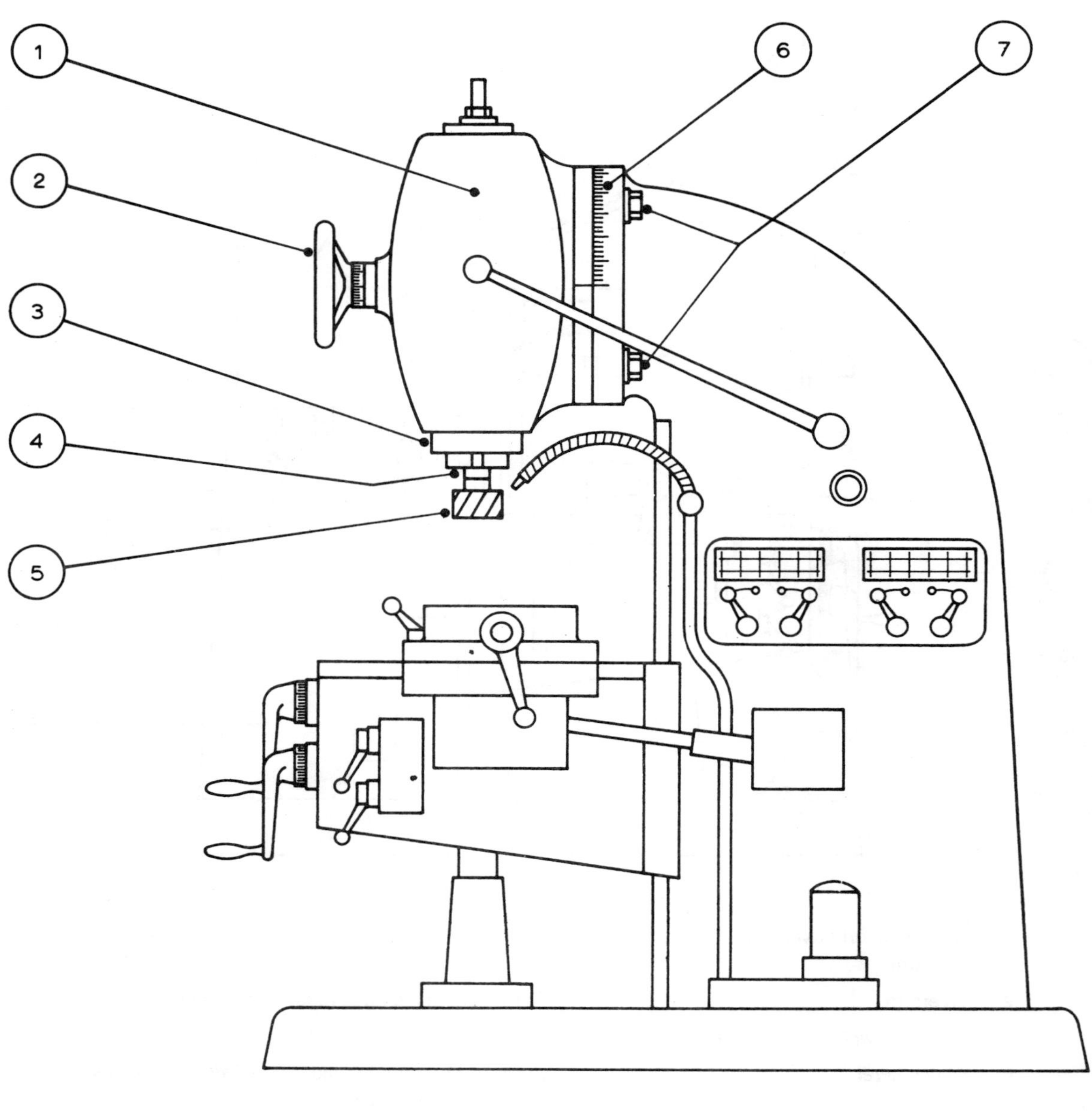

1 Vertical head which tilts
2 Vertical feed handwheel
3 Quill
4 Spindle
5 Cutter
6 Head tilts here
7 Head locking nuts

Note: Remaining parts are similar to the horizontal-type milling machine in Figure 3.13.

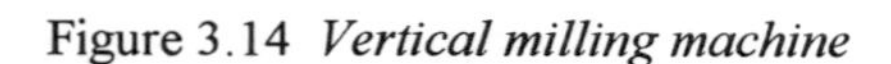

Figure 3.14 *Vertical milling machine*

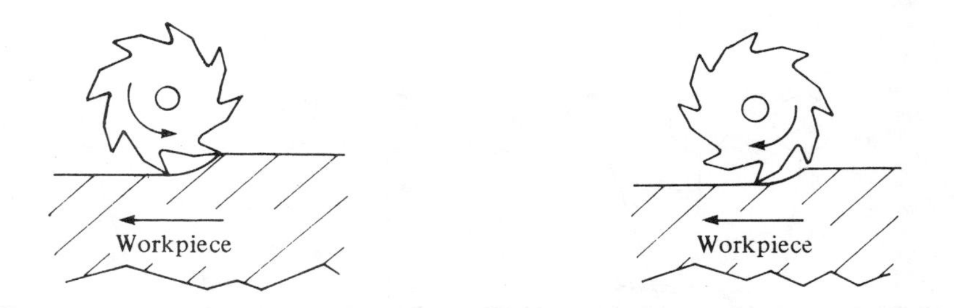

Figure 3.15 *(a) Up milling, (b) down milling*

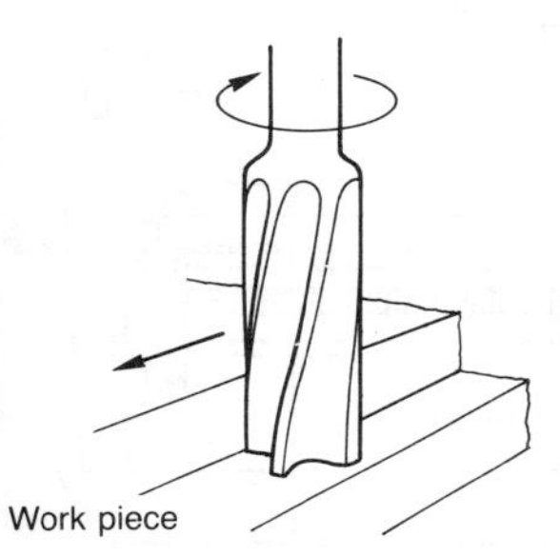

Figure 3.16 *An example of a milling cutter for vertical milling*

A *milling machine* has a cutter with multiple cutting edges, each edge taking its share of the cutting as the work piece is fed past them. As with the hacksaw blade, provision has to be made for chip clearance to avoid clogging. Figure 3.13 shows the basic form of a horizontal milling machine, Figure 3.14 that of a vertical milling machine. With a horizontal milling machine flat surfaces may be generated by the tool action shown in Figure 3.15. Up milling is more commonly used than down milling (often termed climb milling). For a vertical milling machines flat surfaces may be generated by the tool action shown in Figure 3.16.

Milling can be used, with appropriate tools, to produce a wide variety of shapes. For example it can be used to:

Plane surfaces, parallel or at right angles to the base face
Plane surfaces at an angle relative to the base surface
Cut keyways or slots
Cut helical flutes and grooves
Cut circular forms
Cut holes
Cut thread forms
Cut gear teeth

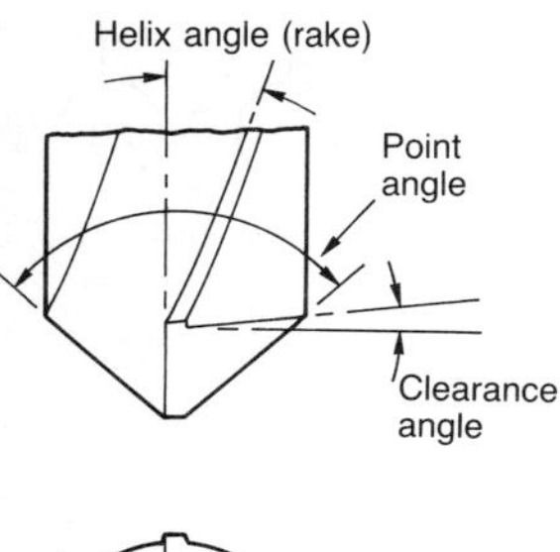

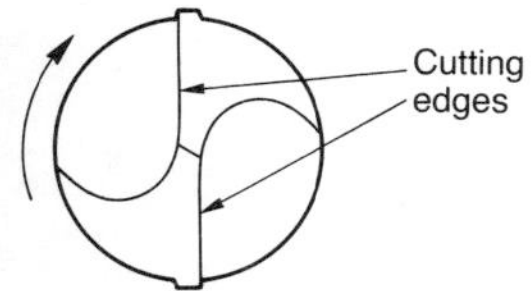

Figure 3.17 *Twist drill*

A *twist drill bit* (Figure 3.17) is another example of a tool. The helix angle represents the rake angle at the outer edge of the lip of the drill. For drilling a soft material such as aluminium, a helix angle of about 40° with a point angle of 90° is used. For a harder material, such as mild steel, the helix angle might be 25° and the point angle 118°. For a hard material, such as cast iron, the helix angle might be 0 to 10° and the point angle 118°.

Turning involves a tool being pressed against a rotating work piece. The machine tool used for turning is the *lathe*. There are several types of lathe. The centre lathe (Figure 3.18) is a versatile machine tool but not ideal for rapid mass production because of the time required for changing and setting tools and making measurements on the work piece. Turret and capstan lathes are more appropriate for mass production. With the centre lathe, the work piece is located and rotated about its axis by the spindle of the lathe. The rotation moves the work piece against the tool. The tool is set to give the required depth of cut by moving it along an axis at 90° to the

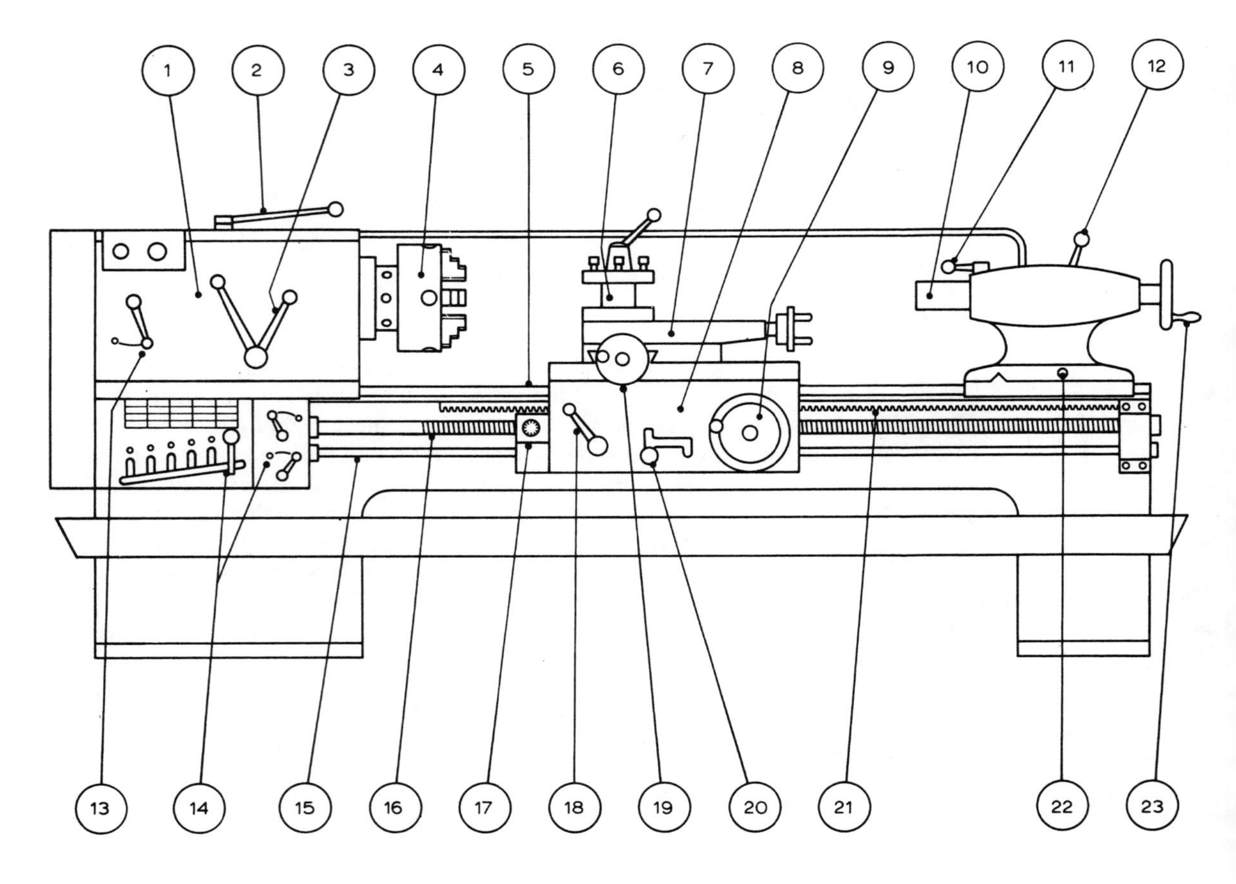

1 All-geared headstock
2 Clutch lever
3 Speed change levers
4 Chuck
5 Bed slideways
6 Four-way turret
7 Compound slide
8 Apron
9 Sliding handwheel
10 Tailstock sleeve
11 Sleeve locking lever
12 Tailstock clamping lever
13 Feed reverse lever
14 Quick-change gearbox
15 Feedshaft
16 Leadscrew
17 Screwcutting dial
18 Screwcutting lever
19 Cross-slide handwheel
20 Feed engage lever
21 Rack
22 Tailstock offset screw
23 Tailstock handwheel

Figure 3.18 *Centre lathe*

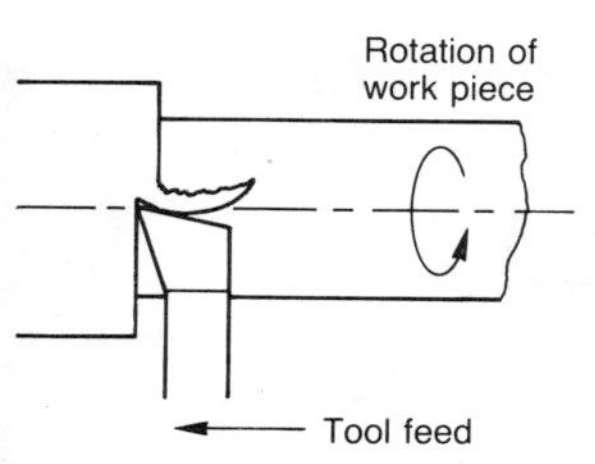

Figure 3.19 *Turning*

work piece axis (Figure 3.19). With parallel turning, the tool is fed along a path which is parallel to the axis of rotation of the work piece. For taper turning, the tool moves along a path which is at an angle to the axis of rotation of the work piece.

In carrying out a cutting process with a machine tool, it is common practice to use a *cutting fluid*. This is a suitable liquid which acts as a coolant and a lubricant for the tool against the work piece. It also can carry away the chips and other waste matter from the cutting area. When the cooling aspect is the most important then a fluid based on 2 to 10% of oil in soft water might be used, where lubrication is important then a straight oil might be used.

3.2.5 Chemical machining

The term *chemical machining* is used when material is removed from a work piece by chemical action. The metal which is to be removed is exposed to a suitable chemical, the metal which is not to be removed is masked by a chemical-resistant material. A common method used is to coat the work piece with a light-sensitive emulsion. A photographic negative of the master pattern is then placed against the work piece and it is exposed to light. The light which passes through the negative affects the light-sensitive emulsion and when the work piece is immersed in a developing agent, those areas that were not exposed to the light are removed, leaving a mask over the exposed part of the work piece. When the work piece is exposed to the chemical, those areas that are not masked have metal removed, but the masked areas are not affected. This method is used in the production of electronic circuit boards to leave conducting tracks, and with other components which are often very complex, small and rather thin, and would present problems if tackled in any other way.

3.3 Shaping polymers

The following are brief details of the common polymer forming processes of casting, moulding, extrusion and forming.

3.3.1 Casting

For polymers, casting is not like the casting of metals where hot liquid metal is poured into a mould. One form of polymer casting involves mixing, in a mould, the appropriate ingredients so that a chemical reaction occurs which leads to the solid polymer being produced. This process is used for encapsulating small electrical components, producing tubes, rods and sheets and can be used with both thermoplastics and thermosets. The term *cold-setting* is used.

3.3.1 Moulding

A very widely used process used with thermoplastics, and to some extent with thermosets, is *injection moulding* (Figure 3.20) In this process, the polymer is fed into the cold end of the injection cylinder. A rotating screw,

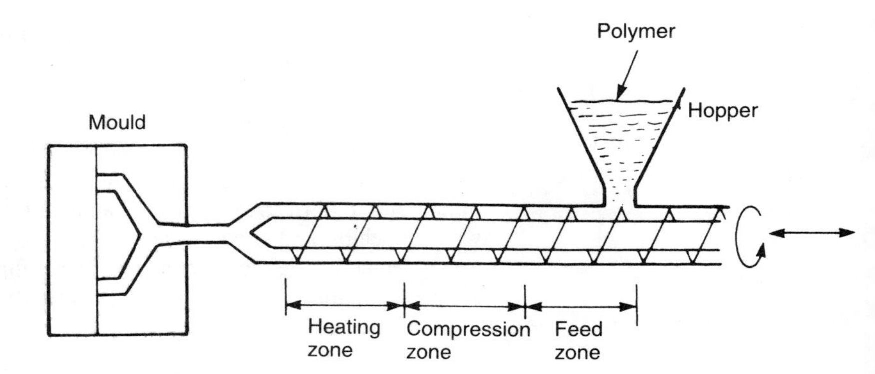

Figure 3.20 *Injection moulding*

or a ram, then compresses the material and passes it through a heated section before injecting it into the mould. When the material has sufficiently cooled, the component is ejected from the mould and the process can be repeated. Complex shapes with inserts, holes, threads, etc. can be produced but enclosed hollow shapes are not possible. Typical products are beer or milk bottle crates, toys, control knobs for electronic equipment, tool handles and pipe fittings.

3.3.2 Extrusion

Extrusion involves the forcing of liquid polymer, a thermoplastic, through a die. The process is comparable with the squeezing of toothpaste out of its tube. Figure 3.21 shows the basic form of the extrusion process. The polymer is fed into a screw mechanism which takes it through the heated zone and forces it out through the die. The operation is continuous, a steady source of liquid polymer being forced through the die. The process is used with thermoplastics for the production of pipes and various sections such as curtain rails, sealing strips and skirting boards.

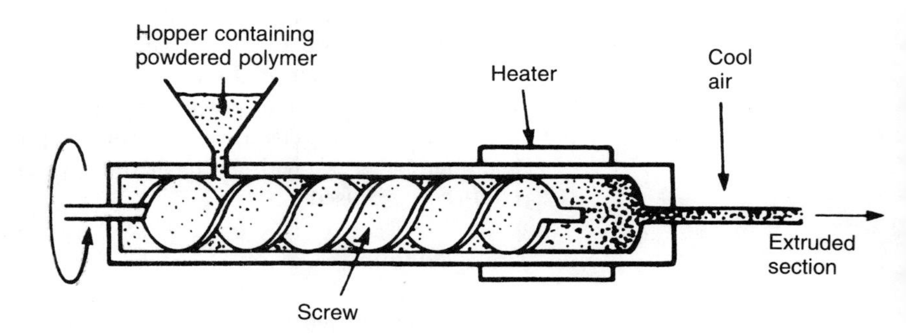

Figure 3.21 *Extrusion*

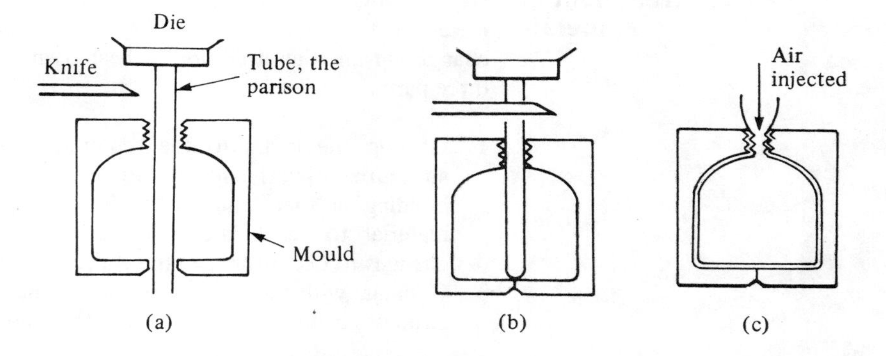

Figure 3.22 *Blow moulding: (a) tube extruded, (b) mould closed up and tube cut off, (c) air injected into the parison forces it to fill the mould*

An important extrusion process is *extrusion blow moulding*. This process is used for the production of hollow articles. The operation involves extruding a hollow tube. A mould is then placed round the hot tube, known as a parison, and closed up on it (Figure 3.22). The mould seals off the lower end of the tube and the top end is cut off with a knife. Air is then injected into the parison and causes the still hot parison to fill the mould. After hardening, the resulting shape is ejected when the mould is opened.

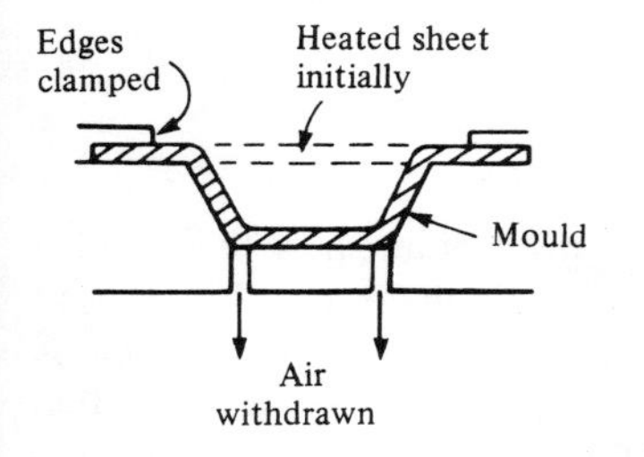

Figure 3.23 *Vacuum forming*

3.3.3 Forming

Forming processes are used to form articles from sheet thermoplastic polymer. The heated sheet is pressed into or around a mould. The term *thermoforming* is often used. The sheet may be pressed against the mould by the application of pressure on the sheet, this method being called thermoforming. Alternatively it can be by the production of a drop in pressure between the sheet and the mould, as illustrated in Figure 3.23. This method is called *vacuum forming*.

Example

Suggest a suitable manufacturing method that could be used to produce (a) a plastic kitchen sink, (b) a length of plastic tubing, (c) control knobs to screw onto threaded metal.

(a) Thermoforming the shape from a sheet of thermoplastic is an obvious method for such a large item.
(b) Long lengths of plastic tubing can be manufactured by estrusion, using an appropriate shaped die.
(c) Injection moulding can be used to produce such items.

3.4 Heat treatment of metals

Heat treatment can be defined as the controlled heating and cooling of metals in the solid state for the purpose of altering their properties by changes in internal structure. A heat treatment cycle consists normally of three parts:

1. Heating the metal to the required temperature for the changes in structure within the material to occur.
2. Holding at that temperature for a long enough time for the entire material to reach the required temperature and for the structural changes to occur through the entire material.
3. Cooling, with the rate of cooling being controlled since it affects the structure and hence properties of the material.

The basic heat treatment processes used with carbon steels are annealing, normalising, hardening and tempering.

3.4.1 Annealing and normalising

Annealing is the heat treatment used to make a metal softer and more ductile. It involves heating the metal to about 30 to 50°C above a temperature, called the *upper critical point*, at which the internal arrangement of the atoms change. This is then followed by very slow cooling, e.g. in the cooling furnace. *Normalising* is when the cooling takes place normally in the atmosphere. This is a faster rate of cooling than with annealing. With normalising the material is harder and less ductile than with annealing. The upper critical point ranges from about 700°C to 900°C, depending on the percentage of carbon present in the steel.

3.4.2 Hardening and tempering

If, after heating a steel to above the upper critical point, it is then cooled very rapidly, for example by being dropped into cold water or oil, the result is a hard, brittle material. This form of heat treatment is called *quenching*. A problem with the severe cooling occurring in quenching with water is that cracks can occur. These are a result of the distortion produced by the structural changes and also differential expansion as a result of different parts of a product cooling at different rates. Oil cools the steel a little more slowly and helps to prevent this happening.

In the quenched state steels have such a low ductility as to be very difficult to use. The process known as *tempering* can however be used to improve the ductility without losing all the hardness gained by the quenching. Tempering involves heating the metal to a temperature typically between about 200°C and 300°C. The higher the tempering temperature used the greater the amount of hardness and brittleness removed from the hardened steel. Table 3.1 indicates some of the temperatures used for various uses of hardened steel. The temperature can be estimated by looking, in the shade and not in bright light, at the colour of the heated steel.

Table 3.1 Tempering temperatures

Product	Temperature °C	Colour of steel
Scribers	200	Faint straw
Hacksaw blades	220	
Planing tools	230	Pale straw
Drills, milling cutters	240	Dark straw
Taps, dies	250	Brown
Punches, reamers	260	Brownish purple
Axes	270	Purple
Cold chisels, wood chisels	280	Dark purple
Screwdriver blades	290	
Springs	300	Blue

Example

List the sequence of processes that might be used to give a screwdriver blade the right degree of hardness.

The steps are likely to be: heating to about 30 to 50°C above the upper critical temperature, quenching in water or oil, then tempering by heating to about 290°C.

3.5 Surface finishes

The steel used for the bumper of a car is likely to have begun as a cast slab which is then rolled to sheet, coated with zinc to give corrosion protection, then formed to the required shape before finally being painted. All this is to give the required shape product with the required surface finish. A large proportion of products are given a form of decorative or protective surface treatment to complete their manufacturing process. Surface finishing procedures may involve:

1 Mechanical surface treatment involving the use of abrasives. The term *polishing* or *buffing* is often used for the process of reducing the roughness of a surface. The buffing process involves the work piece being brought into contact with a revolving buffing wheel that has been charged with a fine abrasive. The abrasive removes minute amounts of metal and so produces a smoother surface. In the case of soft metals, the action may cause some of the metal from peaks to be smeared out into troughs in the surface.
2 Modification of the surface layers of the material by heat treatment. This may by quenching in which the surface layers cool more rapidly than the core of the material and so produce a harder surface than core.
3 Chemical modification of the surface. This can involve diffusing carbon into the surface of a steel to produce a material with a higher carbon

surface content than in the core. The result is a harder surface layer. This treatment is termed *carburization*. Other treatments involve treating the surface oxide layers on steel by heating them and dipping them in oil, the term *blueing* being used for this process.

4 Applying surface coatings. This might be paint or metallic coatings. The paint may be air drying (e.g. cellulose, acrylic), or ones that dry by polymerisation, i.e. forming long polymer chains, or a combination of polymerisation and adsorption of oxygen. Metallic coating might be applied by electroplating. *Electroplating* involves making the work piece the negative electrode, i.e. the so-termed cathode, in a suitable electrolyte with the metal to be used for the coating as the positive electrode, i.e. the anode (Figure 3.24). When a current passes through the electrolyte, metal is removed from anode and deposited on the cathode. Tin, cadmium, chromium, copper, gold, silver and zinc are examples of metals that are used to electroplate products. In some cases, more than one material may be used to coat the work piece. Nickel is often used as an undercoating to gold and silver.

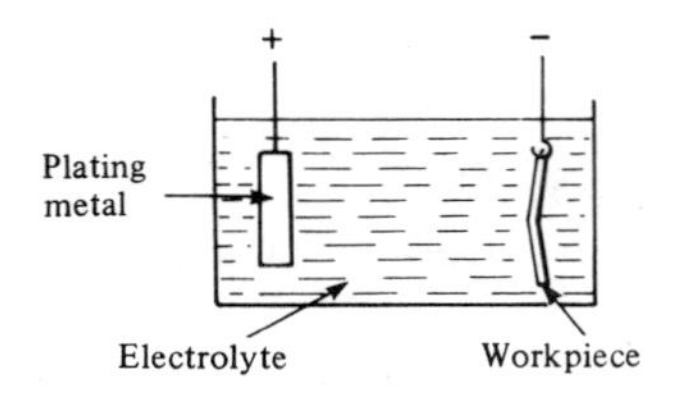

Figure 3.24 *Electroplating*

3.6 Joining and assembly

The main processes for assembly can be summarised as:

1 *Adhesive bonding* Types of adhesives used include natural adhesives, elastomers, thermoplastics, thermosets and two-polymer types.
2 *Soldering and brazing* A metal joining agent is used which is different from the materials being joined but alloys locally with them.
3 *Welding* Heat or pressure is used to fuse together the two materials being joined.
4 *Fastening systems* Fasteners provide a clamping force between the two pieces of material being joined, e.g. nuts and bolts, rivets.

Mechanical fastening systems were discussed in section 2.2.

3.6.1 Adhesives

The use of adhesives to bond materials together has the advantages over other joining methods of:

1 Dissimilar materials can be joined, e.g. metals to polymers.
2 Jointing can take place over large areas.
3 The bond is generally permanent.
4 Joining can be carried out at room temperature or temperatures close to it.
5 A smooth finish can be obtained.

The disadvantages are:

1 The optimum bond strength is usually not produced immediately but a curing time must be allowed.

2 The bond can be affected by environmental factors such as heat, cold and humidity.

It is essential for adhesives to have good wetting properties, i.e. they should spread over the surfaces being joined like water over a clean surface and not ball up or run off the surface like water over a greasy surface. Surface preparation is thus of importance. Joint surfaces are generally slightly roughened to improve adhesion, grease and oil being removed by the use of solvents. For the maximum strength bond to be realised, the maximum area of bonding should be used. Thus a lap joint should be used in preference to a butt joint (Figure 3.25).

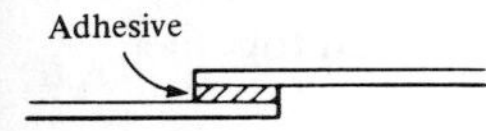

Figure 3.25 *Lap joint*

The term *natural adhesive* is used for vegetable glues made from plant starches, e.g. the type used on postage stamps and envelopes. Such adhesives, however, give bonds with poor strength and are susceptible to fungal attack and are weakened by moisture. They set as a result of solvent evaporation. *Elastomer adhesives* are based on synthetic rubbers and also set as a result of solvent evaporation. The adhesives have poor strength and are mainly used for unstressed joints and flexible bonds with plastics and rubbers. *Thermoplastic adhesives* include a number of different setting types. They consist of a thermoplastic material dissolved in a volatile solvent. Polyamides are widely used with metals, plastics, wood, etc. and have a wide application in rapid assembly work such as furniture assembly and the production of plastic film laminates. Another group are the 'super glues' which set in a matter of seconds. In general, thermoplastic adhesives are generally only used with assemblies subject to low stresses and have poor to good resistance to water but good resistance to oil. Araldite is an example of a *thermoset adhesive*. This adhesive is a two-part adhesive, in that setting only starts as a result of the chemical reaction which starts when the two components of the adhesive are brought together. They will bond almost anything and give strong bonds which are resistant to water, oil and solvents. Thermosets by themselves give brittle joints but combined with a thermoplastic or elastomer, a so-called *two-polymer adhesive*, a more flexible but strong joint can be produced.

3.6.2 Soldering and brazing

Soldering involves heating the joining agent, a metal of low melting point called the solder, together with the materials being joined until the solder melts and alloys with their surfaces. To do this, the surfaces must be clean and free from dirt, grease and surface oxides. A *flux* is used to produce such a clean surface. On cooling, the alloy solidifies and forms a bond between the two materials. In soft soldering, the solder is an alloy of lead and tin and, depending on the proportions of these materials, has a melting point between 183°C and 327°C.

The sequence of steps used in soft soldering is:

1 Apply flux to the joint surfaces.

2 Load the soldering iron bit with solder, i.e. apply the solder to the hot bit so that it melts and leaves a layer of liquid solder on the bit.
3 Transfer the solder to the separate joint surfaces. This means holding the soldering iron in contact with a joint surface until the surface becomes hot enough for alloying to occur with the solder. The surface is then said to be *tinned*.
4 Place the tinned joint surfaces in contact and apply the soldering bit to heat the surfaces so that the tinned surfaces melt and bond together.

There are two main types of flux, corrosive and non-corrosive. A corrosive flux, e.g. killed spirit (zinc chloride), scours the joint surfaces thoroughly and gives a sound joint. However, they leave a corrosive residue which has to be removed. This is done by washing the work piece. For this reason they are not suitable for the assembly of electrical circuits. Non-corrosive fluxes, e.g. resin, are widely used for electrical circuits (see section 3.7.1 for a discussion of soldering with electrical circuits). They do not act as vigorously as the corrosive fluxes and thus more care is needed in the preparation of the joint surfaces. Table 3.2 shows some of the commonly used fluxes.

Table 3.2 Fluxes

Type of flux	Use
Corrosive	
Dilute hydrochloric acid	Zinc galvanised steel
Zinc chloride (killed spirit)	Iron, steel, copper, brass; suitable for most metals
Non-corrosive	
Resin in methylated spirits	Electrical assemblies
Tallow	Lead and plumber's work
Olive oil	Pewter

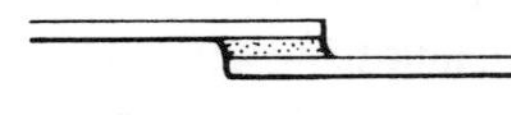

Simple overlap joint
(a)

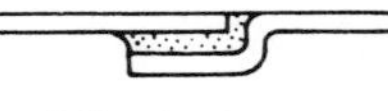

Offset lap joint
(b)

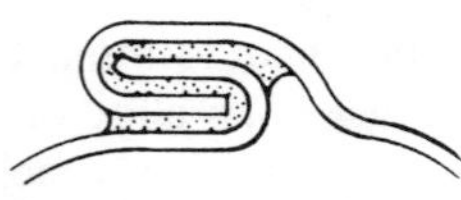

Double-lock joint (used for tin-plate cans)
(c)

Figure 3.26 *Solder joints*

Soft solders are weak structural materials when compared with the metals they join. There is thus a need to ensure that the strength of the soldered joint does no rely on the solder strength. The joints have thus to be designed so that the joining materials interlock or overlap in some way (Figure 3.26). This may involve just a simple overlap joint (Figure 3.26(a)) or, if a smooth surface is required, an offset lap joint (Figure 3.26(b)). Where a higher strength is required, a double-lock joint might be used (Figure 3.26(c)). The joints should be close fitting with a large area of contact and the solder film thickness kept to a minimum.

The term *soft soldering* is used when the temperature at which soldering occurs is below 450°C, the term *hard soldering* being used for soldering at higher temperatures. The term *brazing* is, however, often used

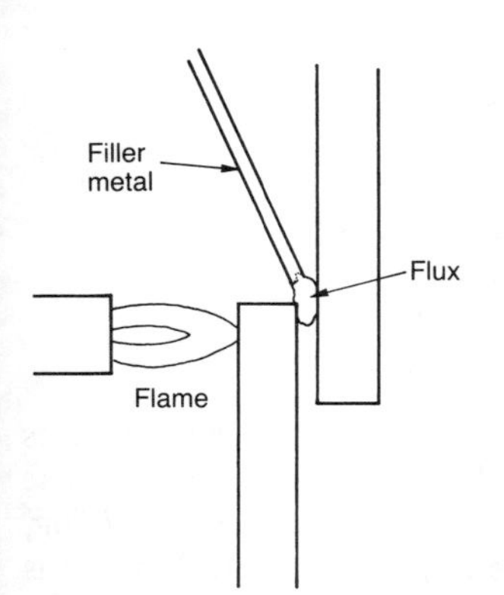

Figure 3.27 *Brazing*

instead of hard soldering. Originally, brazing involved only a copper–zinc alloy, i.e. a brass, as the filler, it having a melting point of between 860 and 890°C. The term 'braze' derives from the use of brass to make the joint. Now-a-days a wide range of materials are used as filler materials. The term silver soldering is sometimes used when the filler is a copper–silver alloy with a melting point of between 650 and 700°C. Hard soldering/brazing involves much higher temperatures than soft soldering. In the case of silver soldering, a blowpipe is used to heat the surfaces being joined. For brazing, a gas torch might be used to heat the surfaces being joined. A brazed joint has a high tensile strength and is tough.

With such processes, as with soft soldering, a flux is used to clean the surfaces and enable the filler material to alloy with the joint surfaces. The surfaces are heated and the flux and filler material drawn into the joints by capillary action (Figure 3.27). Borax is widely used as a flux for brazing fillers which melt about 800°C. Fluoride-based fluxes are often used with lower temperatures fillers. Alkali-halide types fluxes are used with aluminium and aluminium alloys. The purpose of the flux is to remove surface oxides and so clean the joint surfaces.

Table 3.3 shows some of the filler metals used and the joint materials for which they are recommended.

Table 3.3 Filler metals for torch brazing

Filler metal	Melting range °C	Joint materials
Copper–zinc	860–890	Copper, mild steel, carbon steels, alloy steels, stainless steels
Bronze	920–980	Copper, mild steel, carbon steels, alloy steels, stainless steels
Silver–copper	779	Copper, copper alloys,
Nickel–silicon–boron	950–1100	Mild steel, carbon steels, alloy steels, stainless steels
Aluminium–silicon	565–625	Aluminium and certain aluminium alloys

3.6.3 Welding

With brazing and soft soldering, the joint is obtained between two surfaces by inserting a metal filler between them. The metal filler has a lower melting point than that of the materials being joined. With *fusion welding*, the joint is obtained be melting the surfaces of the two parts being joined by the application of heat. Additional molten metal, usually of similar composition to the metals being joined, is added. Figure 3.28 shows the basic stages involved in such welding.

THE LIBRARY
GUILDFORD COLLEGE
of Further and Higher Education

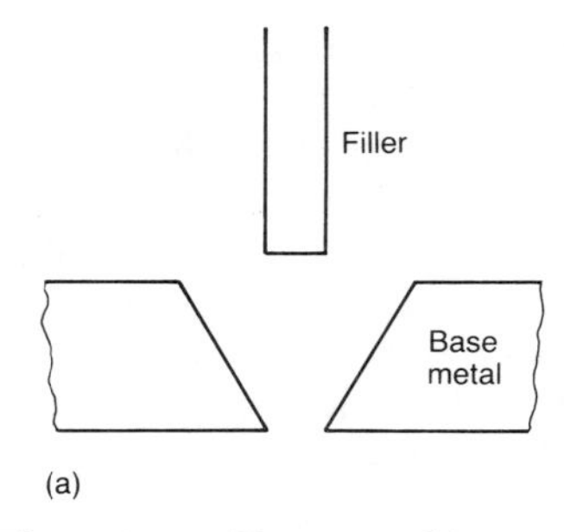

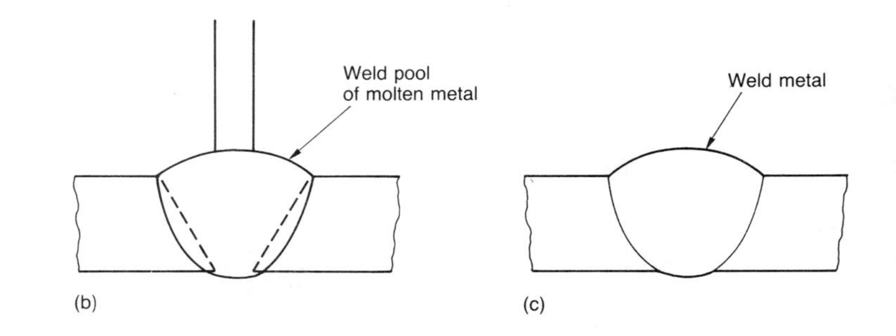

Figure 3.28 *Fusion welding*

In fusion welding, an external heat source is used to melt the interfaces of the joint materials, and the additional metal being added, and so cause the two materials to fuse together when they cool. In *oxyacetylene welding*, the heat is provided by an oxyacetylene flame (Figure 3.29(a)). No flux is used since the molten melting is protected by the products of the combustion of the flame. In *electric arc welding*, the heat is provided by an electric arc (Figure 3.29(b)). A flux is used, the surface surrounding the electrode being coated with it. During the welding, the flux melts. During welding, protective clothing must be worn and goggles or a face mask with coloured filters appropriate to the process being used.

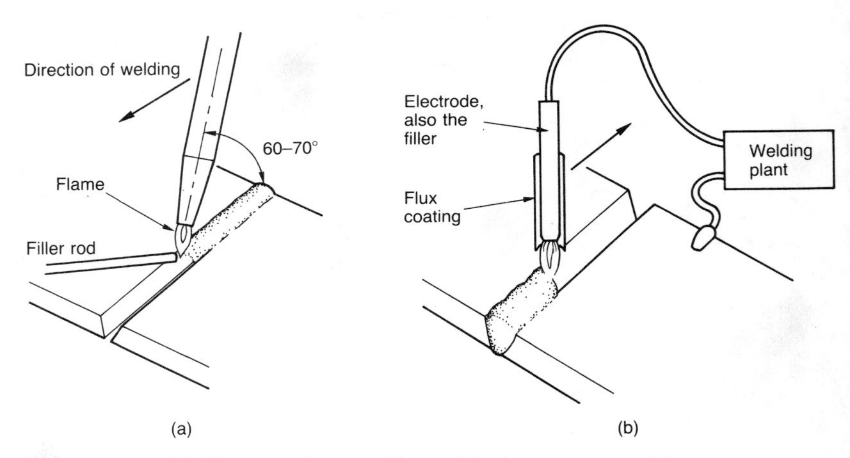

Figure 3.29 *(a) Oxyacetylene welding, (b) electric arc welding*

Figure 3.30 shows some of the forms of joints that can be made with welding. For a butt joint, the edges of the materials may need cutting to form a V with a gap being left between the edges to allow for adequate penetration of the filler metal (Figure 3.30(a)). With a single fillet weld (Figure 3.30(b)), the edge of the right-angled element is bevelled to allow for adequate penetration of the filler metal. With a double fillet weld (Figure 3.30(c)), the filler material is put both sides of the right-angle element. With a corner joint (Figure 3.30(d)) a corner gap is left for adequate penetration of the filler metal.

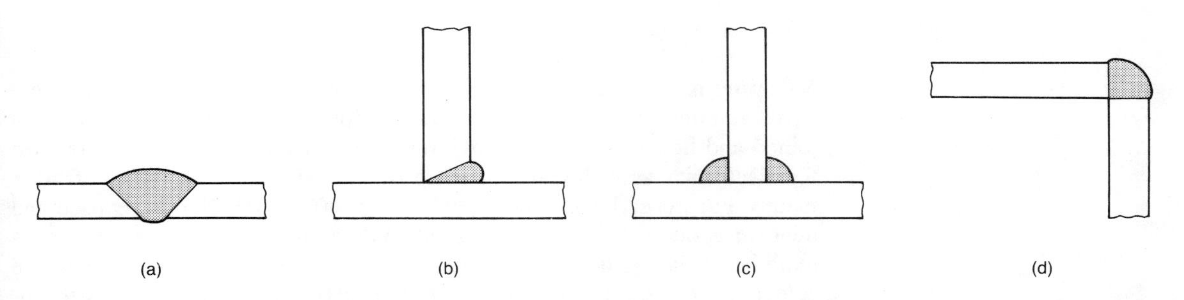

Figure 3.30 *Welding joints: (a) single Vee butt joint, (b) single fillet weld, (c) double fillet weld, (d) corner joint*

Welding is capable of high strength joints. However, because of the high temperatures involved in the welding process, there may be detrimental changes in the materials being joined. There may also be local distortions due to uneven thermal expansion, or internal stress changes, or structural changes in the metals.

3.7 Electrical wiring

An electronic appliance might consist of an integrated circuit, within which connections are made between individual components by means of etched tracks, connections by fine wires between the integrated circuit and the terminals of the package in which it is mounted, mechanical connections as a result of the integrated circuit block terminals being pushed into sockets, tracks on the printed circuit board to interconnect the components mounted on it, connections between the board and perhaps other boards via cables, connections by cables between boards/components and the controls and switches on a cabinet in which the circuit is mounted, connections between the mains supply and the cabinet via a cable and a plug and socket, and then perhaps connections between a number of cabinets, e.g. a hi-fi system where the record player is connected to an amplifier. The above illustrates some of the diversity of connections that have to be made with electronic circuits.

A number of methods are used in electronics for making electrical joints, i.e. joints which enable both a component to be fixed in a particular location and also give electrical conduction paths. These include:

1 *Soldering*
2 *Crimping*
3 *Wire wrapping*
4 *Mechanical fasteners*

The following are some of the common ways electrical connections are made.

3.7.1 Soldering

Soldering is used to make permanent connections. It is a process where a small amount of a metal, the solder, is run between the surfaces being joined and heated so that it alloys with each of the surfaces. Solders used for electronics assembly are made up of tin and lead, with traces of other metals. For general use, a 60:40 solder, i.e. 60% tin 40% lead, is used and melts at about 188°C. To make good soldered joints it is necessary for the materials being joined to be ones that are solderable. This requires the solder to come into contact with the bare metal and alloy with it. Aluminium presents problems in that, in air, the metal is coated with an oxide layer which prevents good bonding. While other materials, such as copper and brass, also oxidise in air, their oxide layers can be easily removed to enable good soldered joints to be made. A good soldered joint requires the surfaces which are to be joined to be free of oxide layers, grease and other surface contamination. This can be obtained by the use of a *flux*. This chemically removes the oxide layer and contamination and enables the flux to flow easily over the surface, the term used is 'wet' the surface. Fluxes used with electrical work tend to be resin based materials. For soldering by hand, solder wire with integral cores of flux is used (Figure 3.31). This enables the flux and the solder to be applied simultaneously. With many electronic components, the surfaces to be soldered may already have been pre-coated with solder or gold so that the surfaces are free of oxide layers and will readily bond. Such surfaces are said to have been *tinned*.

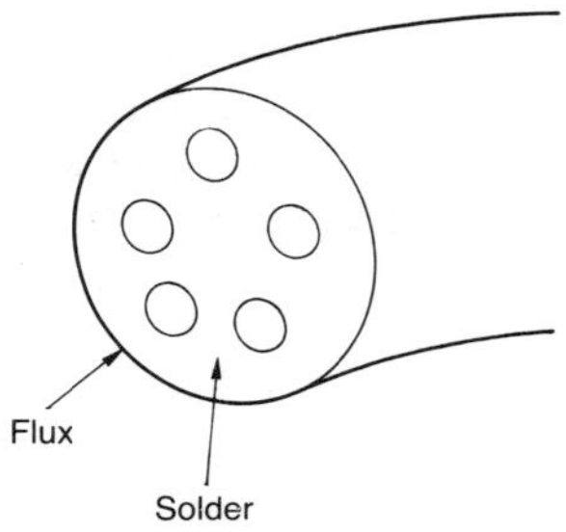

Flux

Solder

Figure 3.31 *Flux cored solder wire*

The sequence of operations used to make a soldered joint using a soldering iron are:

1 Wipe the surfaces clean.
2 Apply the hot soldering iron and fluxed solder wire to the joint so that the joint surfaces and the solder both come up to the melting point of the solder. The soldering iron should not be removed from the joint until the solder has flowed into the joint.
3 Remove the soldering iron and allow the solder to solidify without disturbance.

A bad joint, termed a 'dry joint', is likely to have poor electrical contact across the joint, with possibly a high electrical resistance, and give a poor mechanical joint, possibly falling apart with time. Dry joints occur because the surfaces being joined were dirty or covered with an oxide layer, too little flux was used in the cleaning of the surfaces, or the junction was not heated up to a high enough temperature for alloying to occur between the solder and the surfaces.

Solder is not a very strong material. Thus to improve the mechanical strength of a soldered junction, the wire is usually bent on passing through a connecting hole in a circuit board or a tag to form a hook shape. Figure 3.32 illustrates this for a circuit board. In (a) the connection is too weak, in (b) the connection is difficult to replace, (c), however, is the recommended form of junction.

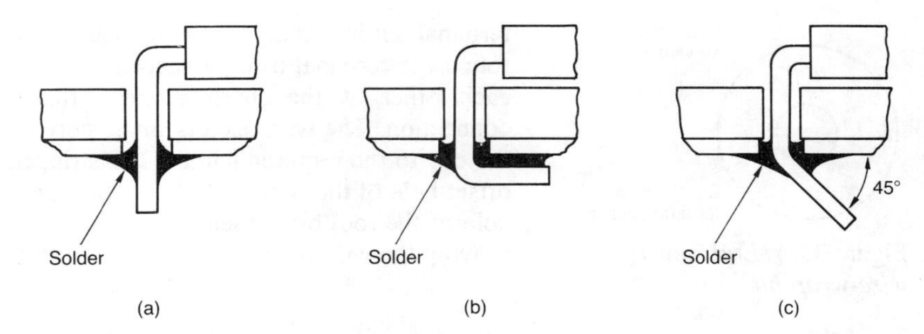

Figure 3.32 *Soldered junctions with circuit boards*

Electronic components tend to be damaged by heat. Thus applying a soldering iron to the connecting leads of a component presents a real danger that the component may be damaged. Thus, when components are being soldered into a circuit, the heat from the soldering iron must be diverted or 'shunted' away from the component. This can be done by placing the nose of a pair of pliers or a crocodile clip on the wire between the joint and the component.

Example

An electrical component is said to have tinned connecting wires. What does this mean?

It means that the connecting wires have already been coated with a layer of solder, or perhaps, gold. This means that good soldered joints are more readily obtained since surfaces oxide layers are not present.

3.7.2 Crimping

Crimping is a jointing technique used in electronics and electrical engineering in which the electrical contact between the joined surfaces relies on the mechanical junction between them. A crimping connection is made by crushing a special terminal onto the connecting wire using a purpose-made tool. The wire is then held by the clamping action of the terminal on the wire. It is a rapid way of making electrical connections. The technique is widely used in the car industry for joining wires in the car electrical wiring. Unsolderable wires can be joined in this way.

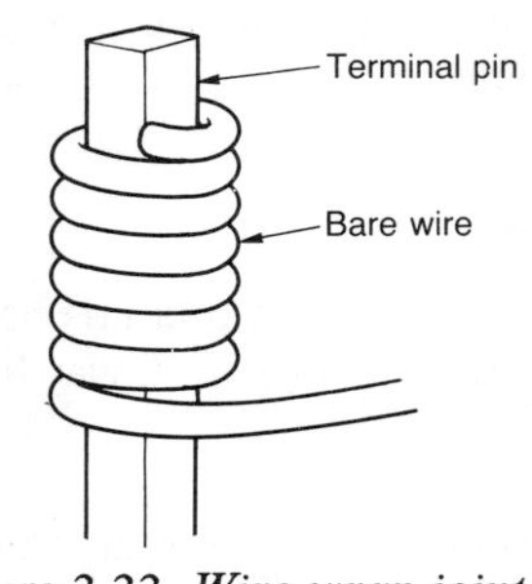

Figure 3.33 *Wire-wrap joint*

3.7.3 Wire wrapping

Wire wrapping is a faster and cheaper way of making electrical connections than soldering. It requires no heat and can be used to make connections to terminals located very close together. It also has the advantage that a wire can be readily disconnected using a special tool which unwraps the joint. The joint consists of about seven turns of bare wire and one or two turns of insulated wire wrapped round a special terminal pin (Figure 3.33). The

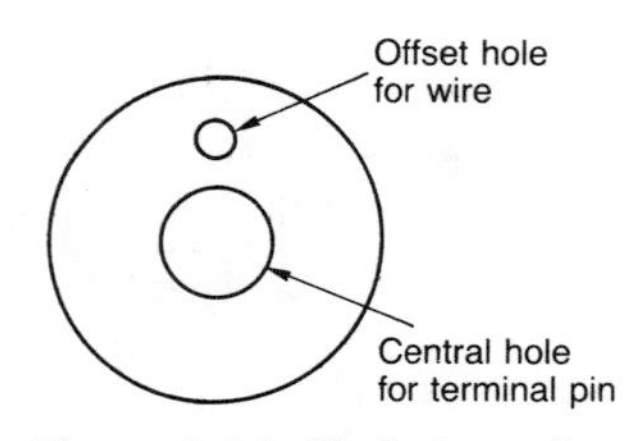

Figure 3.34 *End view of wire-wrap bit*

terminal pin is rectangular with sharp corners. The wire is wrapped tightly round the terminal using a special tool. The wire and the terminal cut into each other at the sharp corners, this making a good metal-to-metal connection. The wire used is single core and has the insulation stripped off the end to the required length. The stripped wire is then inserted into in the offset hole of the wire-wrap bit of the special tool (Figure 3.34). The centre hole of the tool bit is then slipped over the terminal pin and rotated evenly to wrap the wire round the pin.

3.7.4 Mechanical fasteners

Mechanical fasteners include such items as terminals to which wires can be connected by wrapping a wire round them and screwing the cap of the terminal down on to the wire. Plugs and sockets are other examples of mechanical connections. Such methods of connection can be used where demountable, rather than permanent, connections are required.

3.8 Products and processes

Consider a simple screwdriver. It has a shaft and a blade which is made of a high-carbon steel. The steel is used because the material is required to have a high stiffness, high hardness, high toughness and not easily bend. Such an item can be made by forging. The handle of the screwdriver was traditionally made of wood but is now made of a thermoplastic polymer. The polymer is preferred to wood because of the ease with which it can be made to the required shape. A thermoplastic material becomes soft when hot and can be readily moulded to the required shape. Injection moulding can be used.

Consider some of the components used in the average car engine. The cylinder block might be aluminium which is die cast to give the required shape. Aluminium has the advantage of being low density and has adequate strength for what is a comparitively low stressed part. The piston rubs against the cylinder block and wear is a problem. When aluminium rubs on aluminium the wear can be quite significant. So an aluminium piston in an aluminium cylinder block would present problems. To minimize wear, there must be a significant difference in hardness between the piston and cylinder block surfaces. One way of achieving this is to chromium plate an aluminium piston. The pistons are either cast or forged. The sump is likely to have been made from a low-carbon steel sheet which has been pressed into the required shape. The crankshaft might be forged steel or cast iron.

3.8.1 Production sequences

The manufacture of a product is frequently likely to consist of a number of production stages. Consider an example of the manufacture of railway lines. The sequence is likely to be:

1 Casting of the steel into large *blooms*, this being the term for large bars.
2 The blooms are then cooled very slowly to allow gas to escape.

3 The blooms are then reheated and rolled to give the finished rail profile. The hot rails are then sawn to the required lengths.
4 The rails are then either cooled normally in air or subject to faster cooling, depending on the type of strength required of the finsihed product.
5 The rails are then passed through a roller-straightening machine in order to give rails with a high degree of straightness.
6 The rails are then subject to testing to determine whether they are free of flaws.
7 The rail ends are then cold sawn to give exactly the required lengths and finish to the ends.

A sequence involving marking out and the use of cutting tools is given in the next section.

3.9 Basic tools

The following is a brief consideration of common workshop hand tools, their use and maintenance.

3.9.1 Marking out

Marking out is used to provide guide lines on material for a machinist to work to regarding size, features, hole positions, etc. Such lines may be drawn, scribed or centre marked. Drawn lines using a pencil might be used when the line must not cut through a protective coating on the metal, e.g. galvanised sheet steel or tin plate. Such lines are, however, not permanent and easily removed. Scribed lines give clear, fine lines which cannot easily be removed. A *scriber* is a tool with a hard sharp point and is used in conjunction with a straight edge for straight lines or an engineer's square where lines are required to be at right angles to an edge (Figure 3.35). The surface to be marked may be brushed with a suitable *marking out medium* in order that the scribed line shows up clearly. Such a medium might be whitewash in the case of castings and forgings with a dark scale covering. Bright surfaces might be coated with a cellulose lacquer or treated with acidulated copper sulphate, this giving a thin layer of copper on the surface.

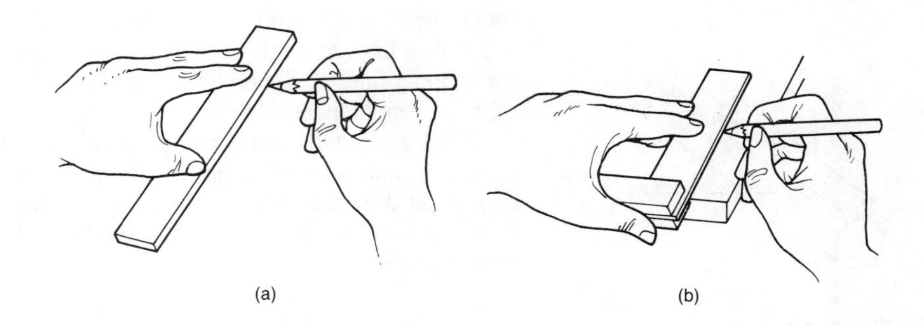

Figure 3.35 *Scribing with (a) a straight edge, (b) an engineer's square*

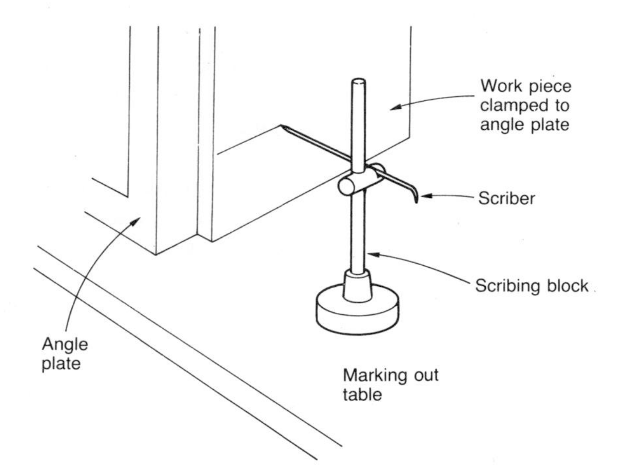

Figure 3.36 *Using a marking out table*

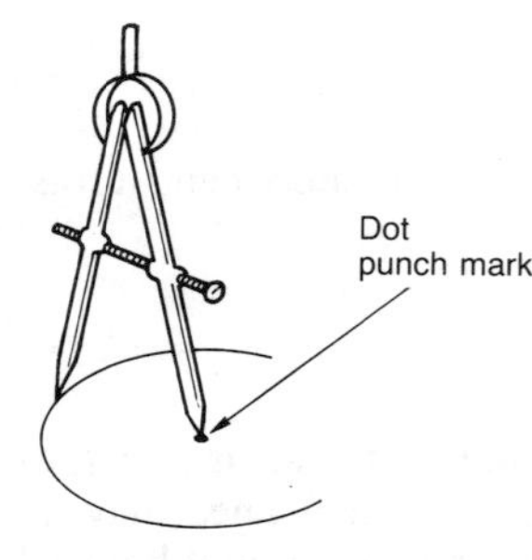

Figure 3.37 *Using dividers*

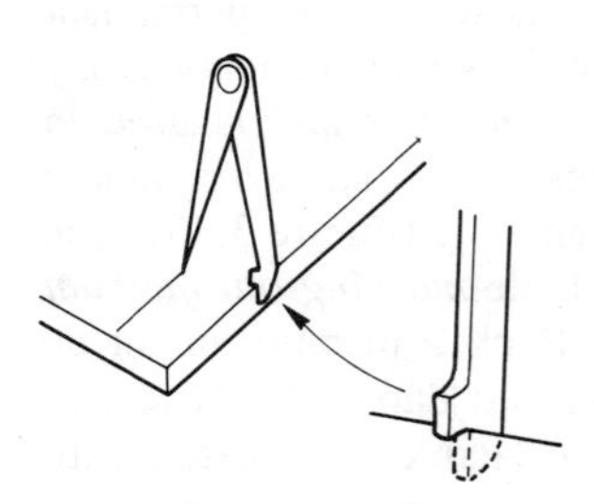

Figure 3.38 *Odd leg calipers*

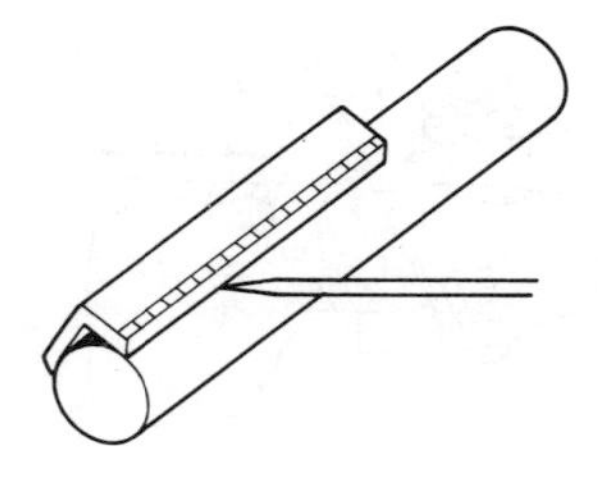

Figure 3.39 *Using a box square*

To scribe straight lines, the minimum equipment required is a scriber and a straight edge. A scriber might be used, for straight lines, in conjunction with a *marking out table* (Figure 3.36). This is a table with rigid legs and a a smooth, flat surface. An angle plate with clamps might be used to support the work piece being scribed.

For arcs and circles, *dividers* can be used (Figure 3.37). Dividers have to be set to the required radii for circles or arcs and this is done by positioning their points in the engraved lines of a rule. An engineer's rule is engraved for this purpose. A *dot punch* might be used to locate the leg of the dividers. A dot punch or centre punch, a *centre punch* is heavier and has a less accurate point, is also used for making a mark to guide the point of a drill or to place a number of marks along a scribed line to give a more permanent indication of its position.

Other marking out equipment includes oddleg calipers, box squares and combination sets. *Oddleg calipers* can be used for finding the centre of a bar or for scribing a line parallel to an edge (Figure 3.38). A *box square* can be used for marking off lines on round bars or tubes (Figure 3.39). A *combination set* is three tools in one. There is a rule and protractor head, a square head, and a centre head. The protractor head can be used to measure and scribe lines at angles other than right angles. The square head is used for scribing lines at right angles and the centre head for locating the centres of, for example, bars or shafts.

Marking out equipment is precision equipment that must be carefully used and maintained. The points of scribing instruments must be kept sharp by regular dressing with an oil slip stone. Dot and centre punches must be kept sharp by grinding. All such equipment should be cleaned after use and returned to their cases.

3.9.2 Basic hand tools

Basic hand tools used for cutting operations are files, chisels, scrapers and hacksaws.

Files are available in a number of grades:

1 Rough, for rapid removal of large amounts of material.
2 Bastard, for rapid removal of smaller amounts of material.
3 Second cut, for removing smaller amounts of material before finishing.
4 Smooth, for giving a smooth surface finish.
5 Dead smooth, for special work that requires a very smooth accurate finish.

The correct file should be chosen for the job. After filing in one direction with the file being moved in the direction of its length, the work should be filed again with strokes at right angles to the original ones. Once the surface has been filed down almost to size, draw filing can be used. This involves filing in a direction at right angles to the direction of the file length. To avoid blunting the teeth on files and breaking files, they should be stored in a tool box or on a rack. They should be kept clean and free from filings by regular brushing with a file card in the direction of the cut.

Chisels (see Figure 3.11 and the associated text) should always be used in a direction away from the body. Chisels that have become mushroomed must never be used as pieces may break off when the chisel is struck with a hammer. When chiselling, goggles and a chipping screen should be used. The chisel should be held firmly at an angle of 30 to 40° to the cutting plane.

Scrapers are used to produce a bearing surface from one which has been filed or machined. Scrapers are classified by the shape of the blade:

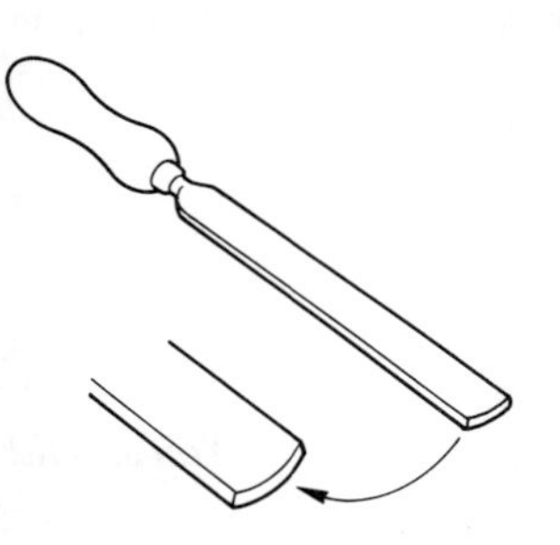

Figure 3.40 *The flat scraper*

1 The flat scraper (Figure 3.40) is used for removing slight irregularities on a flat surface. After scraping in one direction, the scraper should then be used in a direction at right angles to the first one.
2 The half-round scraper is used for internal cylindrical surfaces.
3 The three-square (or bearing) scraper is used for internal surfaces such as bearings.

Hacksaws (see Figure 3.12 and the associated text) should have blades chosen for the job concerned. The greater the thickness of the material to be cut the fewer the number of teeth per 25 mm of blade. When using a saw the angle of cutting should be such as to ensure at least 3 teeth involved at any one time in the cutting.

3.9.3 Production sequence

The sequence of operations that might occur for, say, the production of a rectangular part 10 mm by 20 mm and length 25 mm with a hole of diameter 6 mm drilled in the larger face along its centre line and 12 mm from one end (Figure 3.41) is:

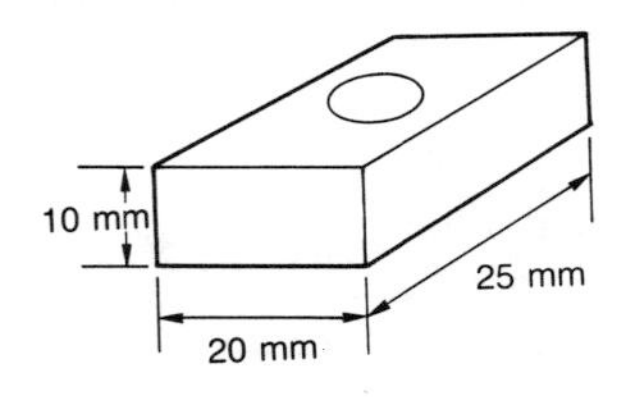

Figure 3.41 *Part*

1 Check using a ruler that the material supplied is the right size for the job. It is assumed that the material supplied is only slightly oversize.
2 File two datum edges flat and square to each other. An engineer's square can be used to determine the angle.
3 Mark a line parallel to the long datum edge and at 10 mm from it. This might be by using marking dye, a marking out table, a scribe and a rule.
4 Using rough and smooth files, file the material down to this line.
5 Mark a line at right angles to the above line and parallel to the long datum edge and at 20 mm from it. This might be by using marking dye, a marking out table, a scribe and a rule.
6 Using rough and smooth files, file the material down to this line.
7 Mark off the overall length of 25 mm from the short datum edge. A hacksaw might then be used to cut just outside this line and then rough and smooth files used to obtain the required length.
8 Mark off the centre for the hole by a centre punch. This might be done by first determining the location of the centre using a marking out table, a scribe and a rule or oddleg calipers.
9 Drill the hole using a bench drilling machine, the work piece being clamped on the work table of the machine.
10 De-burr the hole.
11 Check that all the dimensions are correct.

Problems

Questions 1 to 30 have four answer options: A, B, C and D. You should choose the correct answer from the answer options.

1 Decide whether each of the following statements is TRUE (T) or FALSE (F).

Casting rather than a manipulative process such as forging or extrusion is likely to be the optimum method for manufacturing a product if:
(i) The product is required in long lengths.
(ii) The product has an internal cavity.

A (i) T (ii) T
B (i) T (ii) F
C (i) F (ii) T
D (i) F (ii) F

2 Decide whether each of the following statements is TRUE (T) or FALSE (F).

Sand casting is likely to be a cheaper process than die casting when:
(i) Small numbers have to be manufactured.
(ii) A good surface finish is required.

A (i) T (ii) T
B (i) T (ii) F
C (i) F (ii) T
D (i) F (ii) F

3 Decide whether each of the following statements is TRUE (T) or FALSE (F).

Extrusion is a good process for the production of:
(i) Long continuous lengths of product.
(ii) Products with good surface finish.

A (i) T (ii) T
B (i) T (ii) F
C (i) F (ii) T
D (i) F (ii) F

4 Decide whether each of the following statements is TRUE (T) or FALSE (F).

Deep drawing can be used:
(i) With brittle materials.
(ii) To produce objects such as seamless cans.

A (i) T (ii) T
B (i) T (ii) F
C (i) F (ii) T
D (i) F (ii) F

5 Decide whether each of the following statements is TRUE (T) or FALSE (F).

For a fixed clearance angle, increasing the rake angle of a tool:
(i) Increases the strength of the cutting edge.
(ii) Increases its cutting efficiency with ductile materials..

A (i) T (ii) T
B (i) T (ii) F
C (i) F (ii) T
D (i) F (ii) F

6 Decide whether each of the following statements is TRUE (T) or FALSE (F).

A cutting tool for use with a ductile material:
(i) Should have a large rake angle.
(ii) Will produce granular chips.

A (i) T (ii) T
B (i) T (ii) F
C (i) F (ii) T
D (i) F (ii) F

Questions 7 to 9 relate to the following processes that can be used to manufacture a product:

A Casting
B Forging
C Extrusion
D Deep drawing

7 Which process could be used to manufacture a complex shape with an internal cavity?
8 Which process could be used to manufacture a long length of uniform section?
9 Which process could be used to manufacture a seamless can?

Questions 10 to 12 refer to the following values of rake angles for tools:

A Negative
B 0°
C 10°
D 25°

10 Which is the optimum rake angle for a tool to be used for cutting a ductile material, such as mild steel?
11 Which is the optimum rake angle for a tool to be used for cutting a brittle material, such as cast iron?
12 Which is the optimum rake angle for a hard and brittle tool to be used for cutting a high-strength alloy?

13 When the tool is moved parallel to the axis of the rotating work piece with a centre lathe, the work piece surface produced is:

A Flat
B Cylindrical
C Tapered
D Conical

14 The rake angle of a twist drill is determined by:

A The point angle
B The drill diameter
C The helix angle
D The clearance angle

Questions 15 to 17 refer to the following manufacturing processes that can be used with plastics:

A Casting
B Injection moulding
C Extrusion
D Thermoforming

15 Which method is likely to be used for encapsulating electronic components in plastic?

16 Which method is likely to be used for the production of continuous lengths of plastic curtain rail tracks.

17 Which method is likely to be used for the production of milk bottle crates?

18 Decide whether each of the following statements is TRUE (T) or FALSE (F).

A flux is used with soft soldering:
(i) To clean the joint surfaces.
(ii) To bond with the joint surfaces.

A (i) T (ii) T
B (i) T (ii) F
C (i) F (ii) T
D (i) F (ii) F

19 Decide whether each of the following statements is TRUE (T) or FALSE (F).

Thermoplastic adhesives:
(i) Consist of a thermoplastic material dissolved in a volatile solvent.
(ii) Are not suitable for highly stressed joints.

A (i) T (ii) T
B (i) T (ii) F
C (i) F (ii) T
D (i) F (ii) F

20 Decide whether each of the following statements is TRUE (T) or FALSE (F).

With fusion welding the joint is provided by:
(i) The joint edges of the parent metals melting.
(ii) The filler metal melting.

A (i) T (ii) T
B (i) T (ii) F
C (i) F (ii) T
D (i) F (ii) F

21 Decide whether each of the following statements is TRUE (T) or FALSE (F).

With electric arc welding a flux-coated electrode is used. This is because:
(i) It reduces the amount of electricity used.
(ii) It protects the weld pool from impurities in the filler metal.

A (i) T (ii) T
B (i) T (ii) F
C (i) F (ii) T
D (i) F (ii) F

22 Decide whether each of the following statements is TRUE (T) or FALSE (F).

With oxyacetylene welding, no flux is used because:
(i) The products of combustion protect the weld pool.
(ii) The temperature of the joint is not high enough for oxidation to occur.

A (i) T (ii) T
B (i) T (ii) F
C (i) F (ii) T
D (i) F (ii) F

23 An electrical component is specified as having tinned connecting wires. This means that:

A They are made of tin.
B They have been given a layer of solder.
C They have been cleaned ready for soldering.
D They have already been soldered to a circuit board.

24 Decide whether each of the following statements is TRUE (T) or FALSE (F).

To obtain a good soldered joint in an electrical circuit, the connecting surfaces being joined:
(i) Should be clear of oxide layers.
(ii) Should be heated to the temperature at which the solder melts .

A (i) T (ii) T
B (i) T (ii) F
C (i) F (ii) T
D (i) F (ii) F

25 Decide whether each of the following statements is TRUE (T) or FALSE (F).

Electrical connections are generally made in car electrics by means of crimping rather than soldering. This is because:
(i) Crimping makes better electrical connections.
(ii) Crimping is faster and cheaper.

A (i) T (ii) T
B (i) T (ii) F
C (i) F (ii) T
D (i) F (ii) F

26 A scribed line parallel to a datum edge of a rectangular bar are best made with:

A Scriber and an engineer's square
B Scriber and a straight edge
C Dividers
D A box square

27 Decide whether each of the following statements is TRUE (T) or FALSE (F).

Odd leg calipers can be used to:
(i) Find the centre of a bar.
(ii) Scribe a line parallel to an edge of a bar.

A (i) T (ii) T
B (i) T (ii) F
C (i) F (ii) T
D (i) F (ii) F

28 A scribed line is best protected by:

A Spraying it with lacquer
B Coating it with acidulated copper sulphate
C Making a series of dot punch marks along it
D Making a series of centre punch marks along it

29 Decide whether each of the following statements is TRUE (T) or FALSE (F).

Marking out is to provide:
(i) Guide lines to work to.
(ii) Identification marks for use in assembly.

A (i) T (ii) T
B (i) T (ii) F
C (i) F (ii) T
D (i) F (ii) F

30 Decide whether each of the following statements is TRUE (T) or FALSE (F).

Th marking out of a circle or arc is usually carried out using:
(i) Odd-leg calipers
(ii) Dividers

A (i) T (ii) T
B (i) T (ii) F
C (i) F (ii) T
D (i) F (ii) F

31 Indicate the processes that might be used to manufacture three of the following products, giving reasons for the selection:

(a) Mass production of a plastic box.
(b) Mass production of a mild steel I-section beam for use in building construction.
(c) Mass production of the plastic blades for a hover mower.
(d) Mass production of the electrical circuit of a cheap, portable, radio.
(e) Mass production of small metal (zinc alloy) components, such as a model car steering wheel, for toys.
(f) A one-off production of a rod with varying taper angles along its length.
(g) A one-off production of a large watertight metal box.

4 Safety

4.1 Health and Safety at Work

Safety in a company is everyone's responsibility. While there may be some employees, such as safety officers or safety representatives, who have special responsibilities all the employees and the employers have responsibilities. These responsibilities are a legal requirement under the Health and Safety at Work Act 1974. This chapter gives a brief overview of health and safety at work. For a detailed consideration the reader is referred to the many publications of the Health and Safety Executive; a booklet is available from them listing the publications available. An overview is given in their booklet *Essentials of Health and Safety at Work (HMSO)*.

The *Health and Safety at Work Act 1974* provides a comprehensive and integrated system of law in relation to the health, safety, and welfare of people at work and for the protection of the general public against risks to health and safety arising out of, or in connection with, the activities of persons at work. The Act consists of four parts, Part I being concerned with health, safety and welfare in connection with work and the control of dangerous substances and certain emissions into the atmosphere. Part II is concerned with the Employment Medical Advisory Service. Part III is concerned with building regulations. Part IV deals with miscellaneous and general matters. The Act did not immediately repeal earlier legislation and regulations but allowed them to remain current until revoked and replaced by new regulations or Codes of Practice issued under the act.

Part I is of particular relevance in connection with the topic of this book and has four basic objectives:

1 To secure the health, safety and welfare of people at work.
2 To protect persons other than persons at work against risks to health or safety arising out of or in connection with the activities of those at work.
3 To control the keeping and use of explosive or highly flammable or otherwise dangerous substances and generally prevent the unlawful acquisition, possession and use of such substances.
4 To control the emission into the atmosphere of noxious or offensive substances.

4.1.1 Employers' responsibilities

Under Part I of the Act, general duties are laid down for employers. These are to ensure, as far as is reasonably practicable, the health, safety and welfare at work of all employees. This extends to:

1 *Plant and equipment* This requires the provision and maintenance of plant and systems of work that are, so far as is reasonably practicable, safe and without risk to health. Thus, for example, machinery must be

regularly maintained and fitted with proper guards to ensure safe operation.

2 *Handling, storage and transport* This means ensuring that the arrangements, so far as is reasonably practicable, are safe and there is an absence of risks to health in connection with the use, handling, storage and transport of articles and substances. Thus, for example, materials in store must be safely stacked so that they cannot slip and fall whilst in the stack or being removed from it.

3 *Information, training and supervision* This is the provision of such information, instruction, training and supervision as is necessary to ensure, so far as is reasonably practicable, the health and safety at work of employees. As a consequence this requires employers to identify potential hazards. Employers must ensure that all employees are properly trained in the use of machines and in the relevant codes of practice to ensure that they work in a safe manner.

4 *Safe premises* So far as is reasonably practicable, this requires the maintenance of the place of work in a condition that is safe and without risks to health and the provision and maintenance of means of access and egress from it that are safe and without such risks. Thus, for example, the premises must not represent a fire hazard. They must be in good repair.

5 *Safe working environment* This requires the provision and maintenance of a working environment for employees that is, so far as is reasonably practicable, without risks to health and adequate as regards facilities and arrangements for employee welfare at work. Thus, for example, there must be adequate lighting, heating and ventilation. There must be safe passage for pedestrians and vehicles, handrails on stairs, surfaces which are not slippery, etc. Some manufacturing processes give off harmful fumes, there must therefore be proper extraction for these from the working environment.

In addition, the employer must:

1 Provide, if five or more people are employed, a written statement of the organisation's general policy with respect to the health and safety of the employees and how it will be carried out.

2 Carry out a risk assessment if five or more people are employed. A hazard is anything that can cause harm, e.g. electricity, and risk is the chance of harm actually being done. The risks to workers, visitors to the premises, and the public all need to be considered. Some of the activities which might result in accidents and harm being done are materials having to be lifted or carried, stores with materials falling from shelves or stacks, collisions between people and materials when moving,

exposure to toxic materials, electricity when using hand tools, spillages, etc.

3 Notify the local Health and Safety inspector of occupation of the premises. Health and safety laws are enforced by an inspector from the Health and Safety Executive or the local council. Inspectors can visit workplaces without notice.

4 Consult any appointed union safety representatives of issues affecting health and safety and any information and training which has to be provided.

5 Display the Health and Safety Law poster for employees or give out the leaflet.

6 Notify the inspectors of certain types of injuries, occupational diseases and events occurring.

4.1.2 Employees' responsibilities

Under the Act, employees also have duties with respect to health and safety at work. While at work, employees should:

1 Take reasonable care for the health and safety of him or herself and of other persons which may be affected by his/her acts or omissions at work.

2 Cooperate with the employer as regards any duty or requirement imposed on the employer or any other person as a consequence of statutory provisions in order to enable the duty or requirement to be complied with.

In addition no person shall intentionally or recklessly interfere with or misuse anything provided in the interests of health, safety or welfare as a consequence of statutory provisions. For example, an employee removing the safety guards from a machine could be prosecuted under the Act.

4.1.3 Health and Safety Commission and Executive

Two bodies were set up by the Act: the *Health and Safety Commission* and the *Health and Safety Executive*. The Commission is charged with the task of making arrangements for carrying out research, the provision of training and information, advisory services and the development of regulations. The Executive exercises, on behalf of the Commission, such of its functions as the Commission directs it to exercise. The tasks of enforcing the statutory provisions is given to the Executive, in cooperation with local authorities and other enforcement bodies.

Every enforcing authority may appoint inspectors. Such inspectors may visit workplaces without notice. They may want to examine the safety, health and welfare situation in the company, or perhaps investigate an accident or complaint. They have the right to talk to employees and safety representatives, take such samples and photographs as they consider relevant and in some situations impound dangerous equipment. If, as a result of such investigations an inspector considers a situation to be contravening relevant statutory provisions, he or she may serve an improvement notice requiring the situation to be remedied within a specified time. If there is a risk of personal injury, the inspector can issue a prohibition notice. Such a notice requires the activity concerned to cease until the situation has been remedied. Failure to comply with notices can result in prosecution and fines and/or imprisonment.

4.2 Safe work systems

Employers have the duty that they should provide systems of work that are, so far as is reasonably practicable, safe and without risks to health. This requires the establishment of safe work systems and a consideration of such factors as:

1 *Health* This, for example, means considering the hazards to health of exposure to certain chemicals, or what happens when chemicals are spilt, etc.

2 *Safety* This means, for example, a consideration of the hazards that can occur when a machine or its guard fails, or an operator chooses to do a job in a different way, etc.

3 *Permits to work* The issuing of a safe written procedure may be adequate for some jobs, but where the risks are very high there can be the need to initiate a formal permit system. This permit states exactly what work is to be done and when and if the work is assessed and checked by a responsible person. Those doing the job sign the permit to show that the hazards and precautions necessary are understood.

4 *The needs of individuals* This may include protective clothing and equipment, seating and working space being considered in relation to the needs of each individual. There are also such considerations as whether they are able to understand safety instructions and are able to work safely if under medication, have some handicap, etc.

5 *Maintenance* In order to ensure continuing safety, equipment, buildings and plant need proper maintenance.

6 *Monitoring* The system needs checking to see that rules and precautions are being followed and continue to deal with the risks. New hazards may be found to be introduced by changes in staff, materials, equipment, etc.

4.3 Plant and machinery

The employer is charged with the provision and maintenance of a working environment for the employees that is, as far as is practicable, safe, without risks to health, and adequate as regards facilities and arrangement for their welfare at work. The employer has thus to be concerned with such facilities as hygiene and welfare; cleanliness; floors and gangways being kept clean, dry and not slippery; seats, machine controls, instruments and tools being designed for the best control, use and posture; the place of work being safe with adequate space for easy movement and safe machine adjustment, no tripping hazards, emergency provisions, etc.; lighting that gives good general illumination with no glare, adequate emergency lighting, etc.; and a comfortable environment with a suitable working temperature, good ventilation, acceptable noise levels, etc.

The following are more details about some of the above points in relation to manufacturing.

4.3.1 Machinery safety

Machinery is a cause of many serious accidents in the workplace. Such accidents can arise as a result of parts of the body becoming drawn in or trapped in rollers or belts, hair or clothing becoming entangled in rotating parts, the skin being punctured by sharply pointed parts, abrasion from coming into contact with rough surfaces, people being struck by moving parts of machinery, burns of scalds from hot parts, electrical shock from electrical parts, etc.

All machine operators must be trained and, if necessary, given protective clothing. Some 'prescribed dangerous machines' can only be used by those under 18, and above school leaving age, if they have had full instruction and sufficient training and under close supervision. A guillotine is an example of such a machine. Children must never be allowed to operate or to help at machines. Adequate lighting must be provided for all machines.

If practical, fixed guards should be used to enclose the dangerous parts of machines. They might be fixed in place with screws or nuts and bolts. If fixed guards are not practical the possibility should be considered of interlocking the guard so that the machine cannot start before the guard is in position and cannot be opened while the machine is running. The materials used for guards might be plastic, because it can be easily seen through, or wire mesh, because plastic might be too easily damaged. Figure 4.1 shows examples of some guards. In the maintenance of machines it is necessary to ensure that the guards and safety devices are checked and in working order.

When an operator comes to use a machine he or she must check that:

1. Before starting the machine, he or she knows how to stop it.
2. All the guards are in position.
3. The area around the machine is clean, tidy and free from obstruction.
4. He or she is wearing the appropriate protective clothing and equipment.

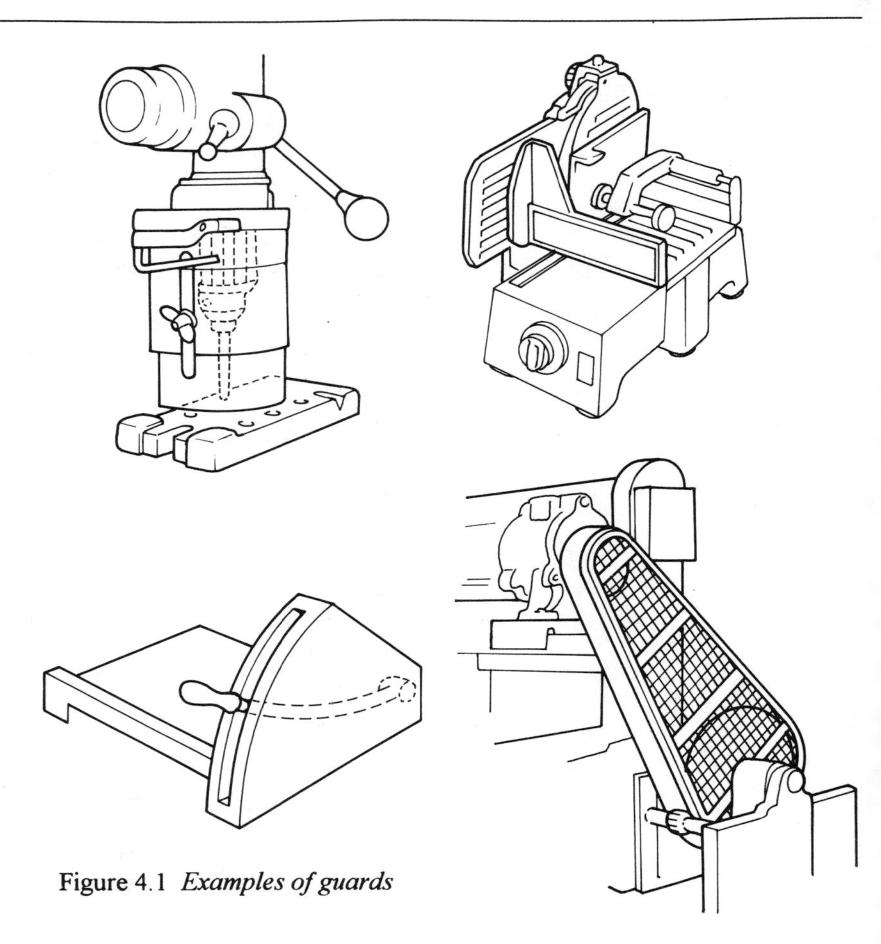

Figure 4.1 *Examples of guards*

If the machine is not working properly or the guards are faulty, the supervisor should be immediately told. A machine must not be used unless the operator is authorised and trained to do so. No attempt should be made to clean a machine in motion. It should be first switched off and unplugged or locked off. The operator should not wear dangling chains, loose clothing, gloves, rings or have free long hair which can get caught up in moving parts of machines.

4.3.2 Plant maintenance

Maintenance is carried out on plant to prevent problems arising and to put faults right. It may arise because a machine has broken down, the term *breakdown maintenance* is used, or as part of a planned programme, the term *preventive maintenance* is used. Breakdown maintenance involves waiting until the plant fails and then repairing it. Preventive maintenance involves anticipating failure or a problem and replacing or making adjustments before failure or the problem occurs. Think of a car where planned maintenance means that the engine oil is changed at regular intervals rather than waiting for problems to occur because of dirty oil.

With regard to items such as hand tools, maintenance should ensure that hammers have heads properly secured to handles, there should be no worn or chipped heads or broken shafts; chisels should have the cutting edge sharpened to the correct angle; screwdrivers should not have split handles; spanners should not have splayed jaws, etc.

Hazards can occur during maintenance. For example, there may be accidental start up of machinery while it is being worked on, confined spaces may be entered where there may be toxic materials or a lack of air, etc. Plant and equipment should be made safe before maintenance starts and safe working areas provided. Maintenance staff should have appropriate protective clothing and equipment.

4.3.3 Handling and transporting

The manual handling of loads is a common source of injuries in factories. This may occur from frequent forced or awkward movements of the body and lead to back injuries and pains in hand, wrist, arm or neck. Injuries to the arms, hands or neck can arise from repetitive handling, e.g. repeated squeezing or pressing, awkward arm or hand movements. Such risks can be minimized by reducing the levels of force required, reducing the amount of repetitive movements, getting rid of awkward positions, etc.

Care should be taken when lifting or moving heavy objects manually. The rules for correctly lifting a load (Figure 4.2) are:

1. Stop and think. Plan the lift and consider whether you need help and the area is free of obstructions. Clear any obstacles.
2. Maintain the back straight and upright, bend the knees with the feet apart and the leading leg forward.
3. Get a firm grip, keeping the arms straight and close to the body.
4. Lift, but do not jerk, avoid sudden movements and twisting of the spine.

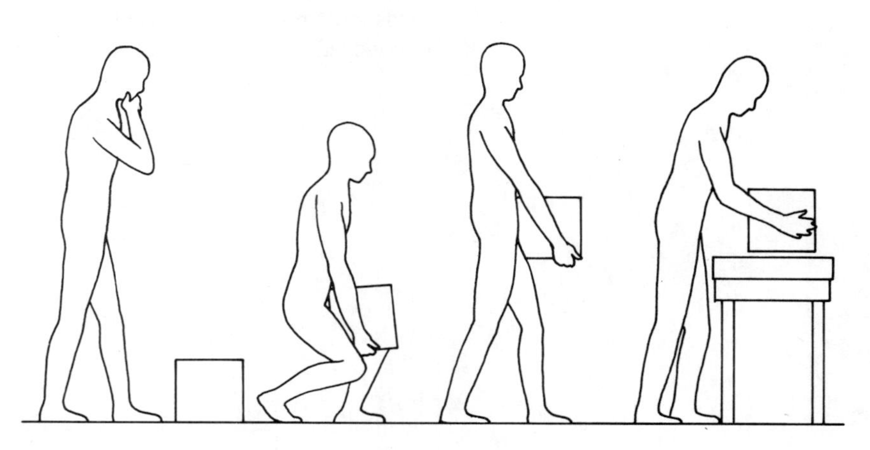

Figure 4.2 *Lifting a load*

5 Move the feet, don't twist the body.
6 Keep the load close to your body.
7 Put the load down and then adjust its position.

Whenever possible, where there is a risk of injury, manual handling should be avoided. Moving loads mechanically can also present hazards. For example, people may be struck by loads falling from a lifting or moving device. For safe lifting by machine the following points must be considered:

1 The form of the load, its weight and position of centre of gravity.
2 How the load is to be attached to the lifting machine.
3 The hook must be positioned above the centre of gravity of the load so as to maintain an even balance and avoid the load tilting and slipping. If in doubt, a trial can be conducted with the load raised just a few centimetres off the ground to see if it tilts. There will be little harm if it drops from such a height.
3 The safe working load (SWL) marked on a lifting machine and on such accessories as slings must not be exceeded. The load on the legs of a multi-sling increases as the angle between the legs increases (Figure 4.3). For example, for a load of 100 kN and a 2-leg sling, when θ is 30° the load on the legs is 52 kN, when 90° it is 70 kN, when 150° it is 200 kN and when 180° it is infinite.
4 Only certified lifting equipment, which is marked with its safe working load, should be used and which is not overdue for examination.
5 Unsuitable equipment should not be used, e.g. badly worn chains shortened with knots, kinked or twisted wire ropes, frayed or rotted fibre ropes.
6 Avoid snatch or sudden loads.
7 The attachment of slings to a load and a crane hook is a skilled job and should only be done by a trained and experienced slinger. He/she alone should give the lift and movement signals to the crane operator.
8 Loads must not be transported over the heads of people working below or loads left hanging without someone in attendance.

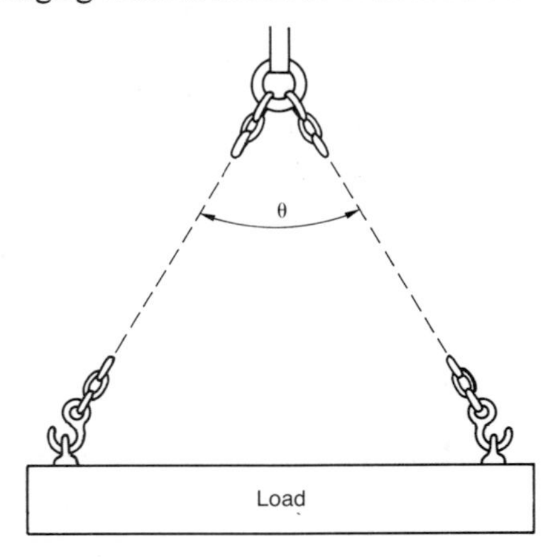

Figure 4.3 *Two-leg sling*

In connection with transport of goods and materials:

1 The workplace should be laid out so that pedestrians are safe from vehicles. The roadways should be properly maintained and adequately lit. Wherever possible, vehicles and pedestrians should be separated. Safe crossing places for pedestrians should be clearly marked.
2 Ensure that drivers can see clearly and pedestrians be seen.
3 Drivers should be trained. Unauthorised people should not drive the vehicles.
4 Vehicles should be checked daily and faults promptly rectified.
5 When vehicles are not in use, the vehicles should be left in a safe state, the hand brake on, the engine switched off and the keys removed, the wheels being chocked if necessary. The keys should be kept secure.
6 Vehicles should be loaded and unloaded safely, materials being safely secured.

Materials and goods should be stored and stacked so they are not likely to fall and cause injury. This means:

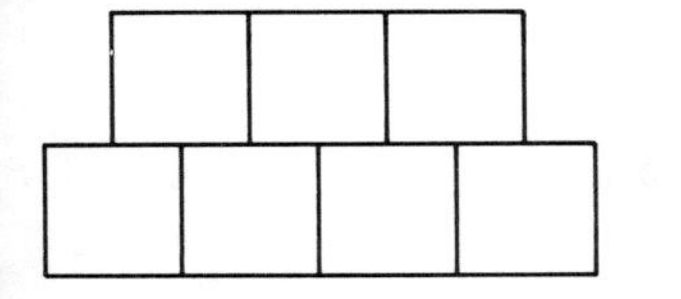

Figure 4.4 *Keyed items*

Figure 4.5 *Chocked items*

1 Stacking on a firm level base.
2 Using a properly constructed rack when needed and securing it to the floor of wall if possible.
3 'Key' stacked items of a uniform size like a brick wall so that no tier is independent of another (Figure 4.4).
4 Use chocks for objects which might roll (Figure 4.5).
5 Keep heavy articles near floor level.
6 Not exceeding the safe load of racks, shelves or floors.
7 Not allowing items to stick out from stacks or bins into gangways.
8 Not climbing racks to reach upper shelves, but using a ladder or steps.
9 Not leaning heavy stacks against walls, the walls might not be able to withstand the load.
10 Not de-stacking by throwing items down from the top or pulling them out from the bottom.

4.3.4 Noise and vibration

Loud noise can cause irreversible hearing damage, accelerating the normal hearing loss which occurs as we grow older and causing other problems. It can also interfere with communication and cause stress. The *Noise at Work Regulations 1989* are intended to reduce hearing damage caused by loud noise. These regulations lay down three action levels at which employers are required to take action. The first action level is when noise reaches 85 dB(A), the second level when it reaches 90 dB(A) and the third, peak, level when it reaches 140 dB(A). Noise is measured in units called decibels, written as dB(A). The following are some examples (taken from *Essentials of Health and Safety at Work, 4th impression 19992, HMSO)* of the levels of noise typically given by various operations:

85 dB(A)	Plastic injection moulding machine
87 dB(A)	Electric drill
93 dB(A)	Sheet metal shop
96 dB(A)	Power press operator
99 dB(A)	Circular saw
108 dB(A)	Hand grinding of metal

For comparison, a loud radio is about 60 to 80 dB(A), a heavy lorry about 7 m away about 100 dB(A) and a jet aircraft 25 m away about 140 dB(A).

At the first action level employers should:

1 Get the noise assessed by a competent person.
2 Inform workers about the risks to their hearing and the precautions.
3 Make ear protectors freely available to those who want it.
4 If anyone thinks their hearing is being affected, suggest they take medical advice.

At the second level, employers must:

1 Control noise exposure, doing all that is feasible to reduce exposure. This might, for example, mean rotating jobs or providing a noise refuge at the machine control point.
2 Mark zones where noise reaches the second level with recognised signs and restrict entry to those zones. Those entering those zones must wear hearing protection.

The control of noise is best achieved by controlling it at source. This can involve:

1 Choosing machines or processes that are quiet. Suppliers of machines should supply noise level data for the machine operators' position.
2 Enclose noisy machines with sound-insulating panels or put them in separate rooms.
3 Fit silencers to exhaust systems.

Vibration is often associated with noise. Operators can be exposed to vibration when using hand-held power tools and machinery. Hazardous vibrations result in tingling of numbness in the hand after a few minutes exposure. Excessive exposure to vibration can cause hand-arm vibration syndrome, a painful condition affecting blood circulation, nerves, muscles and bones in the hands and arms. Such vibration can be reduced by using tools designed to cut down vibration, ensuring that tools are kept sharp, and perhaps altering the form of the job to reduce the grip or pressure that has to be exerted by the operator's hand on the tool. Whole body vibration mainly affects drivers of vehicles and can cause pain in the lower part of the back and spinal damage.

4.3.5 Electricity

The hazards associated with electrical equipment are:

1 Electric shock through contact with live parts. Note that the normal mains voltage, 240 V a.c., can kill.
2 Fire due to overheating of cables, connections and appliances.
3 Explosions due to electrical apparatus or static electricity igniting flammable vapours or dusts.

In working with electricity, the following points need to be taken account of:

1 There should be a switch or isolator near each fixed machine to cut off the electrical power in an emergency.
2 The mains switches are readily accessible and identifiable.
3 There should be enough socket outlets so that overloading by the use of an adapter is not necessary. Such overloading can lead to fires.
4 When equipment is used outdoors, or in damp or corrosive environments, the socket outlets should be of a special design.
5 Fuses and other circuit breakers should be set so that the circuit is broken if the normal, safe, current is exceeded. For example, a 1 kW electric fire connected to the mains supply of 240 V will draw a current of 4.2 A (power = IV). Thus a fuse should be used which will blow if this current is significantly exceeded. A fuse rated at 13 A is too high a rating and would permit the normal current to be exceeded by too great a factor with the possibility of fire as a result of overheating.
6 Power cables to machines must be insulated and protected from damage. This means sheathed and armoured cable or cable installed in conduits.
7 Frayed and damaged cables should be replaced.
8 Lengths of cable should only be joined by using proper connectors or cable couplers.
9 Electric light bulbs. or other items which might easily be damaged, to be protected (Figure 4.6).
10 Tools and power sockets should be switched off before plugging in or unplugging.
11 Appliances should be unplugged before cleaning or making adjustments.
12 Faulty equipment should be taken out of use, labelled 'Do not use' and kept secure until it can be checked by a competent person.
13 Special protection must be used when electrical equipment is used in flammable or dusty environments.

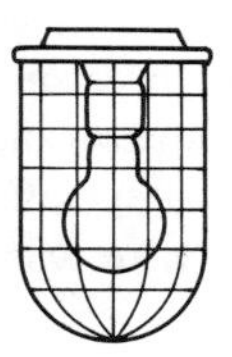

Figure 4.6 *Protected lamp*

In the mains supply, one of the conductors, for safety reasons, is connected to earth. The colours used for the insulator coverings of such cables are brown for the live cable, blue for the neutral cable and yellow/green for the earth cable. If a fault occurred which resulted in the live conductor coming into contact with the metal case of an electrical appliance, current could flow through the case to earth via the body of any person touching it

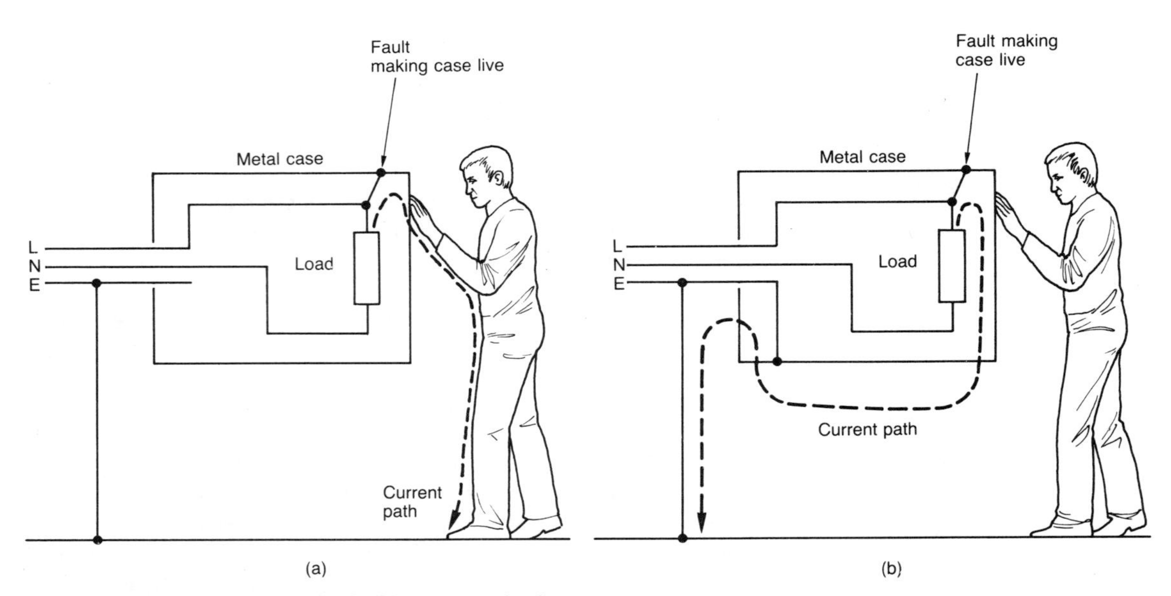

Figure 4.7 *(a) Case not earthed, (b) case earthed*

(Figure 4.7) and so result in an electric shock. If, however, the case is earthed, then since the resistance of the human body is very much greater than that of the earth wire, the current will flow through the earth wire rather than the human.

Where there is a chance that metalwork can become live as a result of a fault, not only should it be earthed but the circuit concerned should be protected by a *residual current device* (RCD). This is a mechanical switching device intended to cause the opening of contacts and so the switching off of current when the residual current attains a specified value. The current in the live wire and the current in the neutral wire will be equal unless there is a fault and some of the current from the live wire passes to earth. The term *residual current* is used for the difference between the live and neutral currents. If this value exceeds the rating of the RCD it will trip, indicting that there is a fault in the circuit. Typical trip values are a few milliamperes. All residual current detectors should be regularly checked by operating the test button.

The procedure that should be adopted if someone is suffering from electric shock is:

1 Switch off the power.
2 If the power cannot be switched off immediately, pull or push the casualty clear with a broom or a wooden chair.
3 Unless the power is switched off, do not touch the person's body with your bare hands otherwise you could receive a shock.
4 Unless wearing thick, rubber-soled boots, stand on lino, rubber or wood. This is to prevent there being a low resistance conducting path through you to earth.

5 If the casualty is breathing, place in the recovery position or on their side and then get them to hospital.
6 If the casualty is not breathing, give mouth to mouth resuscitation. Check the pulse and if absent, apply chest compressions.

4.3.6 Radiations

Hazards can arise from radiations, the term *radiation* being taken to cover not only visible light but radio waves, microwaves, infra-red radiation, ultraviolet radiation and ionising radiations, i.e. radiations such as X-rays and gamma rays. High intensity visible radiation can cause heating and destruction of tissue of the eye or skin. Radio waves and microwaves can cause excessive heating of exposed parts of the body. Infrared radiation can cause reddening of the skin and burns. Ultraviolet radiation can cause sunburn, skin cancer and damage to eyes. Ionising radiations can cause burns, dermatitis, cancer, cell damage and blood changes.

The following are a few points about how such hazards can occur in the workplace and precautions that should be taken. Ultraviolet light can occur with welding. Thus during welding special goggles or a face screen should be worn. Infrared radiation can occur from hot bodies such as molten metal. Protective clothing may be needed to reduce the warming, burning and irritation of the skin and eye protection, with suitable filters, worn to avoid discomfort.

4.4 Harmful substances

Health hazards can arise from many substances if they get into the body. The Control of Substances Hazardous to Health Regulations 1988 deals with such hazards.

The Control of Substances Hazardous to Health Regulations 1988 (COSHH) imposes duties on employers to protect employees and other persons who may be exposed to substances hazardous to health and also imposes certain duties on employees. The term 'substances hazardous to health' covers substances:

1 listed in an approved list as dangerous and for which the general indication of the nature of risk is specified as very toxic, toxic, harmful, corrosive or irritant;
2 for which a maximum exposure limit has been specified or for which there is an occupational exposure standard;
3 any micro-organism hazardous to health;
4 dust of any kind when present at a substantial concentration in air;
5 any other substance which could create a hazard to health comparable to those listed above.

Employers are required to:

1 Assess the risk of exposure to hazardous substances and the precautions which should be taken to prevent exposure, or if this is not reasonably practicable, to achieve adequate control. This will mean reading the

labels and safety sheets issued by suppliers of such substances, considering the possible routes into the body (breathed in, swallowed, taken through the skin) and the worst result that can occur, the concentrations or conditions likely to cause problems, who and how many people might become exposed, how often and how long they work with the substance, the first symptoms of over-exposure.
2 Introduce appropriate control measures to prevent or control the risk. This might involve putting the harmful substances in a separate room or building or outside, reducing the amount used and the number of people exposed, using closed transfer and handling systems, using local exhaust ventilation which extracts dust or vapour as close as possible to its source, keeping dangerous chemicals locked away, clearing up spillages quickly and safely, having smooth working surfaces to allow easy cleaning. It also means not allowing smoking, eating or drinking in chemical handling areas, the use of protective clothing, washing hands before smoking, eating or drinking, not siphoning or pipetting hazardous chemicals by mouth but using a pump or hand-operated siphon.
3 Ensure that the control measures are used.
4 Ensure that the control measures are properly maintained and periodically examined. Thus all ventilation equipment must be regularly examined and tested by a competent person.
5 Where necessary, the exposure of workers should be monitored.
6 Where necessary, carry out health surveillance of workers.
7 Inform, instruct and train employees and their representatives of the risks involved, the precautions to be taken and the results of monitoring.

Some substances present particular problems and there are special regulations governing their use. Thus lead and its compounds are covered by the *Control of Lead at Work Regulations 1980* and an *Approved Code of Practice*; asbestos by the *Control of Asbestos at Work Regulations 1987*, the *Asbestos (Licensing) Regulations 1983* and the *Asbestos (Prohibitions) Regulations 1992*.

As an indication of the types of hazards needing to be controlled, consider the processing of plastics. Moulding and extrusion machines for the manufacture of plastic products have automatic alarm systems to limit overheating of the polymers used. These may not always be correctly set. Thus problems may occur since heating thermoplastic materials to beyond their processing range causes them to decompose and produce fumes. The normal processing of polymers may also give rise to fumes. In some case these fumes can be toxic. Some moulding and extrusion machines are fitted with vents and extraction may be required at these points, e.g. extraction hoods within a semi-enclosed area. In addition there may be dust hazards during the supply of dry materials to the feed units of the machines. The hazards posed by such fumes and dust needs to be assessed and guarded against. Thus there should be provision of a good standard of general ventilation in the workroom to ensure that the concentration of fumes in the work areas does not reach hazardous levels. This might involve having the air changed six time every hour.

Some substances may constitute fire hazards. The supplier's data sheets should be consulted on how best to handle such materials. Precautions can involve reducing the amounts kept on site and storing them in areas separate from process areas in safe, well-ventilated, isolated, buildings. Full and empty gas cylinders should, where possible, be stored in a secure outside compound. They should be stored with their valves uppermost, protected from damage and not stored below ground level or near drains or basements since most gases are heavier than air. Cylinders should be changed away from sources of ignition in a well-ventilated place. At the end of each day's work, the cylinder valves should be turned off. To avoid 'flash-back' into hoses and cylinders when lighting up welding flames, operators must be trained to adopt the correct lighting up and working procedures and non-return valves and flame arresters fitted.

4.5 Protective clothing and equipment

Wherever possible, employers should eliminate or control risks so that protective clothing and equipment do not have to be used. Some jobs, however, require by law specified protective clothing or equipment. The following is an indication of some of the hazards and protection that might be used:

Eyes

Hazards: chemical or metal splashes, dust, protectiles, gas, radiation.

Protection: spectacles, goggles, face screens, helmets.

Head and neck

Hazards: falling objects, bumping head, hair entanglement, chemicals.

Protection: helmets, bump caps, hairnets, caps, skull-caps.

Hearing

Hazards: impact noise, high levels of sound, high and low fequency sound.

Protection: Earplugs, muffs.

Hands and arms

Hazards: abrasion, temperature extremes, cuts, chemicals, electric shock, skin infections, vibration.

Protection: gloves, gauntlets, wrist cuffs, armlets.

Feet and legs

Hazards: wet, slipping, cuts, falling objects, heavy pressures, metal and chemical splashes, abrasion.

Protection: safety boots, gaiters, leggings.

Respiratory protection

Hazards: toxic and harmful dusts, gases and vapours, micro organisms.	Protection: disposable respirators, mask respirators, fresh air hose equipment.

The body

Hazards: heat, cold, weather, chemical or metal splashes, impact, contaminated dust, entanglement of clothing.	Protection: overalls, boiler suits, warehouse coats, donkey jackets, aprons, specialist protective clothing.

4.6 Accidents and emergencies

There is a need to plan for dealing with emergencies, whether it is a simple accident or a major incident. People need to be told what might happen, how the alarm will be raised, what to do, where to go, who will be in control, and what essential actions need to be taken, e.g. closing down plant. They need training in emergency procedures. Access ways need to be kept clear for emergency services and escape routes. Fire fighting equipment, electrical isolators and shut off valves need to be clearly labelled. Emergency equipment needs testing regularly. The *Management of Health and Safety at Work Regulations 1992* covers emergencies.

The *Health and Safety (First Aid) Regulations 1981* cover the requirements for first aid equipment. There must be:

1 Someone who can take charge in an emergency.
2 A first-aid box
3 Notices telling people where the first-aid box is and who the appointed person is.
4 A trained first aider and a first-aid room if the work gives rise to special hazards.

After an accident or a serious incident, the immediate emergency should be dealt with, any injuries treated and the premises or plant made safe. Any injuries should be recorded in an accident book. If applicable, the incident should be reported to the inspector. The *Reporting of Injuries, Diseases and Dangerous Occurrences Regulations 1985* require an employer to report immediately by phone if as a result of, or in connection with, work someone dies, receives a major injury, is seriously affected by, for example, electric shock or poisoning, or there is a dangerous occurrence (for further information, see the Health and Safety Executive leaflet *Reporting an injury or a dangerous occurrence* (11)). Also, a written report needs to be sent within seven days confirming the telephone report and notifying any injury which stops someone doing their normal job for more than three days, certain diseases suffered by workers, and certain types of events involving flammable gas in domestic or other premises (see the Health and Safety Regulations booklet *Guide to reporting of injuries, diseases and dangerous occurrence regulations 1985* (23)).

4.6.1 Fire precautions

With regard to fire precautions, employers need to:

1. Provide enough exits for everyone to get out easily.
2. Provide fire doors and escape routes which are clearly marked as such and unobstructed. Fire doors should never be wedged open, they are there to stop smoke and fire spreading.
3. Provide escape doors which can be opened easily from the inside.
4. Provide enough fire extinguishers of the right type and properly serviced.
5. Ensure, by regular checks, that the fire alarm is working and can be heard above the background noise.
6. Ensure that everyone knows what to do in the case of fire. Clear instructions should be displayed and there should a periodic fire drill.
7. Ensure that people know how to raise the alarm and use the extinguishers.

A fire certificate may be needed for the premises. This will depend on the type of business and the number of people employed.

Fire extinguishers are colour coded depending on what they contain and the type of fire they can be used on.

1. *Red* is water with CO_2 or soda-acid for use with wood, paper, textiles, etc. In the pressurised water extinguisher, operation involves removing the cap and striking the exposed knob. This punctures a compressed carbon dioxide cartridge and releases gas under high pressure which then forces out the water. The soda-acid extinguisher operates by the plunger being depressed, or the extinguisher being inverted, causing an acid to mix with sodium carbonate solution. The reaction generates carbon dioxide under pressure which forces water out under pressure. It is unsafe for all situations where voltages are concerned.
2. *Blue* is dry powder for use with flammable liquids and low voltages up to 1000 V.
3. *Cream* is foam for use with flammable liquids and is unsafe at all voltages. Removing the cover and striking the plunger, or inverting the extinguisher, causes chemical to react and produce foam
4. *Black* is CO_2 gas and is for use with flammable liquids and is safe at high voltages.
5. *Green* is vaporising liquids and is safe for use with high voltages. The extinguisher is operated by pumping the handle and spraying a jet of the liquid at the seat of the fire. The vapour is, however, highly toxic and so this type should not be used in confined spaces.

With electrical equipment, fire extinguishers that cover the device with a conducting liquid should never be used. Blue, black and green extinguishers can be be used because they are non-conducting.

Fires need heat, oxygen and a combustible liquid or solid. Water from fire buckets, hose reels and fire extinguishers of the red colour employ the cooling down method to reduce the temperature below that required for combustion. Fire blankets, dry powder, foam and carbon dioxide types of extinguisheres prevent oxygen reaching the fire.

In the case of a fire the action that might be advocated is:

For the person discovering the fire

1 Sound the fire alarm.
2 Phone 999 to call the fire brigade.
3 Attack the fire using the fire applicances provided. Do not take risks.

On hearing the alarm

1 Leave the building by following the fire exit signs.
2 Do not return to the building for any reason unless authorised to do so.
3 Close all doors behind you.
4 Do not use lifts.
5 Report to the assembly point.

4.7 Examples of processes and safe working practices

The following are outlines of how two processes might be carried out in a safe manner.

The procedure to be adopted with *manual arc-welding* might be:

1 Know and comply with the prescribed safety precautions and fire prevention procedures.
2 Wear the necessary protective clothing. This can include an adjustable helmet with a welding screen, special cape and sleeves, special apron and gloves.
3 Check that portable screens are in position to avoid unscreened persons seeing the arc.
4 Check that the return earth lead is firmly connected to the bench or work piece and the power source.
5 Check that the welding lead is connected to the power source and that the connection to the electrode holder is tight and sound.
6 Switch the power on.
7 Concentrate on the welding process.
8 Make sure you have full control of the electrode and can hold it steady.
9 Position yourself to avoid stretching and the risk of overbalancing.
10 Support the arm holding the electrode by keeping it near the body.
11 Ensure that the welding screen is in front of your eyes before striking the arc and keep is there until the arc is broken.
12 The arc may be struck by: holding the electrode in a gloved hand. pointing the electrode downwards and away from the body at an angle of about 60 to 70° to the plate surface, lowering the electrode until the striking end just touches the plate, this contact starts the current flow, immediately withdrawing the electrode a slight distance from the plate then establishes the arc. The arc may be broken by withdrawing the electrode.
13 Place the electrode holder in a safe place when not in use.
14 Use goggles when chipping off hot slag.

15 Switch off the power source when the equipment is not in use.
16 Switch off the mains supply to the power source at the end of a work period.
17 Leave the work area in a tidy and orderly manner with all equipment properly stowed.

The procedure to be adopted when using a *pillar drilling machine* for the drilling of holes in a work piece might be:

1 Hair must be short or covered. No loose clothing should be worn.
2 The guard on the drill must be in position and remain in position during the drilling.
3 The hole centres on the work piece should have been accurately marked out and centre punched. The hammer used with centre punching should have a hammer face clear of blemishes and a secure head. The punch should not have a mushroomed head but have a correctly ground point. A pilot drill is necessary when drilling a large diameter hole.
4 The work to be drilled must be properly secured.
5 There must be proper support for the work for when the drill breaks through.
6 The drill should be used at the correct speed and rate of feed. Where appopriate, use a suitable coolant.
7 Swarf should not be allowed to jam in the drill.

Problems

Questions 1 to 20 have four answer options: A, B, C and D. You should choose the correct answer from the answer options.

Questions 1 to 3 relate a company where tubes are welded and then spray painted. The following safety equipment/clothing is available:

A Face masks
B Protective gloves
C Ear plugs
D Respirator

Select the most appropriate equipment/clothing from the above for a worker engaged in the following operations:

1 Welding the tubes.
2 Removing the hot welded tubes from the welding fixtures.
3 Spraying paint onto the tubes.

4 For safe operation of machinery, the operator must check before starting that:

A The working area is clean
B There is enough material for the job
C All the guards on the machine are correctly fitted
D The machine is connected to the power supply

5 According to the Health and Safety at Work Act, safety in the workplace is the responsibility of:

A The safety officer
B The employer alone
C The employee alone
D The employer and employee jointly

6 When an operator comes to operate a machine, he/she finds that the safety guards are faulty. He/she should:

A Start the machine with the faulty guards
B Start the machine without the guards
C Immediately tell the supervisor
D Fit non-faulty guards

7 Decide whether each of the following statements is TRUE (T) or FALSE (F).

Safe manual lifting requires the person performing the lifting to:
(i) Bend their back.
(ii) Lift the load without jerking.

A (i) T (ii) T
B (i) T (ii) F
C (i) F (ii) T
D (i) F (ii) F

8 Decide whether each of the following statements is TRUE (T) or FALSE (F).

In carrying out mechanical lifting with multi-leg slings it is necessary to remember that:
(i) As the sling angle is increased the load acting in the legs increases.
(ii) The safe working load marked on a sling should not be exceeded.

A (i) T (ii) T
B (i) T (ii) F
C (i) F (ii) T
D (i) F (ii) F

9 Decide whether each of the following statements is TRUE (T) or FALSE (F).

The noise level for an operator of a machine is measured as being at the first action level. This means that:
(i) Hearing protection must be worn.
(ii) The operator must be told of the risks and precautions.

A (i) T (ii) T
B (i) T (ii) F
C (i) F (ii) T
D (i) F (ii) F

10 Decide whether each of the following statements is TRUE (T) or FALSE (F).

Safe stacking of materials requires:
(i) A firm, level base.
(ii) Chocks if pipes or drums are stacked.

A (i) T (ii) T
B (i) T (ii) F
C (i) F (ii) T
D (i) F (ii) F

11 The safe working load (SWL) for mechanical lifting gear is:

A The only load for which the equipment should be used.
B The average load for which the equipment is suitable.
C The minimum load for which the equipment can be used.
D The maximum load for which the equipment can be used.

12 A residual current device (RCD) is a device for detecting when there is a:

A Current in the live wire.
B Current in the neutral wire.
C Difference between the currents in the live and neutral wires.
D No current in the live wire.

Questions 13 to 15 relate to the following personal protective equipment:

A Ear muffs
B Gloves
C Respirator
D Goggles

For the following processes, which protective equipment would be needed:

13 Operating a machine where the noise is at the second action level.
14 Operating an ultraviolet lamp.
15 Handling chemicals which might irritate the skin.

16 Electrical equipment in metal cases should have the cases earthed to prevent:

A Fuses repeatedly blowing
B Electric shock to anyone touching the case
C Electrical interference from neighbouring electrical items
D The equipment causing electrical interference with other equipment

17 If someone suffers from electrical shock, the first step that must be taken is:

A Leave the victim and fetch medical help
B Commence artificial respiration
C Pull the victim away from contact with the supply
D Switch off the supply

18 Which of the following types of fire extinguishers must not be used with a fire involving electrical equipment?

A Pressurized water
B Foam
C Carbon dioxide
D Vaporising liquid

19 Decide whether each of the following statements is TRUE (T) or FALSE (F).

When carrying out manual arc welding it is necessary to:
(i) Wear protective clothing.
(ii) Check that portable screens are in place to avoid unscreened persons seeing the arc.

A (i) T (ii) T
B (i) T (ii) F
C (i) F (ii) T
D (i) F (ii) F

20 Decide whether each of the following statements is TRUE (T) or FALSE (F).

When drilling a hole using a pillar drilling machine it is necessary to ensure that:
(i) The hole centre has been centre punched.
(ii) The workpiece must be properly secured.

A (i) T (ii) T
B (i) T (ii) F
C (i) F (ii) T
D (i) F (ii) F

21 What are the general duties imposed by the Health and Safety at Work Act, 1974 on (a) employers, (b) employees?

23 Explain the functions of (a) improvement notices, (b) prohibition notices.

24 Discuss the following situations and consider what actions should be taken or what consequences could occur:

(a) A worker removes the safety guards from a machine because they reduce the number of items he can produce and hence his wages, which are based on piece-rate.
(b) The company is short of storage space for a large consignment of goods that have just arrived and so they temporarily stack them in the passage ways leading to the main exit from the factory floor.
(c) A worker is just about to become married and in celebration fellow workers let off the fire extinguishers.
(d) An inspector wants to go into the tool room but the management tell him that they will not allow it because the work there is commercially highly secret and he might tell their competitors.
(e) An inspector wants to talk to a worker about his machine but the worker refuses to talk to him.
(f) A worker is injured and as a result is off work for a month. The employers pay him/her during that time but take no other action.

25 What protective clothing and equipment might be suitable in the following situations:
(a) chemicals may splash into the eyes,
(b) a press is very noisy during operation,
(c) a storekeeper has to move sheets of metals.

5 Graphical communication

5.1 Communicating engineering information

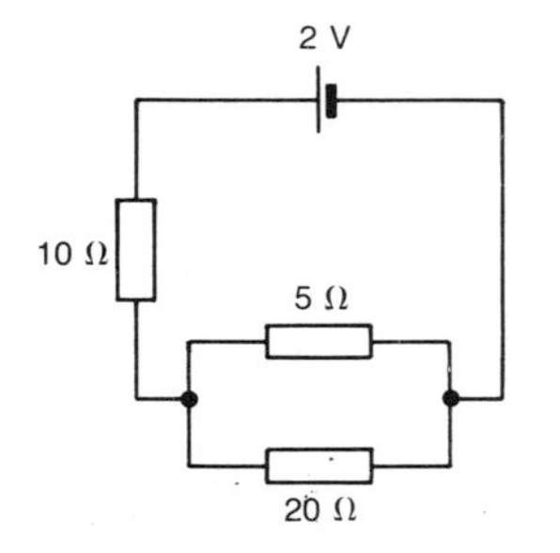

Figure 5.1 *Circuit diagram*

If you look in any textbook, for example this one, you will find that not only are there words but many diagrams. The diagrams illustrate some aspect of a piece of equipment or product. With a relatively few lines, a lot of information can be conveyed to the reader and in a way that is often easier to comprehend, and takes less space, than if words alone had been used. Thus, for example, the circuit diagram in Figure 5.1 shows what components are required and how they are to be connected in the circuit. Describing it in words would take more space and be not so easily comprehended. Though Figure 5.1 is a simple electrical circuit, try describing it in words alone so that there is no ambiguity about what the components are and how they are to be connected. Another example of graphical communication is a road map. The map graphically shows the roads, towns and other features in sufficient details to enable routes to be planned between places. Think of the problem of trying to convey the information contained on a map by words alone.

Graphical communication in engineering is used to convey such information as:

The size and shape of a product.
How a product is to be assembled.
How a component is to be installed.
The overall appearance of a product.
The wiring diagram for an electrical component.
The sequence of operations to be used in some process.
The flow of information through a sequence of units.

and to people such as:

Manufacturing engineers
Maintenance engineers
Sales engineers
Customers

The manufacturing engineer, for example, will require information which will enable the product to made, the maintenance engineer information about a product which will enable maintenance to be carried out, the sales engineer the information necessary to sell the product and the customers the information necessary to install and operate the product. The methods used for this diversity of tasks might be:

Block diagrams
Flow diagrams
Circuit diagrams

Sketches
Detailed technical drawings

This chapter is a general overview of such forms of graphical communications and the criteria for selection of particular types.

Charts and graphs can also be considered to be means of communicating engineering information. They give descriptions of numerical and mathematical data. They are discussed in chapter 8 with examples of their use in engineering science in chapter 9.

Example

What type of information would a machine operator require if he/she had to machine a piece of metal to a particular shape?

The operator would need detailled information about the shape and the dimensions of the item.

Example

What type of information would a computer programmer require if he/she had to write a program for a microprocessor-controlled device to carry out a task?

The programmer would need to know the sequence of the operations that had to be followed for the task to be performed.

5.2 Graphical methods

The following are brief details of graphical communication methods and the criteria which might be used to determine which method to use.

5.2.1 Block diagrams

Block diagrams aim to show systems and how information passes from one system or subsystem to another. A *system* can be defined as an arrangement of parts which work together and around which we can draw an imaginary line and talk of there being some form of output from a specified input or inputs. The block diagram is a useful way of representing such a system. Figure 5.2 shows such a diagram for an electric motor system. Within the boundary described by the box outline is the system. Inputs to this system are shown as arrows entering the box, outputs from the system are shown as arrows leaving the box. Thus in the case of the electric motor system, we can consider there to be just one input – electrical energy, and just one output – mechanical energy. As another example, an air compressor can be considered to be a system which has an input of air at atmospheric pressure and an output of air at a higher pressure. A manufacturing process, such as a machine tool, is a system which has inputs of raw materials, electrical energy, and information about the form of product required with outputs of finished products, waste heat and scrap.

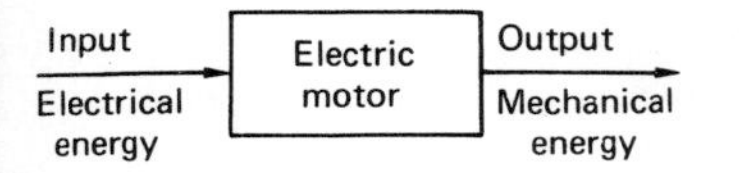

Figure 5.2 *The electric motor*

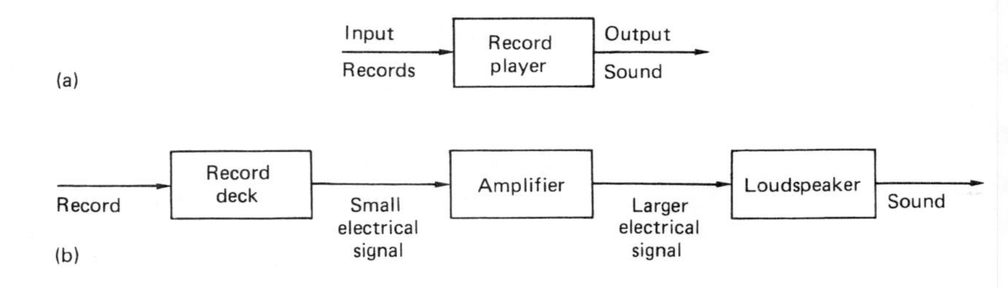

Figure 5.3 *The record player system: (a) as a complete system and (b) as a series of interconnected subsystems*

A record player can be considered as a system which has an input of a record and an output of sound (Figure 5.3(a)). An alternative way of considering the system is as a series of interconnected subsystems (Figure 5.3(b)). The record is thus the input to the record player subsystem. The output from this is a small electrical signal. This is then the input to an amplifier subsystem. The output from the amplifier is a larger electrical signal. This is then the input to the loudspeaker subsystem, with this then giving the final output of sound.

The term *subsystem* is used for self-contained blocks within a larger system. In drawing blocks to represent a system in terms of subsystems, it is necessary to recognise that the lines drawn to connect boxes indicate a flow of information, the direction being indicated by an arrowhead, and not necessarily physical connections. Figure 5.4 shows the basic form of a communication system, e.g. a radio system. The input signal is passed through a modulator system to put it into a suitable form for transmission. The signal is then transmitted through the communication channel, with the possibility of distortion, noise or interference inputs. The received signal is then demodulated to put it into a form suitable for reception.

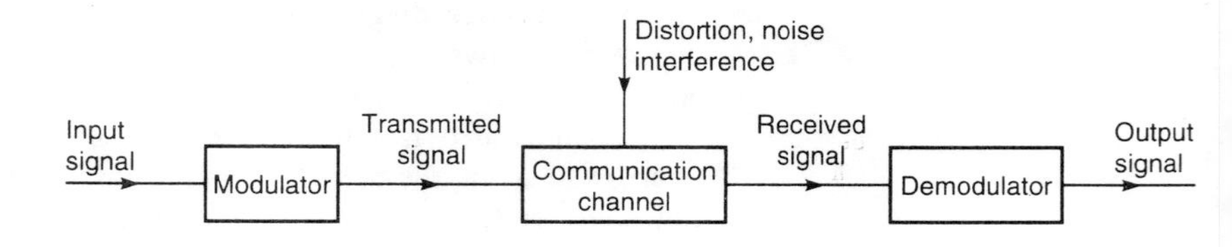

Figure 5.4 *A communication system*

5.2.2 Flow diagrams

Flow diagrams aim to show a sequence of events. They are particularly used in computing to show the sequence of steps that have to be carried out by a computer program. With computer flow diagrams a convention is used of using diamond-shaped boxes to contain items requiring decisions, rectangular boxes when arithmetical operations are involved, parallelogram-shaped boxes when there is an input or output of numbers, a circle to show a start of a sequence, rectangular boxes with rounded ends to show a stop of a sequence of operations, arrow-ended lines to indicate the direction of flow through the chart, etc. Figure 5.5 shows some of these commonly used symbols.

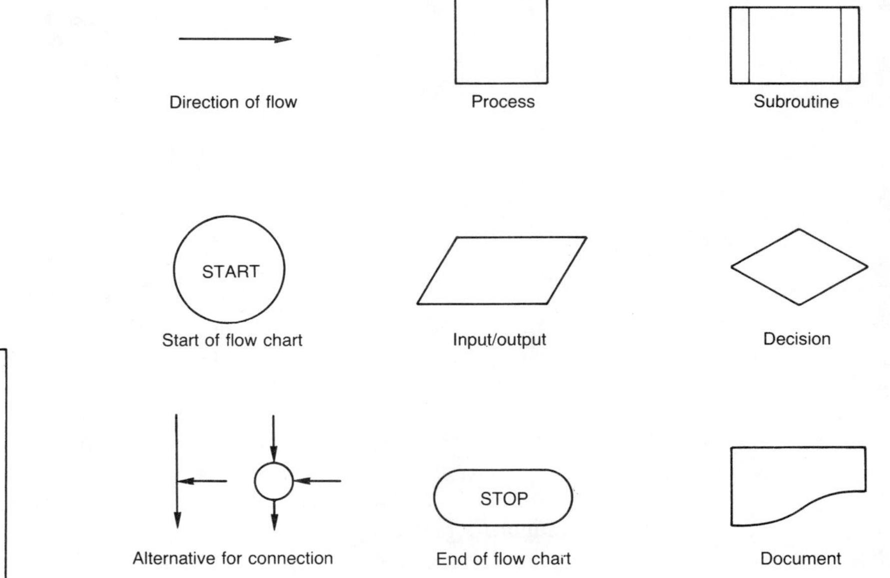

Figure 5.5 *Flowchart symbols*

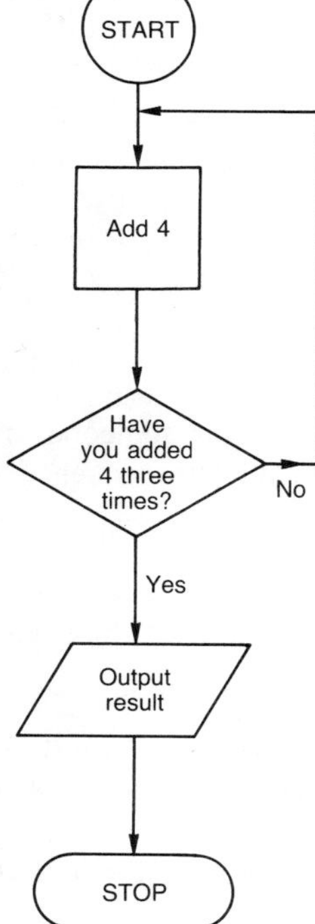

Figure 5.6 *A flowchart for 4×3*

Figure 5.6 shows what a flow chart might look like for carrying out the multiplication of 4 and 3. The sequence of operations are:

1 Start the progam
2 Add 4
3 Have you added 4 three times? If the answer is no then you have to follow the loop back up the flowchart and repeat the operation of adding 4. This repeating occurs until the number 4 has been added three times. Then the answer to the question is yes and you can proceed to the next operation.
4 Output the result, i.e. an output of 12.
5 Stop the program

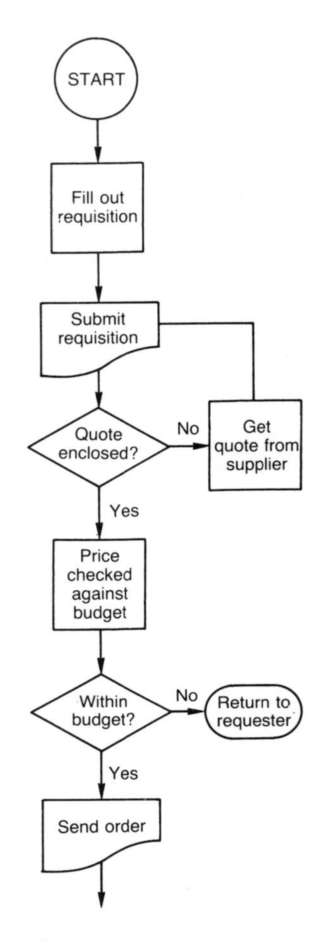

Figure 5.7 *Placing an order*

Figure 5.7 shows the form a flowchart might take for someone in a company wanting to place an order for some item. The sequence of operations are:

1 Start the program.
2 The person wishing to purchase the item, the requester, fills out a requisition form.
3 This is submitted to a superior.
4 If there is no quote for the item attached to the requisition form then it is returned to the requester. If a quote is attached then it proceeds to the next stage.
5 The quoted price is checked against the amount allowed for it in the budget.
6 If it is not within the budget then the order is denied and it is returned to the requester.
7 If it is within budget then the order is sent out.

There is then likely to follow other stages, such as acknowledgement from the supplier, etc. before the program is complete.

5.2.3 Circuit diagrams

Circuit diagrams can take the form of showing by means of standard symbols the parts of an electrical/electronic circuit and the interconnections between them. Figure 5.8(a) shows an example. Chapter 7 gives details of the standard symbols used and the conventions adopted for the drawing of such diagrams. A circuit diagram of this form does not show the physical positions of the parts of the circuit or their physical appearance. It is a means of communication between the circuit designer and those who have to manufacture the circuit or possibly remedy faults developed in the circuit. Another possibility is for a circuit diagram showing the actual layout of the components. Figure 5.8(b) shows a possible layout for the circuit in (a) using a circuit board.

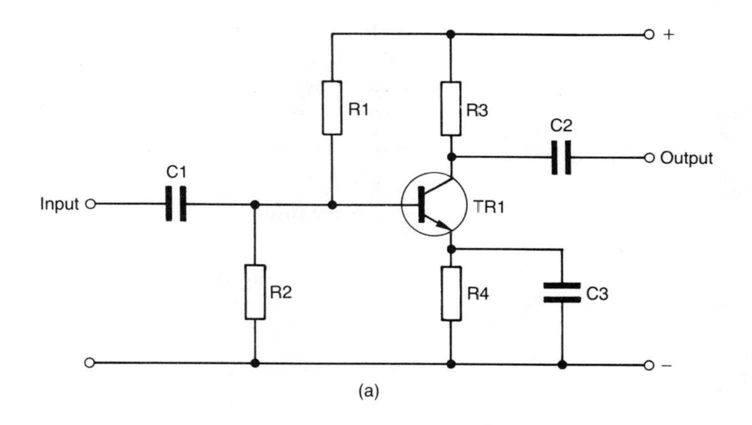

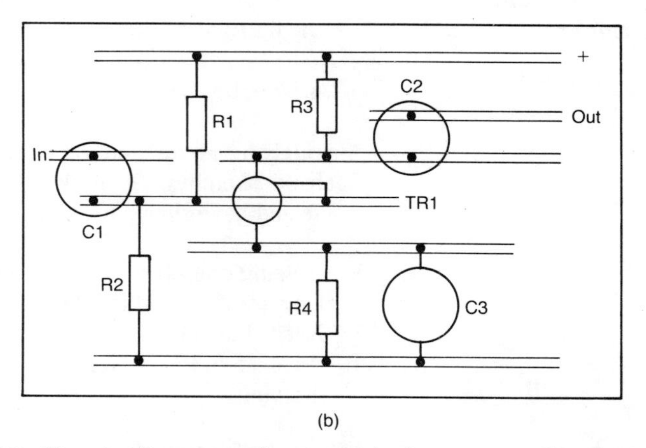

Figure 5.8 *Circuit diagrams showing (a) the circuit, (b) the layout of components in the circuit*

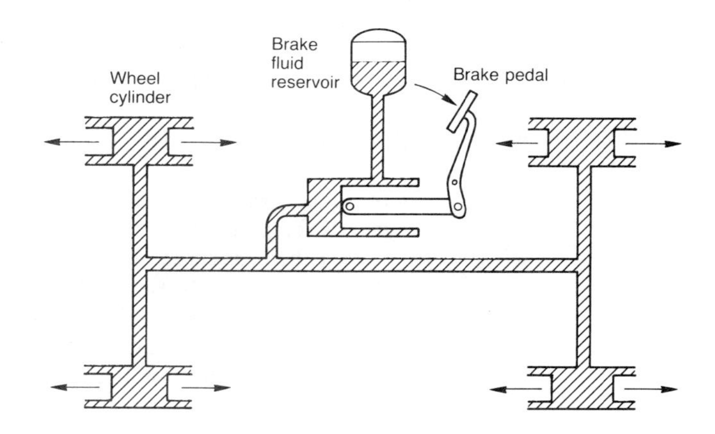

Figure 5.9 *Hydraulic braking system*

It is not only electrical circuits that can be represented by circuit diagrams. Figure 5.9 shows a circuit for the hydraulic braking system of a car. Pressing the brake pedal causes the fluid to be pressurised and this pressure transmitted to the wheel cylinders and hence result in the brakes being applied.

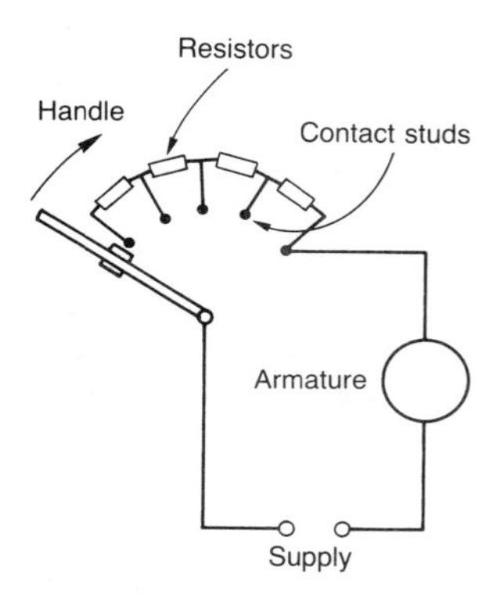

Figure 5.10 *Manually operated starter*

Example

Figure 5.10 shows a manually operated starter for a motor. On the basis of the diagram, explain what happens as the handle is moved from its initial position across the studs to the far right position.

When the handle makes contact with the first stud, all the resistors are in the circuit and so the current to the armature of the motor is kept low. As the handle moves across successive studs, so fewer resistors are in the circuit and so the current increases. Only when the handle reaches the far right position is the full circuit current applied to the motor.

5.2.4 Sketches

A widely used method of communicating engineering information is by freehand sketches. Such sketches are often drawn without using instruments and have such aims as:

1 Helping oneself to visualise ideas or requirements.
2 Convey ideas and information to others
3 Record details of measurements for later use.
4 As a step towards the production of a precisely drawn engineering drawing.

The methods of pictorial projection described in section 6.3 can be used for representing a three-dimensional object by a two-dimensional sketch.

5.2.5 Technical drawings

Detailed technical/engineering drawings are used to represent three-dimensional solids on a two-dimensional sheet of paper. Such drawings might include pictorial views (see section 5.2.4) and, what is termed, *orthographic drawing*. With pictorial drawing the true shapes and sizes are not generally used. Thus for the cube shown in Figure 5.11, though we can recognise it as a cube with all the face angles as 90°, it is obvious that on the figure the angles are not actually 90°. Often some of the lines are likely to be foreshortened in order to give a more realistic view. The aim with a pictorial drawing is to give the three-dimensional appearance of the object and this is not achieved by drawing to scale. With orthographic drawing true shapes and sizes are required and so a scale drawing is required.

A scale drawing means that each dimension on the drawing is proportional to the same dimension on the object. The angles on the object are, however, exactly the same on the drawing. If a drawing is the same size as the item, the scale is referred to as 'scale full size' or scale 1:1. If the drawing is half size the scale is referred to as 'scale half size' or scale 1:2. This means that every dimension on the drawing is half the size of the corresponding dimension on the item. If the drawing is one-fifth the scale is referred to as 1:5.

In order to give sufficient information so that an item can be manufactured, a number of views of the object are given. Thus there might, for example, be a view of the object from the front, a view showing the object from one side and perhaps a plan view showing what it looks like when seen from above. There might also be views showing what the object would be like if sectioned at a particular plane. See chapter 6 for details of how such drawings are made.

There are a number of types of technical drawing. These include:

1 *A single-part drawing* This shows all the dimensions needed to completely define the item so that it can be manufactured. This could include, the form, dimensions, tolerances, material, heat treatment, finish, etc.

2 *A single-part assembly drawing* This is made when several components have to be fitted together to make a complete product. The drawing shows all the information necessary for the assembly. The components are each identified by an item number or by reference to a single part drawing number of the item, and their place in the overall assembly shown. In some cases the components are listed on a separate sheet, a parts list, distinct from the assembly drawing. Figure 5.11 shows an example of an assembly drawing together with a parts list. In other cases the individual parts, the assembly drawing and the parts list might all be combined on a single sheet of paper.

3 *A collective single-part drawing* This is a drawing of similar parts of different sizes (Figure 5.11). The dimensions might be indicated by letters A, B, C, etc. and then a key indicates the values for these for particular versions

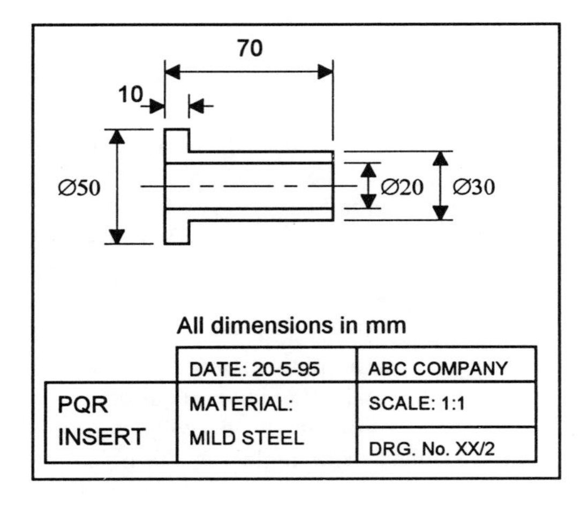

(a)

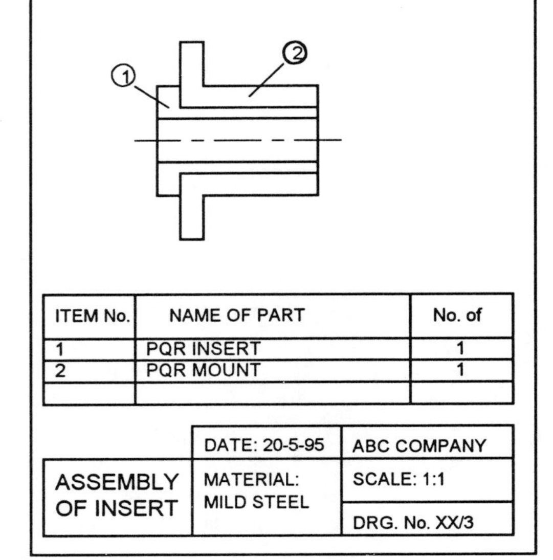

(b)

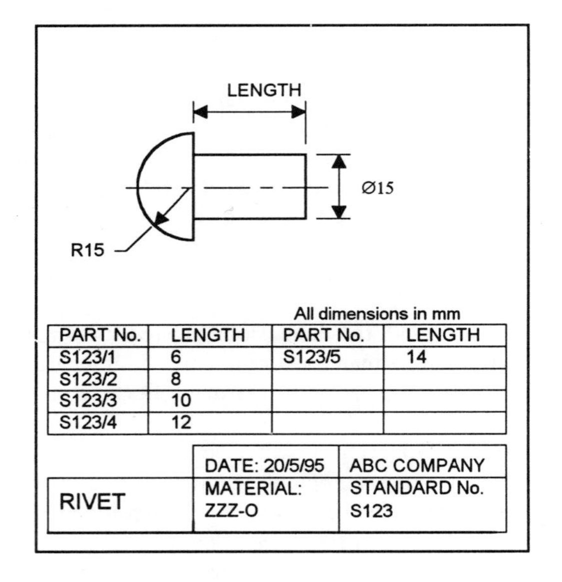

(c)

Figure 5.11 *(a) Single-part drawing, (b) assembly drawing, (c) collective single-part drawing*

of the part. For example, there may be a drawing of a bolt which is, for the same basic form, available in a number of different sizes. Standard parts might be shown in catalogues in this way, i.e. one drawing and then each item with the dimensions specified.

4 *A collective assembly drawing* This is a drawing of products assembled essentially from the same or similar parts. The resulting assemblies differ only slightly from each other. An example of such a drawing might be of a gear box with different ratios.

5.3 Selecting graphical methods

When selecting which graphical method to use to communicate engineering information it is necessary to take into account the audience being addressed and the type of information being communicated. The following three tables illustrate how the various methods could relate to such criteria:

Table 5.1 Engineering information related to graphical methods

	Engineering information				
	Marketing information to customers	Planning production sequences	Workshop manufacturing	Conceptual-ising ideas to colleagues	Costing/ estimating
Block diagrams				√	√
Flow diagrams		√			√
Circuit diagrams- symbolic				√	
Circuit diagram - layout			√		
Sketches	√			√	√
Detail drawing			√		
Assembly drawing			√		

Table 5.2 Engineering information related to purposes and possible communication method

	Purposes				
	Conceptual-ising	Marketing	Costing/ estimating	Planning	Manufacturing
Electrical	Sketch, block/circuit diagram	Sketch	Flow diagram	Flow diagram	Circuit layout
Electronic	Sketch, block/circuit diagram	Sketch	Flow diagram	Flow diagram	Circuit layout
Mechanical	Sketch	Sketch	Flow diagram	Flow diagram	Detail/assembly drawings
Hydraulic	Sketch, block/circuit diagram	Sketch	Flow diagram	Flow diagram	Circuit layout
Pneumatic	Sketch, block/circuit diagram	Sketch	Flow diagram	Flow diagram	Circuit layout

Table 5.3 Type of engineering information related to graphical methods of communication

	Type of engineering information			
	Qualitative, i.e. not involving numbers	Quantitative, i.e. involving numbers	Overall idea	Detailled information
Block diagrams	√		√	
Flow diagrams	√			
Circuit diagrams - symbolic		√		√
Circuit diagram - layout				√
Sketches	√		√	
Detail drawing		√		√
Assembly drawing				√

Example

Select a form of graphical communication to use when (a) a sales person needs to show a customer what a product will look like, (b) a technician who has to wire up a circuit, (c) a computer programmer needs to get a view of the steps required in a computer program, (d) a machine operator has to machine metal to form a product.

The likely answers are: (a) a sketch, (b) a circuit layout diagram, (c) a flow chart, (d) a technical single-part drawing. The criteria, in general, determing the method are: those which most clearly communicate the information, who is being communicated to and the cost.

5.3.1 Standardization

Communication would be made rather difficult if every person who drew a circuit diagram used different symbols for the components. Every time they used a symbol they would have to explain what it mean. Standard symbols for electrical components and the way in which electrical circuits are drawn aids communication. Likewise with technical drawings of components. For example, a standard method of representing a screw thread that everyone can understand helps with communication.

Standardization operates at international, national and company level. The *International Organization for Standardization (ISO)* is the international agency for standardization for all except electrotechnical activities which are internationally controlled by the *International Electrotechnical Commission (IEC)*. The aim of these organizations is to promote the development of standards in the world with a view to facilitating international exchange of goods and services, and to develop mutual cooperation. The *European Committee for Standardization (CEN)* comprises the national standard bodies of the EEC. The electrotechnical counterpart is the *European Committee for Electrotechnical Standardization (CENELEC)*. At the national level there is the *British Standards Institution (BSI)*. At the company level there might be standards laid down by a particular company for use in-company.

The standards used in the next two chapters for graphical communications are *British Standard (BS) 308: Engineering Drawing Practice*, *British Standard (BS) 3939: Graphical symbols for electrical power, telecommunications and electronics diagrams* and *British Standard (BS) 2917 Specification for graphical symbols used on diagrams for fluid power systems and components*.

5.3.2 Graphical techniques

Drawing may involve freehand sketching when perhaps only a pencil is used, or more formal drawing where drawing instruments such as rules, protractors, compasses, set squares, T-squares and stencils are used, or where computer programs are used. Computer software for drawing can vary from simple drawing programs that can be used to produce simple sketches to elaborate programs for producing detailed engineering drawings. The output from computers can be directed to plotters, dot matrix printers, ink/bubble jet printers or laser printers. Dot matrix printers produce only relatively coarse images, while ink/bubble jet printers and laser printers can be used to produce finer images. Ink/bubble jets can be used to produce, relatively cheap, colour drawings. Dot matrix printers, ink/bubble jet printers and laser printers generally only print on paper up to A4 size (see table 6.1), for larger sizes plotters have to be used.

The use of computers for engineering drawings is further discussed in section 6.5.

Problems

Questions 1 to 12 have four answer options: A, B, C and D. You should choose the correct answer from the answer options.

Questions 1 to 4 relate to the following methods that can be used for communicating engineering information:

A Block diagrams
B Circuit diagrams
C Pictorial sketches
D Detailed technical drawings

Select the most appropriate means of communication from the above for communicating:

1 Details of how the individual components in an electrical circuit are to be wired together.
2 A method of explaining how the various parts of a hi-fi system, i.e. record player, amplifier, speakers, etc. interact.
3 Details of an item so that a machinist can manufacture it.

4 Details for a marketing manager showing how a product will appear when manufactured.

5 Decide whether each of the following statements is TRUE (T) or FALSE (F).

A flow diagram is a method of:
(i) Specifying the steps needed to lead to some event occurring.
(ii) Showing how information flows from one system to another.

A (i) T (ii) T
B (i) T (ii) F
C (i) F (ii) T
D (i) F (ii) F

6 Decide whether each of the following statements is TRUE (T) or FALSE (F).

A pictorial sketch is a good method of:
(i) Specifying the product so that it can be manufactured.
(ii) Showing a salesperson what a product will look like.

A (i) T (ii) T
B (i) T (ii) F
C (i) F (ii) T
D (i) F (ii) F

7 Decide whether each of the following statements is TRUE (T) or FALSE (F).

A circuit diagram shows:
(i) How the components are to be connected in an electrical circuit.
(ii) Where the components are to positioned on a circuit board.

A (i) T (ii) T
B (i) T (ii) F
C (i) F (ii) T
D (i) F (ii) F

8 Decide whether each of the following statements is TRUE (T) or FALSE (F).

A block diagram is used to represent a system or sub-system with a system.
(i) Each block will have and input or inputs and output or outputs.
(ii) The lines joining systems or subsystems indicate the paths along which information flows.

A (i) T (ii) T
B (i) T (ii) F
C (i) F (ii) T
D (i) F (ii) F

9 Decide whether each of the following statements is TRUE (T) or FALSE (F).

A single-part technical drawing of a part must have:
(i) The angles the same as on the part.
(ii) The dimensions the same as on the part.

A (i) T (ii) T
B (i) T (ii) F
C (i) F (ii) T
D (i) F (ii) F

10 A fitter having to assemble a machine would need:

A A sketch of the machine
B A block diagram of the systems
C An assembly drawing
D A component drawing

11 Decide whether each of the following statements is TRUE (T) or FALSE (F).

A flow diagram is to be used to indicate the steps involved in carrying out the arithmetic operation of doubling a number. It must include:
(i) The arithmetic box for: add the number.
(ii) The decision box for: have you added the number twice?

A (i) T (ii) T
B (i) T (ii) F
C (i) F (ii) T
D (i) F (ii) F

12 Decide whether each of the following statements is TRUE (T) or FALSE (F).

A simple sketch can be used to:
(i) Present marketing information to customers.
(ii) Conceptualise the initial ideas for colleagues.

A (i) T (ii) T
B (i) T (ii) F
C (i) F (ii) T
D (i) F (ii) F

13 Select graphical methods for communicating the following engineering information, indicating the reasons for your selection:
(a) The general appearance of a product for a sales brochure for the general public.
(b) The sequence of operations needed for the operation of a domestic washing machine.
(c) Details to enable a product to be manufactured.
(d) Details of how the components in an electrical circuit are to be connected.

(e) The sequence of operations needed for a computer program.
(f) An instruction leaflet for the wiring up of a ceiling lamp.
(g) Information to enable assembly of a component from a number of parts.
(h) The sequence of the stages required in a manufacturing process.

6 Engineering drawings

6.1 Drawing conventions and standards

This chapter is about the techniques of engineering drawing. There are certain conventions and standards which are adopted concerning the layout of drawings, types of lines, lettering, methods of projection, methods of representing sections, scales and the representation of common features such as screw threads, bearings, etc. The representation of common features is discussed in more detail in chapter 7, the points regarding layout being discussed in this chapter.

6.1.1 Drawing layout

Drawing sheets in general use have the *standard (ISO) sizes* shown in table 6.1 (note that this is not a complete list, there are other sizes but they are not commonly used). Each of the sizes is twice the size of the one below it. In order to allow for errors of location of the drawing when copies are made, a frame tends to be drawn or pre-printed on the drawing sheet to enclose the drawing area. This should provide a border of at least 10 mm.

Table 6.1 ISO drawing sheet sizes
(Table 1: BS 308: Part 1: 1993)

Designation	Size in mm	Border in mm
A0	841 × 1189	20
A1	594 × 841	20
A2	420 × 594	10
A3	297 × 420	10
A4	210 × 297	10

Drawings should contain a *title block* which gives essential information necessary for the identification, administration and interpretation of the drawing. To save time, a standard title block is generally pre-printed on sheets to a standardized layout for a particular company. The basic information that appears in such a block is:

1 Name of the company.
2 Drawing number. This is a number which uniquely identifies the drawing. A particular drawing may require more than one drawing sheet. Thus there may be Drawing No. 231, Sheet 1 of 2 sheets and Drawing No. 231, Sheet 2 or 2 sheets.
3 Descriptive title of the part or assembly shown. Thus there might be a title such as 'Pulley Assembly'.
4 The scale used, e.g. 1:2 (see section 5.2.5).

5 The date of the drawing.
6 The signature or signatures of who made the drawing and checked it.
7 Information regarding which issue of the drawing it is. See the note below on revisions.
8 A symbol or statement indicating what type of projection has been used. These symbols and the forms of projection are discussed in section 6.2.
9 The unit used for all the measurements given in the drawing. There might thus be a statement 'All dimensions in mm'.
10 Possibly a statement about the drawing being copyright and references to drawing standards adopted.

In addition, the following types of information might be included:

1 The material to be used and its specification. Thus there might be a statement that the material is to be mild steel, or perhaps a more specific specification of the material by its British Standard reference number.
2 The hardness of the material.
3 Heat treatment requirements.
4 Protective treatment requirements. This might be a statement that it is to be given a coat of a rust protecting paint.
5 The finish required. This might be a statement that it is to be 'self finish', i.e. as is left by the manufacturing process, or perhaps that it is to undergo some finishing treatment.
6 The surface texture required.
7 The general tolerances to be adopted. This is an indication of how accurate the dimensions given by the drawing are to be treated.
8 Screw thread forms.
9 Tool references.
10 Warning notes. For example, 'DO NOT SCALE' might be used.

A possible standard title block that might be printed on the drawing sheets used by a particular company might be as shown in Figure 6.1. It is standard practice to put the title block at the bottom of the drawing sheet with the drawing number in the lower right-hand corner. The drawing number might also be repeated in the top right or top left corner.

Since drawings are often revised and it is important to know which revised version is given by a particular drawing, and which versions it replaces, the type of information shown in Figure 6.2 might be included on the drawing sheet. Each revision is entered and so there is a complete record of all revisions that have occurred.

All dimensions in mm

	Materials	Drawn	Company name
Title	Finish	Checked	Scale:
	Projection	Date	Drg. No.

Figure 6.1 *Typical title block*

Revision number	Revision	Signature	Date

Figure 6.2 *Revision information*

It is often useful to know where on a drawing the revision has occurred. To this end the drawing sheet may be divided into a grid. Figure 6.3 shows such a grid. Then marginal grid references can be used to identify the position of the revision and a zone reference indicated in the reference table. Thus, for example, B4 could indicate a particular zone on the drawing sheet in which revision occurs. Where such a grid is used then the block shown in Figure 6.2 would have an extra column for the zone to be entered.

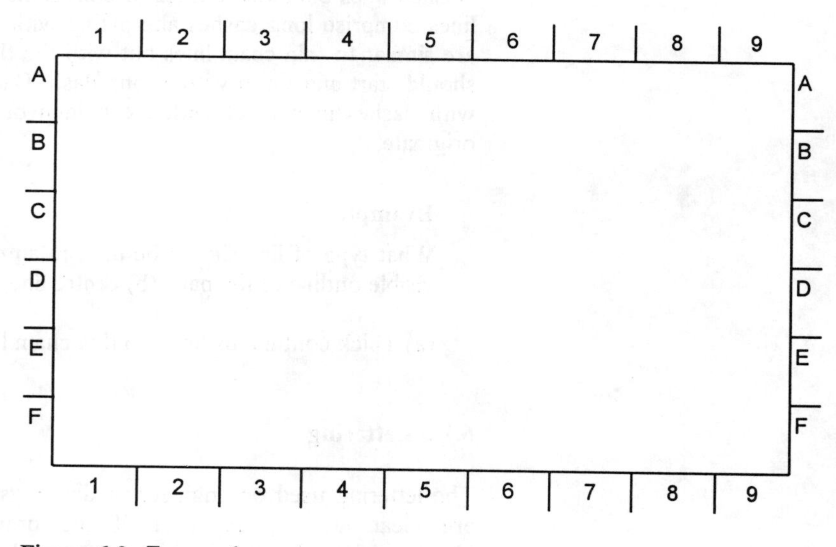

Figure 6.3 *Zone referencing*

In the case of assembly drawings, a block for the parts list may also be included on the drawing sheet. Figure 6.4 shows what such a block might look like with some idea of the types of entries that might be used.

Part no.	Detail ref.	Name of part	Material	No. of
6				
5				
4				
3	Standard	Cap screw	40 × M8 × 1.0	4
2	231/2	Pulley	Mild steel	2
1	231/1	Bracket	Mild steel	1

Figure 6.4 *Parts list*

6.1.2 Lines

The lines on an engineering drawing should be black, dense and bold and of a consistent density. Thus lines on any one drawing sheet should preferably all be in pencil or all in ink. The type of line used will depend on what the line is indicating. Figure 6.5 shows the types of lines and an example illustrating their use. Two thicknesses of line are used, 0.3 mm and 0.7 mm. Dashed lines comprise dashes of consistent length and spacing. Thin chain lines comprise long dashes alternating with short dashes. Thick chain lines are similar to thin chain lines but with the thicker line used. All chain lines should start and finish with a long dash. Dashed lines should start and end with dashes in contact with the hidden or visible lines from which they originate.

Example

What type of line should be used in an engineering drawing for (a) the visible outline of the part, (b) centre lines, (c) hidden outlines?

(a) Thick continuous line, (b) thin chain lines, (c) thin short dashes.

6.1.3 Lettering

The lettering used on engineering drawings needs to have characters that are clear and legible, even if the drawing is reduced in size by photocopying. To this end, characters should not have serifs or other embellishments. The text in this book has characters with serifs, i.e. 'curly bits'. With serifs we have

Aa Bb Cc Dd Ee Ff Gg, etc. and 1 2 3 4 5, etc..

The comparable text without serifs is

Aa Bb Cc Dd Ee Ff Gg, etc. and 1 2 3 4 5, etc.

Types of line

Line	Description	Application
A	Continuous thick	A1 Visible outlines A2 Visible edges
B	Continuous thin	B1 Imaginary lines of intersection B2 Dimension lines B3 Projection lines B4 Leader lines B5 Hatching B6 Outlines of revolved sections B7 Short centre lines
C	Continuous thin irregular	*C1 Limits of partial or interrupted views and sections, if the limit is not an axis
D	Continuous thin straight with zigzags	†D1 Limits of partial or interrupted views and sections, if the limit is not an axis
E	Dashed thick	E1 Hidden outlines E2 Hidden edges
F	Dashed thin‡	F1 Hidden outlines F2 Hidden edges
G	Chain thin	G1 Centre lines G2 Lines of symmetry G3 Trajectories and loci G4 Pitch lines and pitch circles
H	Chain thin, thick at ends and changes of direction	H1 Cutting planes
J	Chain thick	J1 Indication of lines or surfaces to which a special requirement applies (drawn adjacent to surface)
K	Chain thin double dashed	K1 Outlines and edges of adjacent parts K2 Outlines and edges of alternative and extreme positions of movable parts K3 Centroidal lines K4 Initial outlines prior to forming §K5 Parts situated in front of a cutting plane K6 Bend lines on developed blanks or patterns

NOTE. The lengths of the long dashes shown for lines G, H, J and K are not necessarily typical due to the confines of the space available.

†This type of line is suited for production of drawings by machines.

‡The thin F type line is more common in the UK, but on any one drawing or set of drawings only one type of dashed line should be used.

§Included in ISO 128-1982 and used mainly in the building industry.

Figure 6.5 *Types of line (Table 3 and Figure 2: BS 308: Part 1: 1993)*

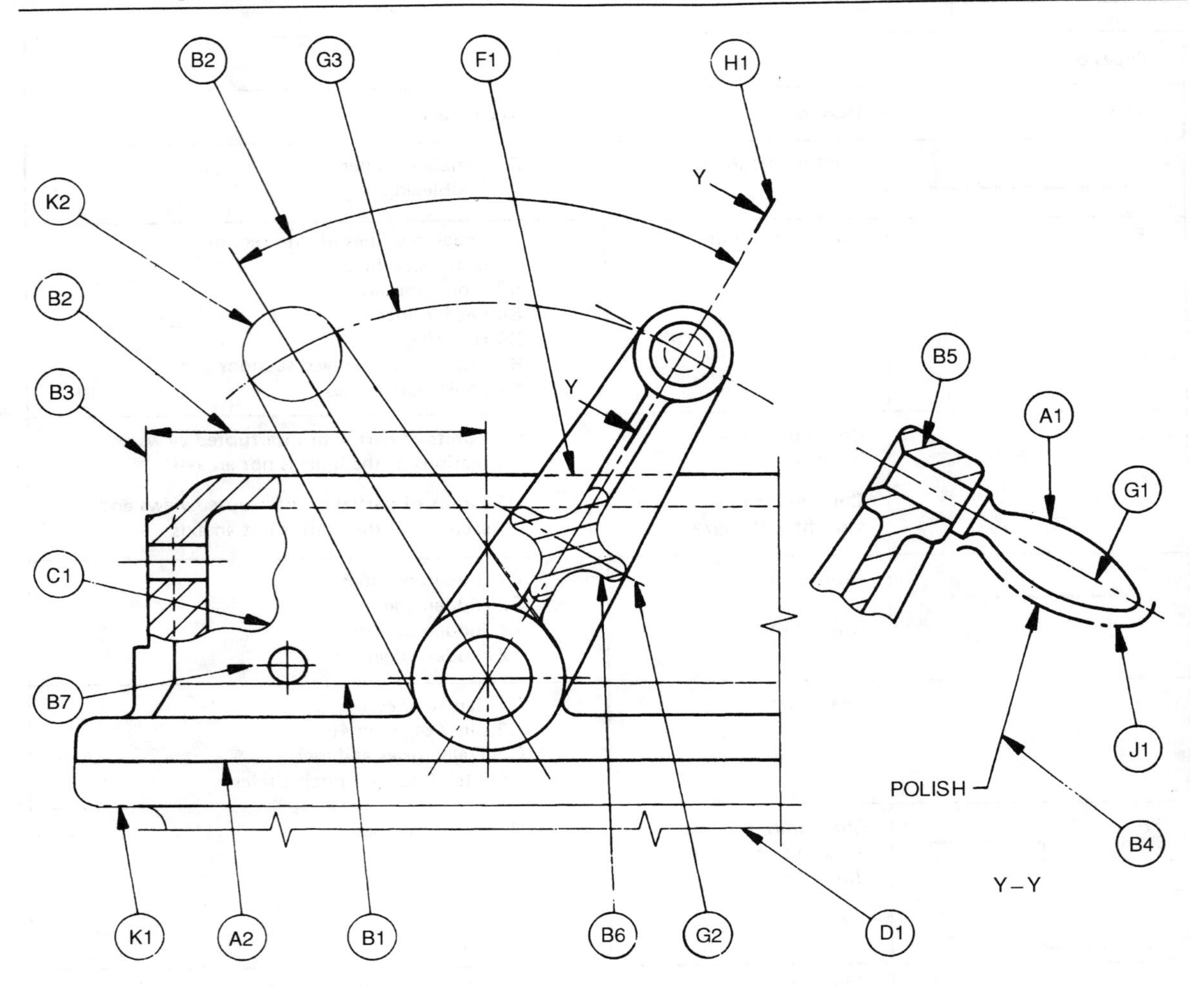

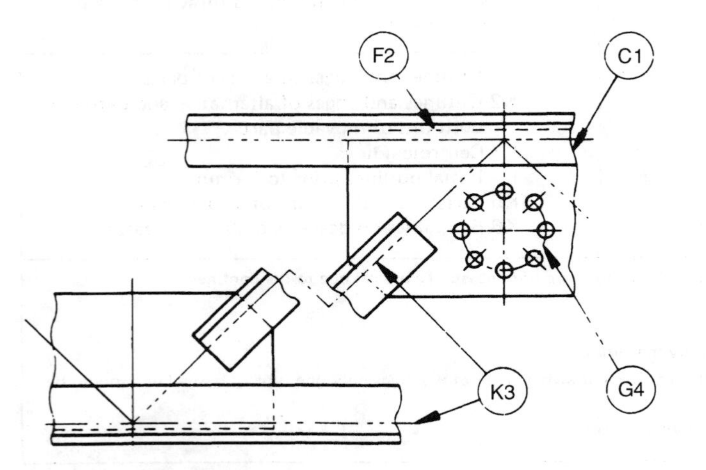

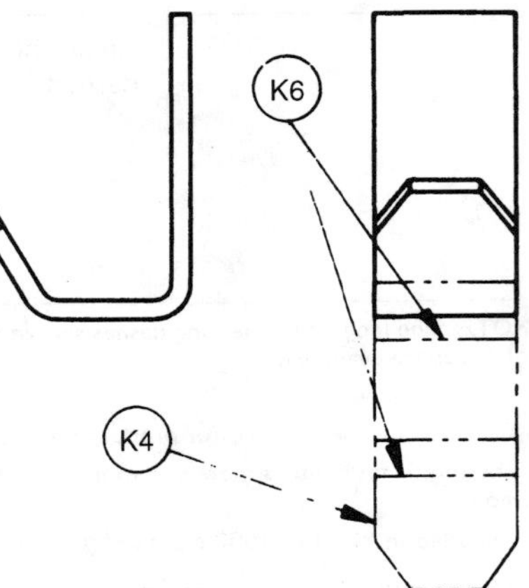

Figure 6.5 *(continued)*

Vertical or sloping characters can be used, the sloping characters, without serif, being

Aa Bb Cc Dd Ee Ff Gg, etc. and *1 2 3 4 5*, etc.

Capital letters are preferred to lower case as they are less likely to be misread. If a note is to be emphasised, it should not be underlined but drawn using larger characters.

6.1.4 Dimensions

A *dimension* is a number on a drawing which, together with the units indicated for that drawing, indicates the size of specific features of the part drawn. Each dimension that is needed for the complete definition of a finished product should be given on the drawing and should appear only once. It should not be necessary for such dimensions to have to be deduced from measurements made on the drawing and using the scale given, or deduced from other dimensions.

Since the most prominent linework on an engineering drawing is the visible outline of the part, such lines are drawn as thick continuous lines while projection and dimension lines are thin continuous lines. *Projection lines* are lines projected from points, lines of surfaces to enable the dimensions to be placed, wherever possible, outside the outline of the part. Where projection lines are extensions of lines of the outline, they should start a little clear of the outline and extend to just a little beyond the dimension line. The arrowheads on dimension lines should be easily seen and normally not less than 3 mm long, the points touching the projection or other limiting lines. Figure 6.6 illustrates the above points.

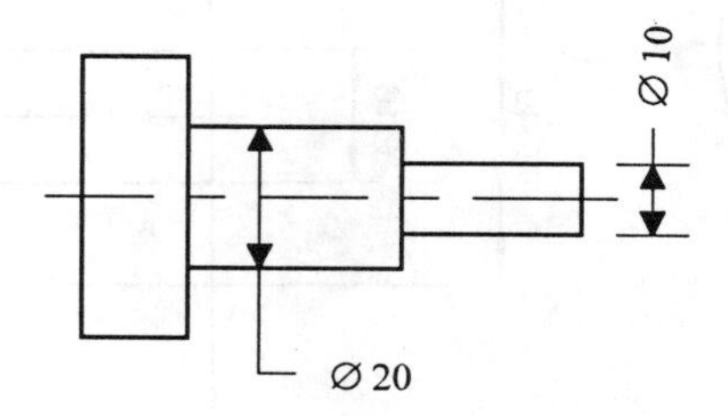

Figure 6.6 *Projection and dimension line*

Where a number of dimensions are to be given from a common datum line, surface or point, the normal method to be used is shown in Figure 6.7.

Some dimensions are marked differently to those indicated in Figure 6.7. In some instances an overall dimension for a part may be given although such a dimension is not necessary if each of the intermediate dimensions are given. Such a dimension is termed an *auxiliary dimension*. Auxiliary dimensions should be included in parentheses, i.e. (...). A dimension indicating the diameter of a circle should be preceded by the symbol Ø. Radii should be

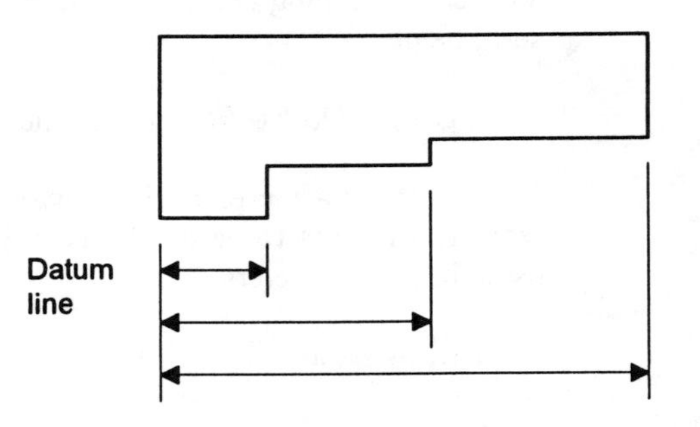

Figure 6.7 *Dimensions from a common datum*

dimensioned by a dimension line which passes through, or is in line with, the centre of the arc. The line should have just one arrowhead at the end toughing the arc. The letter R should precede the dimension. Figure 6.8 shows an examples of a drawing which includes an auxiliary dimension, dimensions of diameters and radii.

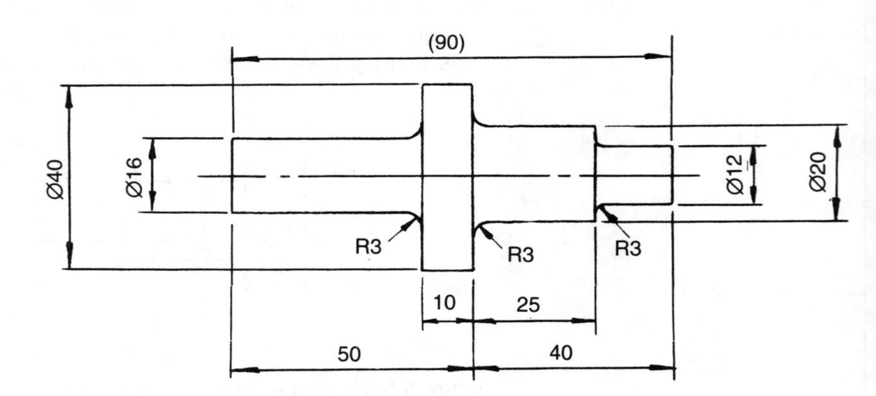

Figure 6.8 *Example of dimensioning*

Figure 6.9 shows methods that can be used for giving angular dimensions.

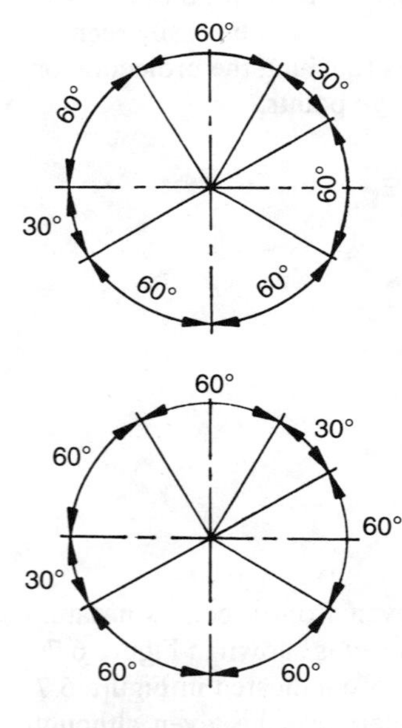

Figure 6.9 *Methods of representing angular dimensions (Figures 16: BS 308: Part 2: 1985(1992))*

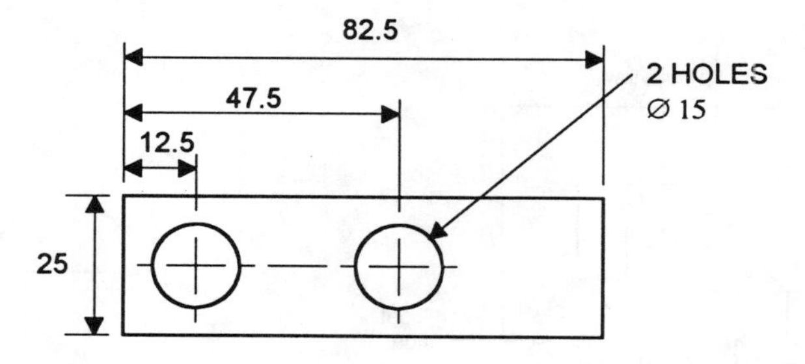

Figure 6.10 *Examples of leader lines*

6.1.5 Leader lines

Leader lines are lines drawn to indicate where dimensions or notes are intended to apply. They should be thin continuous lines terminating in arrowheads when terminating on a line and dots when terminating within the outline of the part. Figure 6.10 shows an example of such a line. To avoid confusion with other lines, leaders that touch lines should be nearly at right angles to the surface and not be parallel to dimension or projection lines.

6.2 Orthographic projection

Orthographic projection is a method used by engineers to represent three-dimensional shapes on a two-dimensional drawing sheet as a number of two-dimensional views. There are two methods that can be used, *first angle projection* and *third angle projection*. With first angle projection, a view is obtained by looking from one side of the object and projecting what is seen through to the other side. With third angle projection, a view is obtained by looking from one side and indicating on that same side what is seen. You can think of the drawing sheet as being folded to form the sides of a box with the views projected on to it. Then the box is unfolded to give the views on the flat sheet. Figure 6.11 illustrates these forms of projection with views being taken from the front, from above and from the right side and the results as they would appear on a drawing sheet. We could also have views taken from the left side and upwards from underneath. With first angle projection the right side view appears on the left of the front view, with third angle projection it appears on the right side. Likewise, with first angle projection the left side view appears on the right of the front view, with third angle projection it appears on the left side. In selecting which views to show, the choice is determined by which features it is essential to show on the drawing. It is necessary to clearly indicate on a drawing which form of projection is being used. This is sometimes done by stating it in words, in other cases symbols are used (Figure 6.12).

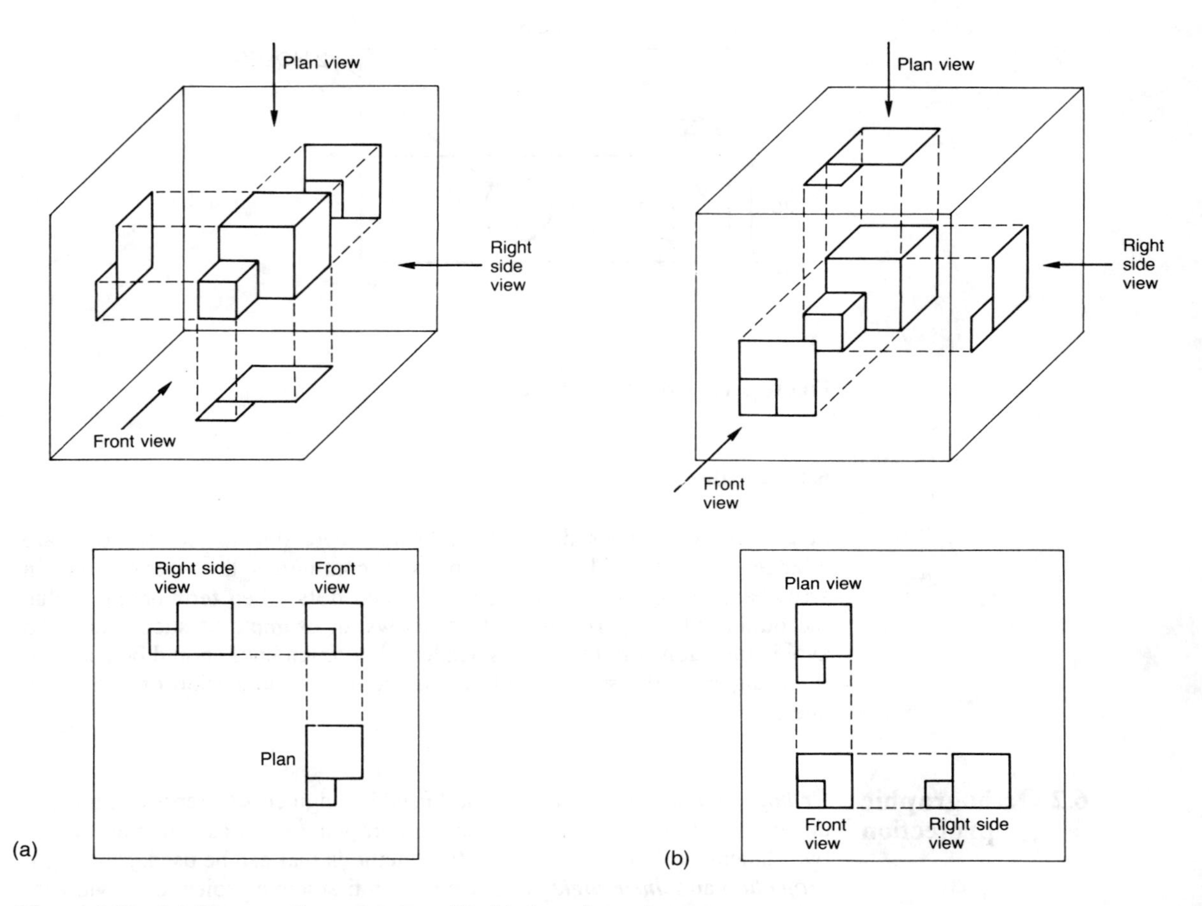

Figure 6.11 *(a) First angle projection, (b) third angle projection*

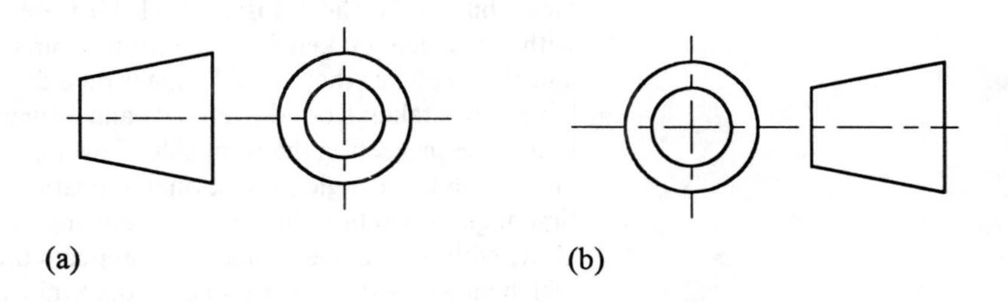

Figure 6.12 *Symbols indicating method of projection (a) first angle, (b) third angle*

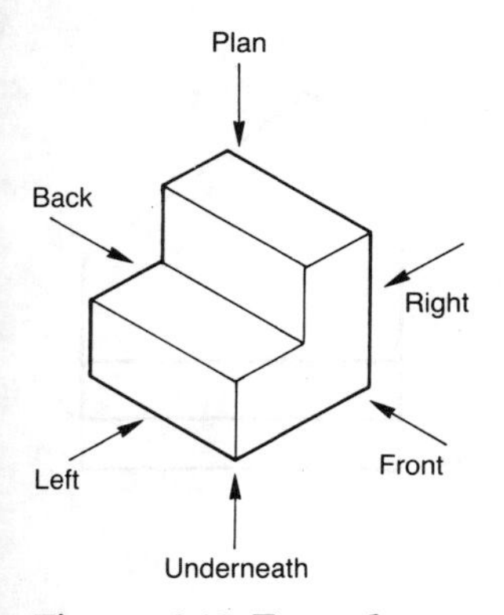

Figure 6.13 *Example*

.Example

Sketch, using first angle projection, all the six views of the part shown in Figure 6.13 and then select the views best able to show the details of the part.

Figure 6.14 shows all the six views. To show that the part has a rectangular base we need either the plan view or that obtained looking upwards at it. The view from above, however, shows the position of the step. A view from the left shows the position of the step, a view from the right giving less information. A view from the front, or the rear, shows the depth of the step. Thus the three views, from above, the left and the front give the required information.

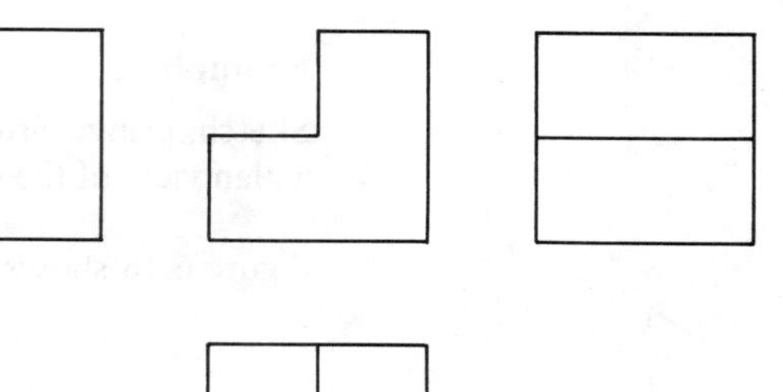

Figure 6.14 *Example*

Example

Sketch, using first angle projection, three views of the part shown in Figure 6.15, selecting the views best able to show the details of the part.

We need to decide what views should be used. A front view will show the base thickness and length, the thickness and position of the vertical wall, the radii at its base and the shape of the lip at the left hand end. A view from the left will show the detail of the lip and that is goes right across the base. A view from the right would not show this information. A plan view would show that the base is rectangular. A view from the underneath would give the same information. Thus the choice would be a front view, a view from the right and a plan view. Figure 6.16 shows the resulting sketch.

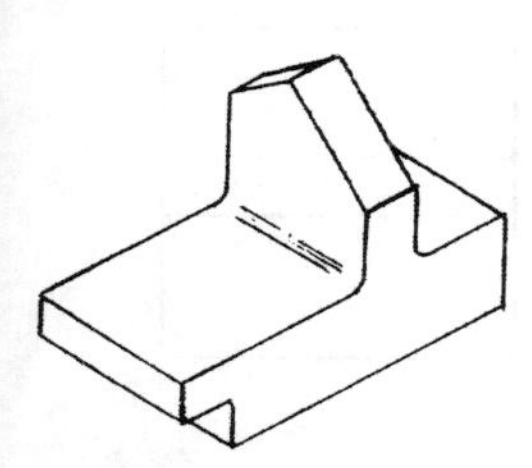

Figure 6.15 *Example*

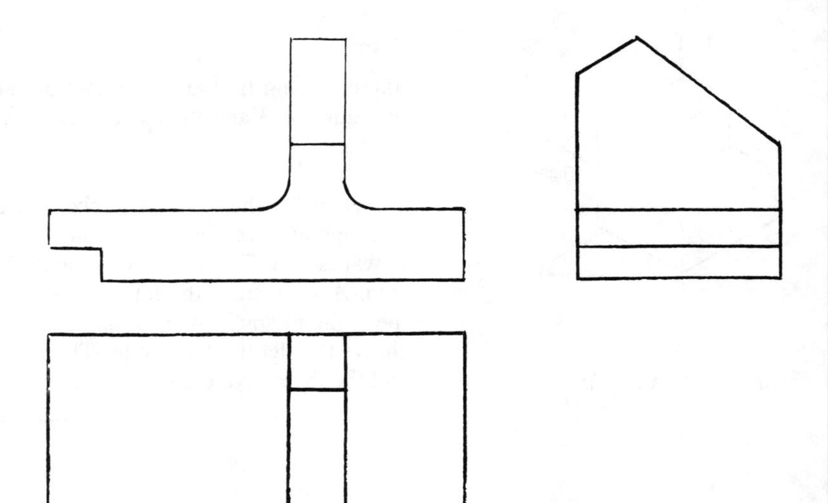

Figure 6.16 *Example*

Example

Sketch, using third angle projection, the front view, both end views and a plan view of the object shown in Figure 6.17.

Figure 6.18 shows the resulting sketches.

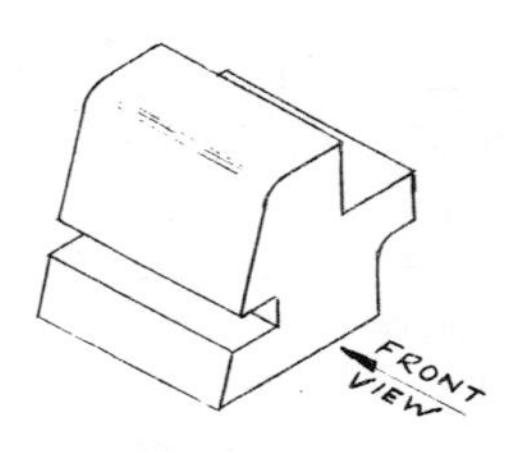

Figure 6.17 *Example*

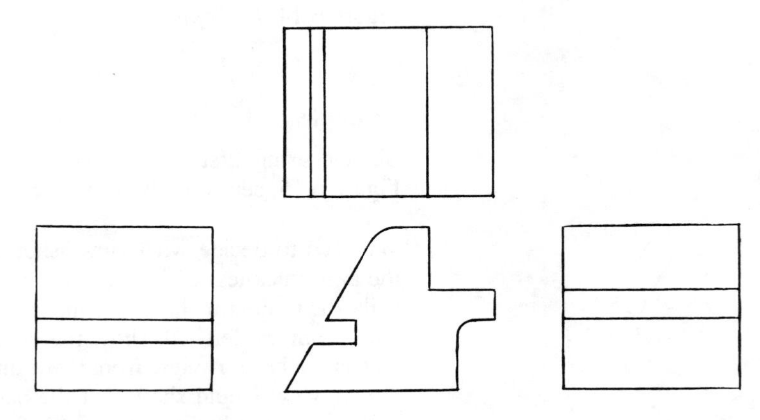

Figure 6.18 *Example*

6.2.1 Projection of views

When making a drawing, dimensions are transferred from one view to another by *projection*. This method is illustrated in Figure 6.19 for a first angle projection drawing. The sequence might be:

1 Using measurements taken from the object, draw the front view.
2 Draw projection lines (thin lines) downwards to indicate the positions on the plan of features.
3 Using width measurements taken from the object, draw the plan.
4 The third view is now obtained from projections from the front and plan views. For the left side view, draw a thin line at 45° to the horizontal to the right of the plan (for the right side view it would slope in the other direction and be to the left of the plan).
5 Project lines from the plan horizontally and where they touch the 45° line project them vertically upwards. The intersection of these projection lines with lines projected horizontally from the front view locates the features on the side elevation.

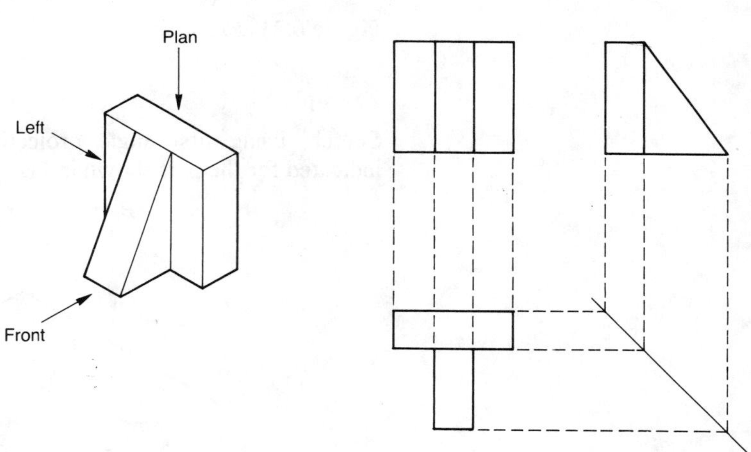

Figure 6.19 *Projection*

6.2.2 Hidden detail

Details that cannot be directly seen from a particular view, i.e. hidden detail can be shown by dashed lines. These hidden details may be internal features or external features.

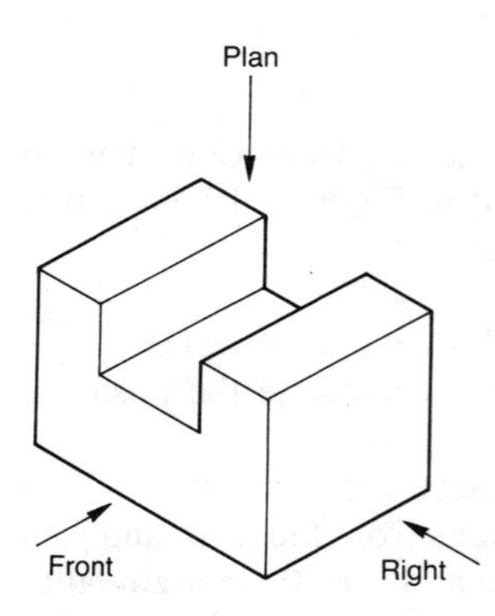

Figure 6.20 *Example*

Example

Sketch, using first angle projection, the three views in the directions indicated for the part shown in Figure 6.20.

Figure 6.21 shows the resulting sketch. For the plan view and the front view there is no hidden details. With the view from the right, there is hidden detail of the groove and this is indicated by the dashed line.

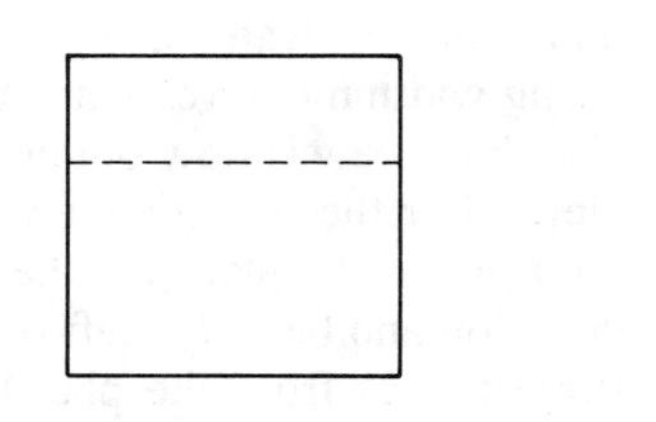

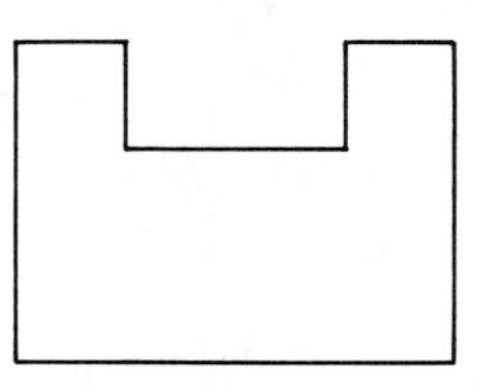

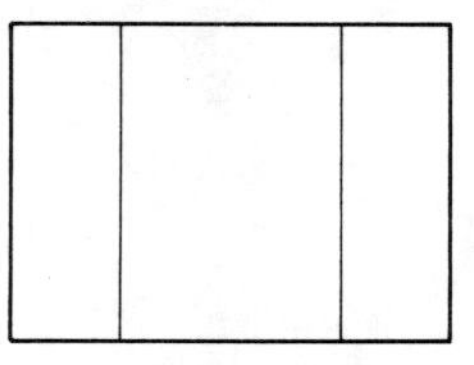

Figure 6.21 *Example*

Example

Sketch, using first angle projection, three views in the directions indicated for the part shown in Figure 6.22.

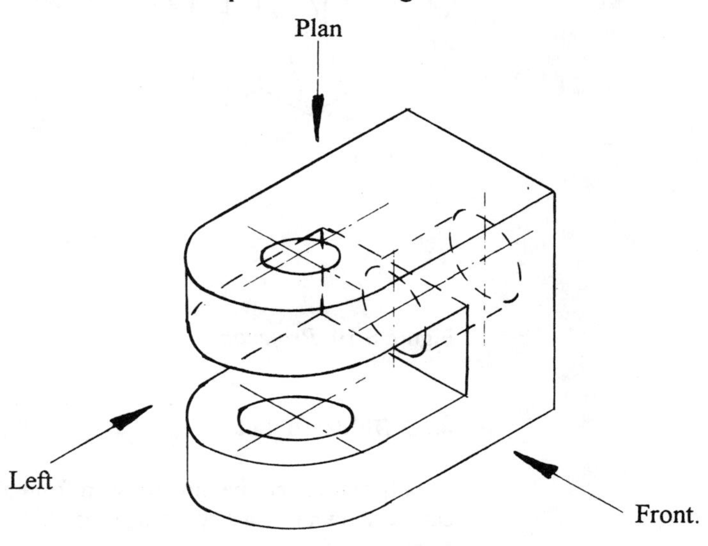

Figure 6.22 *Example*

Figure 6.23 shows the resulting sketches. In the front view all the holes are hidden from sight and can only be indicated as hidden detail by the use of dashes. In the plan view the central hole in the end of the fork and the inner vertical face are hidden and can only be represented by dashed lines. For the left side view the two holes in the fork ends are hidden and can only be represented by dashed lines. Note the use of thin chain lines to indicate the centre lines of the holes.

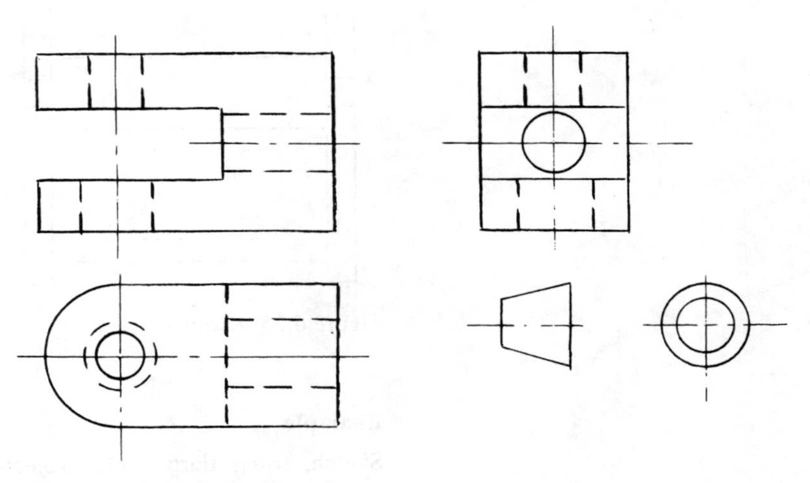

Figure 6.23 *Example*

Example

Figure 6.24 shows a drawing of an object in first angle projection. Hidden details have not been included. Add to the views any hidden details that can be included.

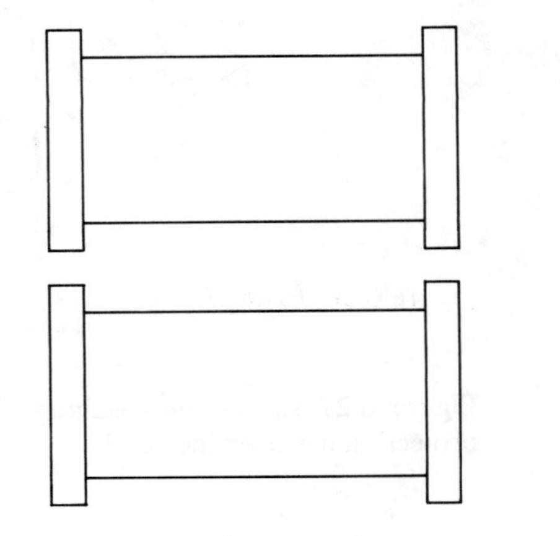

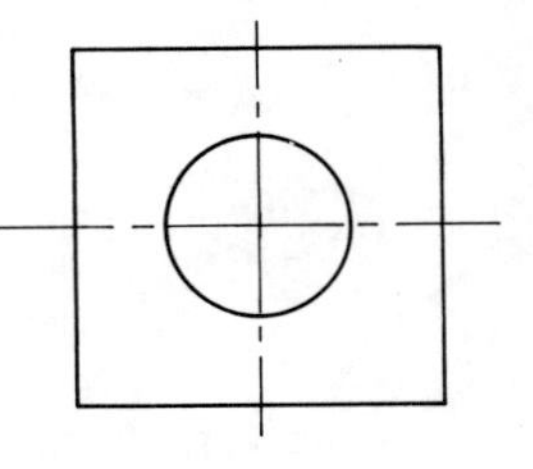

Figure 6.24 *Example*

Figure 6.25 shows the drawing with the hidden details included. Note that it has been assumed that the central portion of the item is circular, it could, however, be square. The symbol for first angle projection has been included.

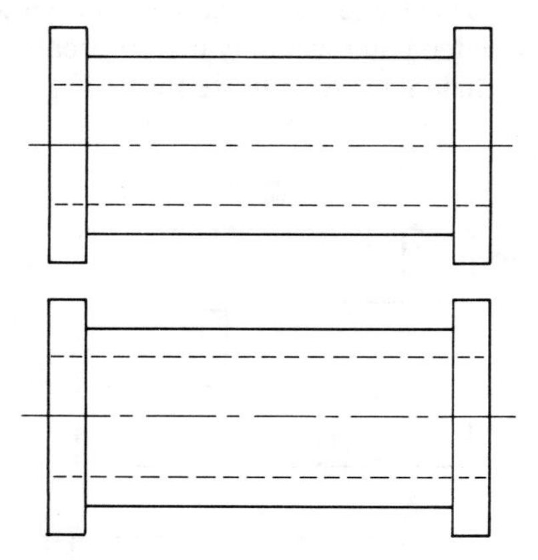

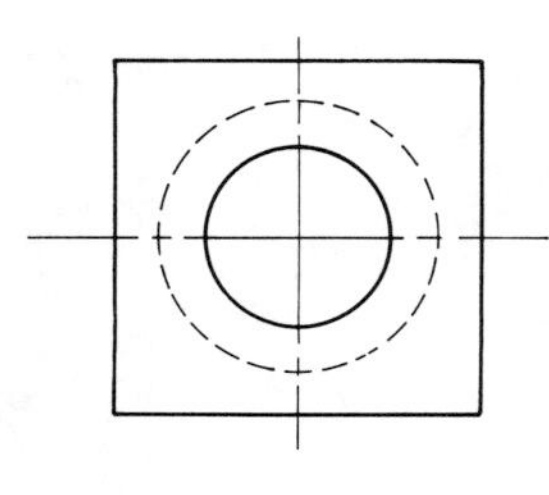

Figure 6.25 *Example*

Example

Sketch, using third angle projection, three views in the directions indicated for the part shown in Figure 6.26.

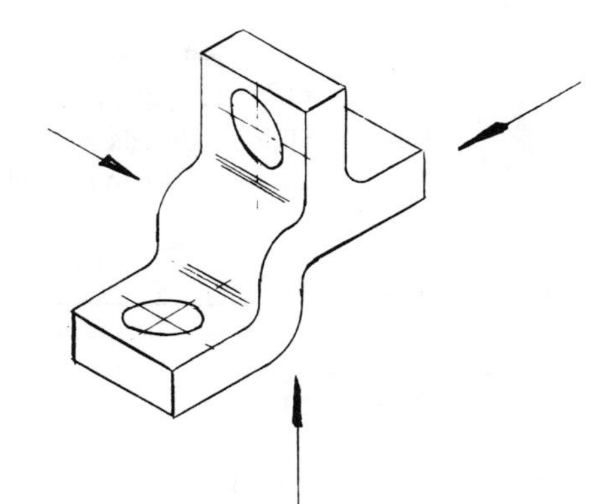

Figure 6.26 *Example*

Figure 6.27 shows the resulting sketch. The symbol for third angle projection has been included.

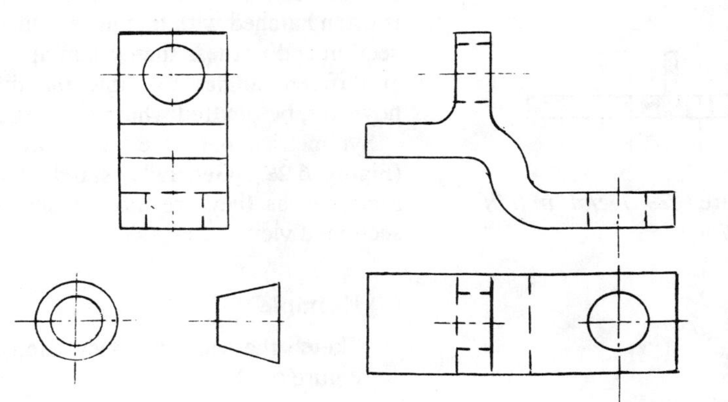

Figure 6.27 *Example*

6.2.3 Sectioned views

With objects having interior features, external views even with hidden details indicated may not be able to adequately show all the details of such features. In such situations sectional views can be drawn. A sectional view is obtained by taking an imaginary cut through the object at that part of the object where some feature needs to be clearly seen. Cutting planes are indicated by chain lines, thickened at the ends and at changes of section, thin elsewhere and designated by capital letters with the directions of viewing being indicated by arrows resting on the cutting line (Figure 6.28).

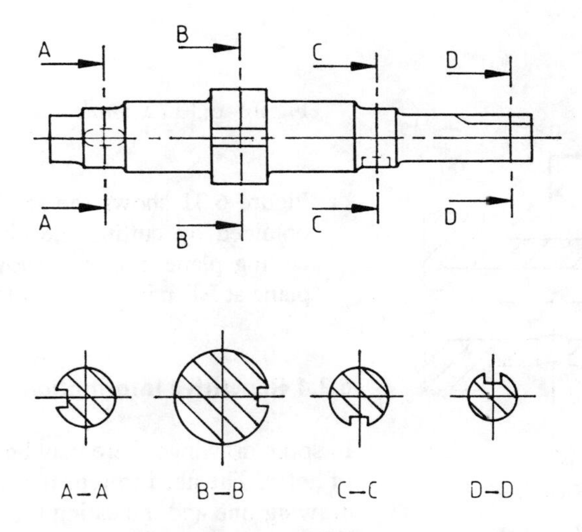

Figure 6.28 *Sections (Figure 35(b): BS 308: Part 1: 1993)*

In Figure 6.28 the sections are shown at a number of cutting planes, namely AA, BB, CC and DD in the directions indicated by the arrows. With a sectional view, the solid material which has been cut by the cutting plane is often hatched with thin lines, usually at 45° to the horizontal or axis of the section and not less than 4 mm apart. Adjacent cut parts should be hatched at different angles to show the differences in the parts. Hatching may, however, be omitted where the meaning of the drawing is clear without it.

Symmetrical parts are often drawn half in outside view and half in section (Figure 6.29). Generally, standard items such as bolts, nuts, etc. are not sectioned as they are more easily recognised by an external view than a sectioned view.

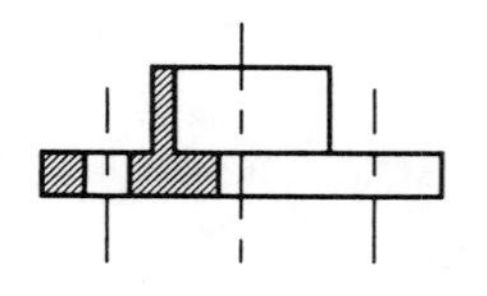

Figure 6.29 *Detail in half section*

Example

Sketch the sectioned view along the cutting plane shown for the item Figure 6.30.

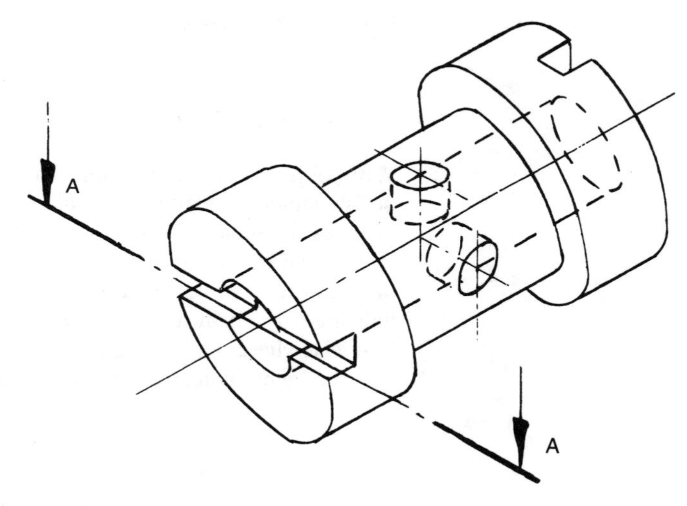

Figure 6.30 *Example*

Figure 6.31 shows the sectioned view obtained by imagining the view obtained by cutting the object at the indicated plane. Note that this cutting plane does not show all the hidden detail and another cutting plane at BB might be used to show more detail.

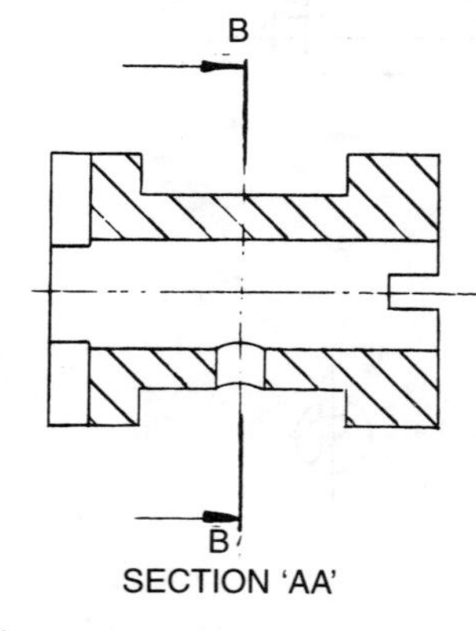

Figure 6.31 *Example*

6.2.4 Repetitive information

In some drawings there may be repetition of identical features, e.g. a series of holes. The need to repeatedly draw the same features may be avoided by drawing one and indicating the positions of the others by their centre lines. Where necessary a note may be added to make this repetition clear.

6.3 Pictorial projection

Sometimes there is a need to show a view of part in three dimensions, i.e. pictorially, instead of the two-dimensional views used with orthographic projection. Such a view enables the part to be more easily visualised. There are two main forms of pictorial projection used by engineers: isometric projection and oblique projection. The following gives details of these methods.

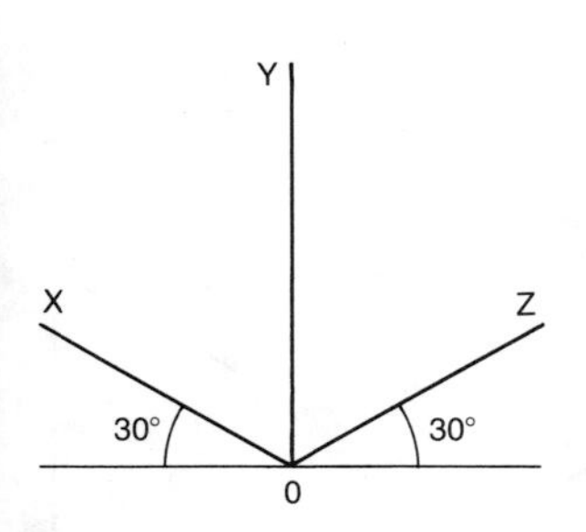

Figure 6.32 *Axes*

6.3.1 Isometric projection

Isometric projection is based on the three axes drawn the way shown in Figure 6.32. OY is the vertical axis, OX and OZ are the other two axes and are drawn from O at 30° to the horizontal. Thus if we take the views from an orthographic projection drawing, we have the views in the positions shown in Figure 6.33.

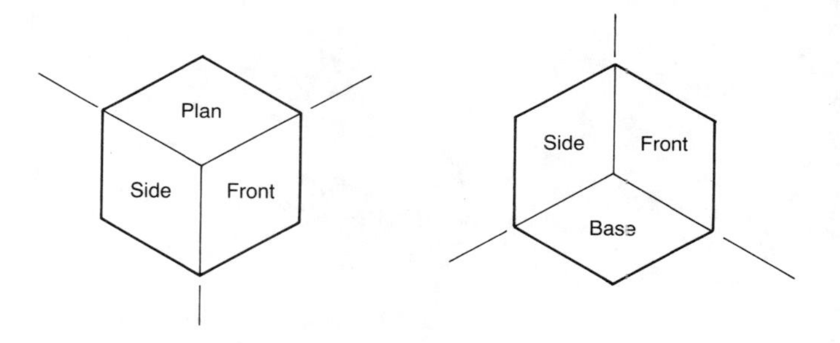

Figure 6.33 *Isometric construction*

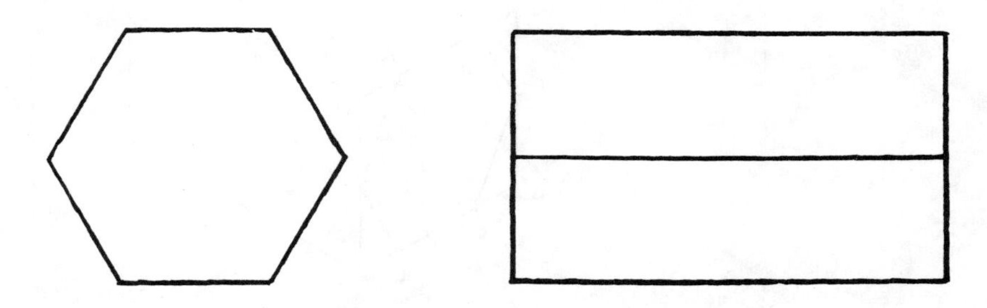

Figure 6.34 *Orthographic projection drawing*

The procedure that can be adopted for obtaining an isometric projection is illustrated by the following example. Consider a hexagonal prism which has the end view and front view shown in Figure 6.34. Consider first the end view. A rectangular frame GHJK is drawn round it to completely enclose it (Figure 6.35(a)) and this box then drawn isometrically (Figure 6.35(b)), i.e. GH and KJ are drawn along parallel axes at 30° to the horizontal, with GK and HJ vertical lines. The length JD is taken from the orthographic end view in (a) and measured out along the isometric JK line

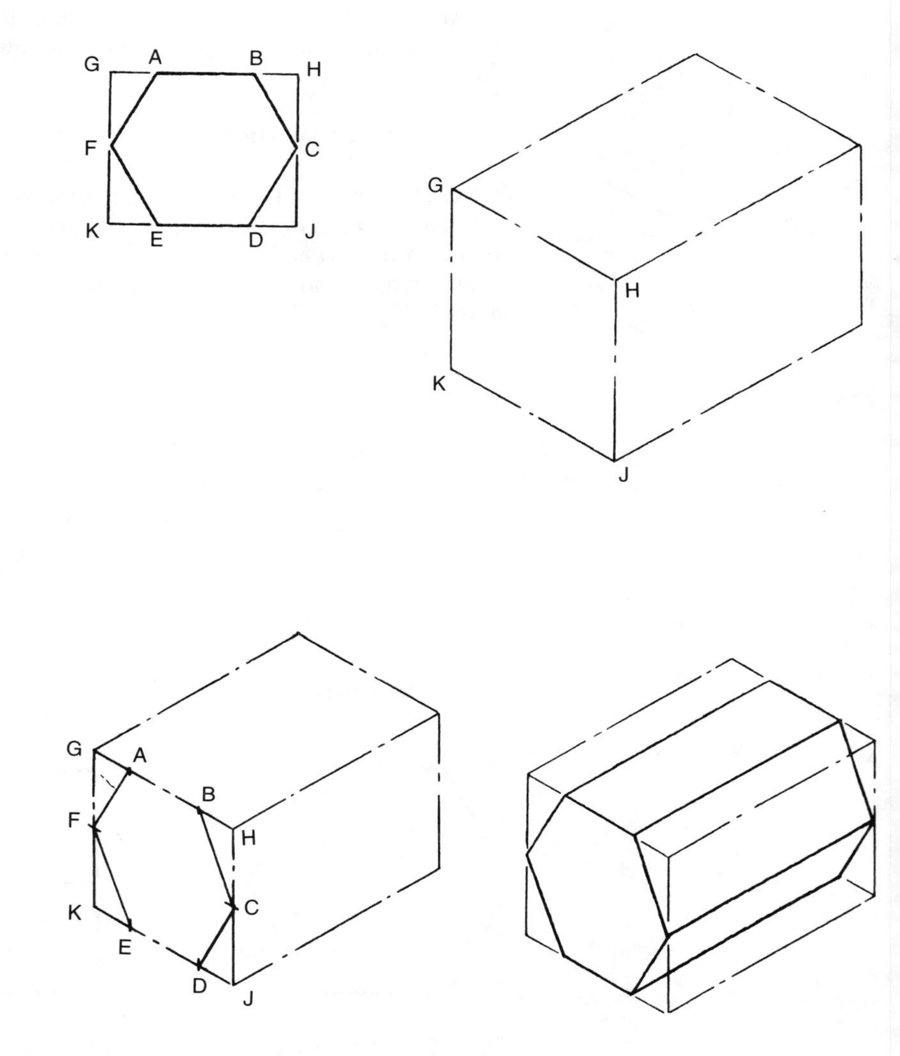

Figure 6.35 *Isometric projection*

to locate the point D. The length JC is taken from the orthographic end view in (a) and measured out along the isometric JH line to locate point C. By repeating this for each of the points we can obtain Figure 6.35(c). A similar operation for the front view enables the isometric projection view shown in Figure 6.35(d) to be obtained.

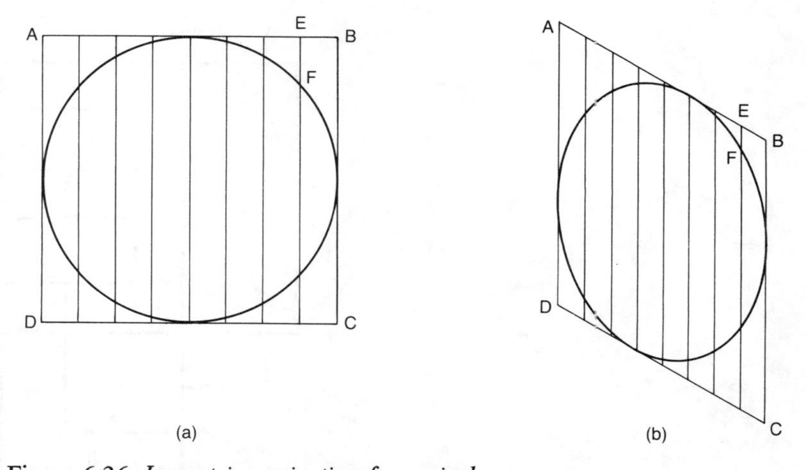

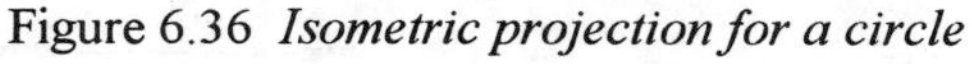

Figure 6.36 *Isometric projection for a circle*

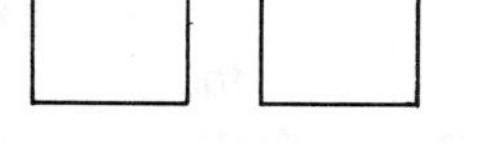

This method of transferring dimensions from the orthographic projection drawing to the isometric projection drawing can be used to make views of objects of any shape. Figure 6.36 illustrates this for a circle, the circle in the orthographic projection being turned into an ellipse in the isometric view. As before, a box ABCD is drawn round the orthographic view (Figure 6.36(a). This box is then drawn isometrically with AD and BC vertical and AB and CD at 30° to the horizontal (Figure 6.36(b)). In order to locate points on the circle, a number of other vertical lines are drawn. Thus for point F on the circle, this is located on the isometric view as being the distance BE out along the BA line and then down the vertical a distance EF. In a similar way, other points round the circle can be transferred to the isometric view.

A pictorial drawing with all the dimensions to the same scale often appears out of proportion. For example, for a cube with the orthographic projection shown in Figure 6.37(a) we can obtain the isometric projection shown in Figure 6.37(b). However, this cube looks bigger than the orthographic drawing would suggest. To overcome this, the sides of the cube are shortened to 0.82 of their actual dimensions to give the isometric projection shown in Figure 6.37(c). This appears a better representation of the cube.

(a)

(b)

(c)

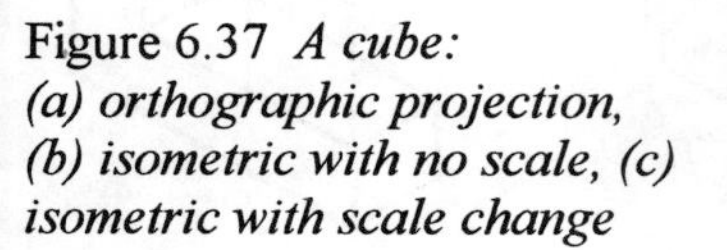

Figure 6.37 *A cube: (a) orthographic projection, (b) isometric with no scale, (c) isometric with scale change*

Example

Sketch the isometric view of the object described by Figure 6.38.

Figure 6.39(a) shows the isometric view box 40 mm × 10 mm × 60 mm in which the object could fit. Figure 6.39(b) shows two end faces which are in contact with the box and for which dimensions can be readily transferred. By transferring other dimensions the isometric view shown in Figure 6.39(c) can be built up.

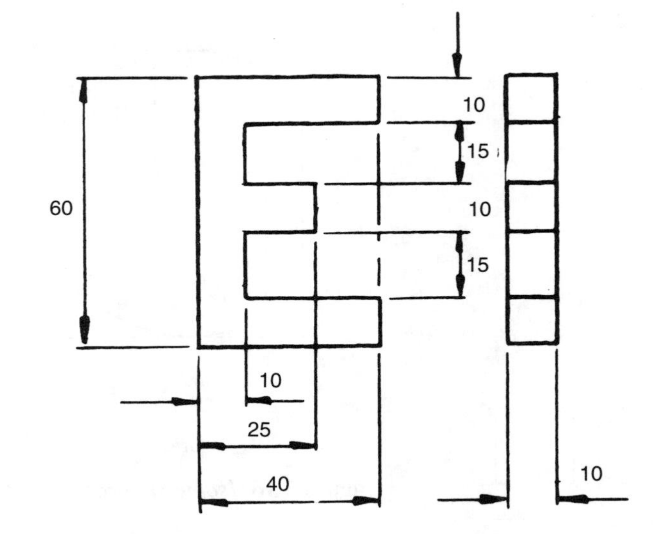

Figure 6.38 *Example*

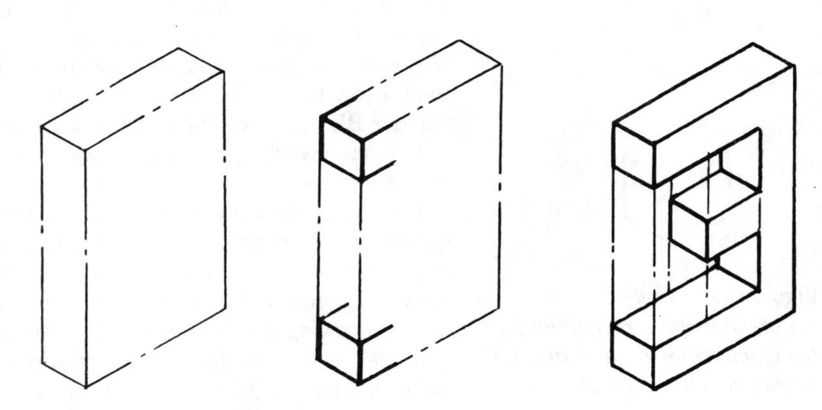
Figure 6.39 *Example*

Example

Sketch the isometric view of the object described by Figure 6.40.

Figure 6.41(a) shows how the end view of the object can be boxed and divided up in order to obtain the end view dimensions for the isometric view. Figure 6.42(b) shows the isometric box and the construction lines that can be used to determine the positions of points in the isometric view.

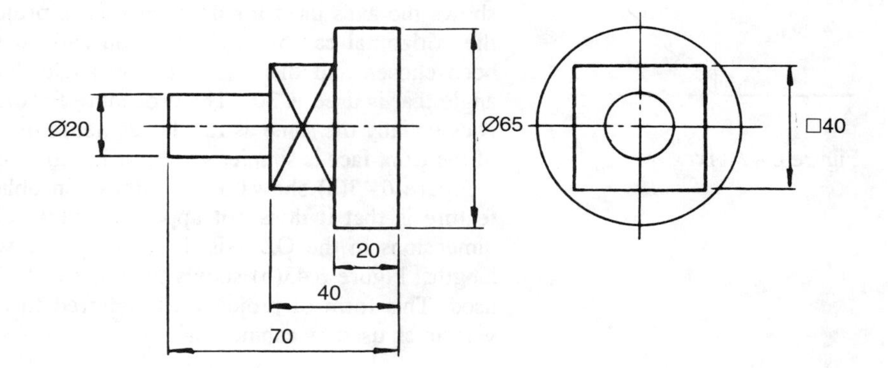

Figure 6.40 *Example*

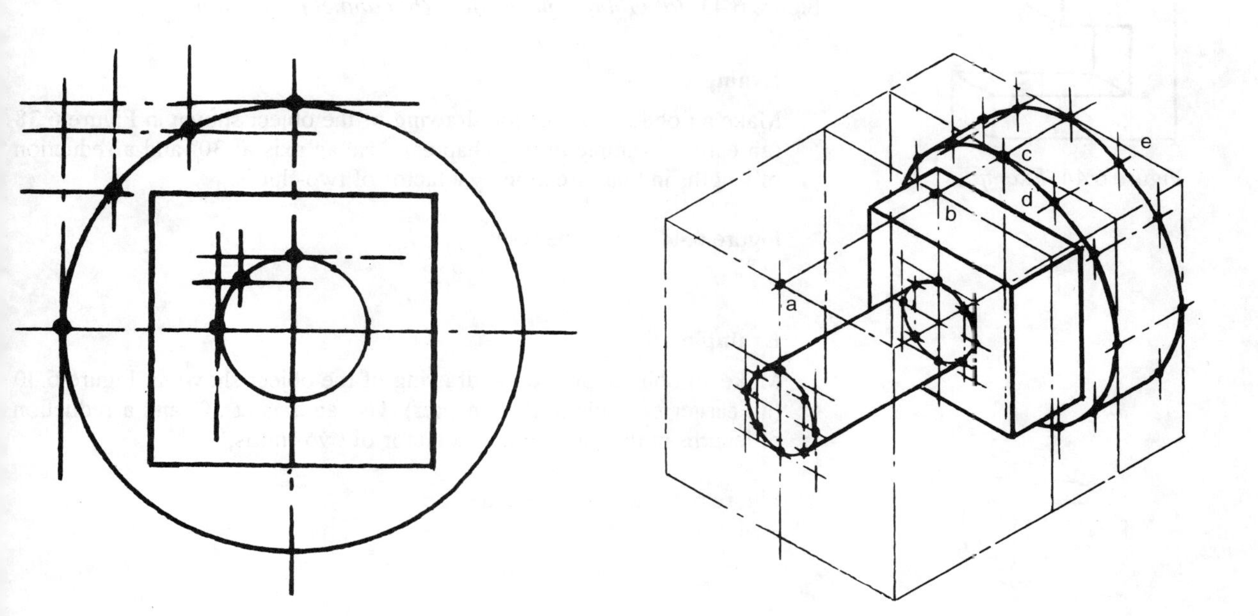

Figure 6.41 *Example*

6.3.2 Oblique projection

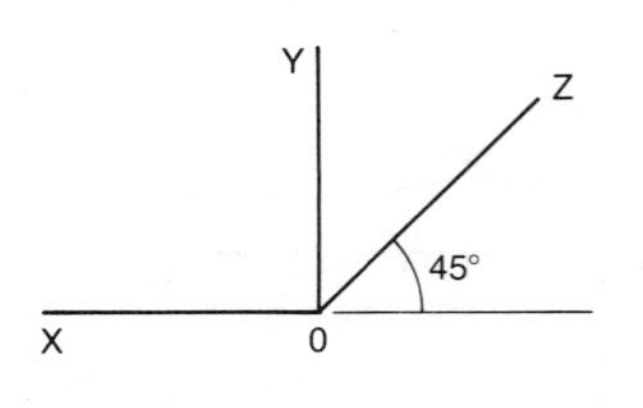

Figure 6.42 *Axes*

Oblique projection has the advantage over isometric projection that the front face of the object is shown face-on and to its true shape. Figure 6.42 shows the axes used for drawing such a projection. The OZ axis angle to the horizontal can be any value you care to select. In the figure, 45° has been chosen and this is a common angle that is used. Another common angle that is used is 30°. The procedure for drawing this form of projection is essentially the same as for the isometric projection, although the drawing of the front face is simpler since it is its true shape.

Figure 6.43(a) shows a cube drawn in oblique projection. A noticeable feature is that it does not appear to be a cube. To avoid this distortion, dimensions in the OZ axis direction are drawn at a fraction of their true lengths. Figure 6.43(b) shows the cube with the factor of two-thirds being used. This form of projection is referred to as *cabinet projection* since it was much used by cabinet makers.

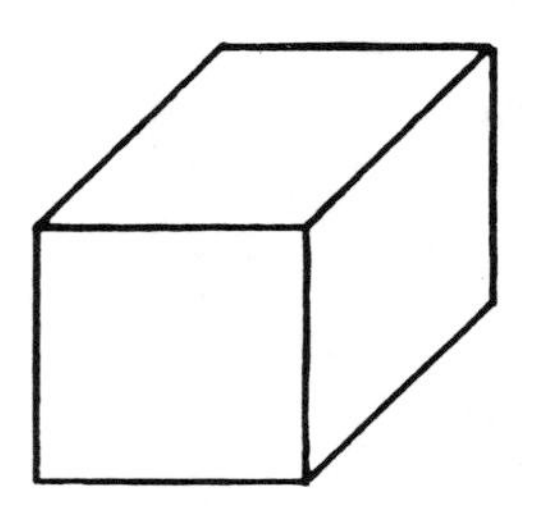

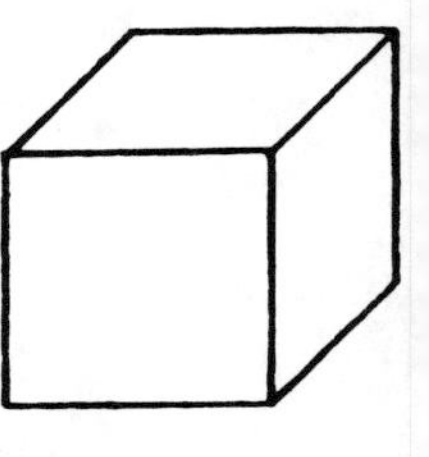

Figure 6.43 *(a) Oblique projection, (b) cabinet projection*

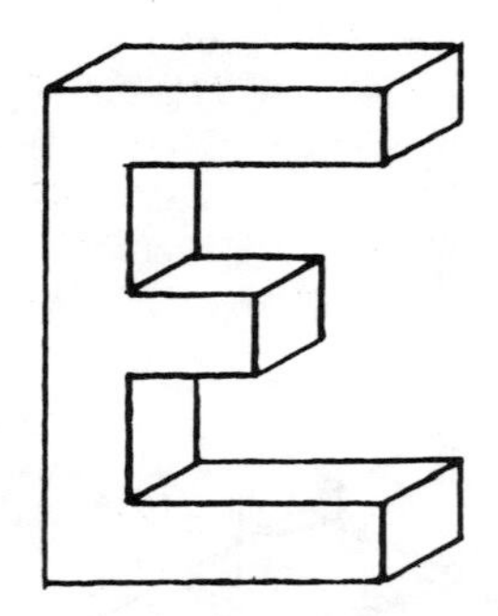

Figure 6.44 *Example*

Example

Make an oblique projection drawing of the object shown in Figure 6.38 (an earlier example in this chapter). Use an axis at 30° and a reduction of lengths in that direction by a factor of two-thirds.

Figure 6.44 shows the result.

Example

Make an oblique projection drawing of the object shown in Figure 6.40 (an earlier example in this chapter). Use an axis at 30° and a reduction of lengths in that direction by a factor of two-thirds.

Figure 6.45 shows the result.

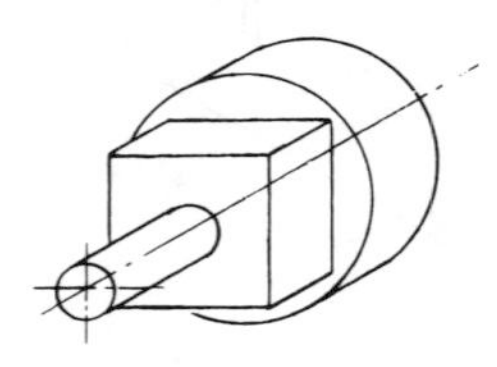

Figure 6.45 *Example*

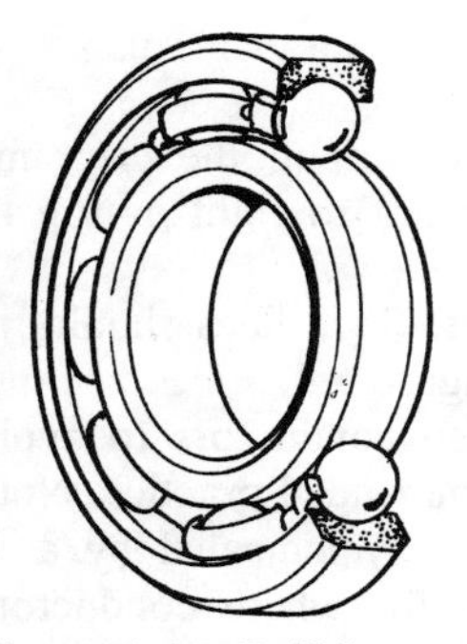

Figure 6.46 *Ball bearings*

6.3.3 Sectioned views

In order to show internal details, sectioned views may be used with isometric or oblique projections. Parts of the item are effectively removed to show internal details. Figure 6.46 illustrates this to show the details of ball bearings.

6.4 Circuit and flow diagrams

Circuit and flow diagrams are concerned with how specified components are connected. The aim is to ensure that the reader is able quickly and easily to understand what components are being used and how they are to be connected. With circuit diagrams, attention has to be paid to the choice of the correct symbol to represent a component, how the symbols are to be arranged on the drawing sheet and the careful routing of the interconnections to avoid misunderstandings. The same basic approach is used to electrical circuit diagrams as to flow diagrams for hydraulic or pneumatic systems. British Standard 5070 gives general information regarding drawing practice for such engineering diagrams, more specific information being given in the standards relating to particular types of diagrams. The following are a few points of relevance to such diagrams.

6.4.1 Electric circuit diagrams

Figure 6.47 shows an example of an electric circuit diagram. The standards for use in electrical drawings are laid down in the British Standard 3939. A more detailed discussion is given in chapter 7 where symbols are discussed.

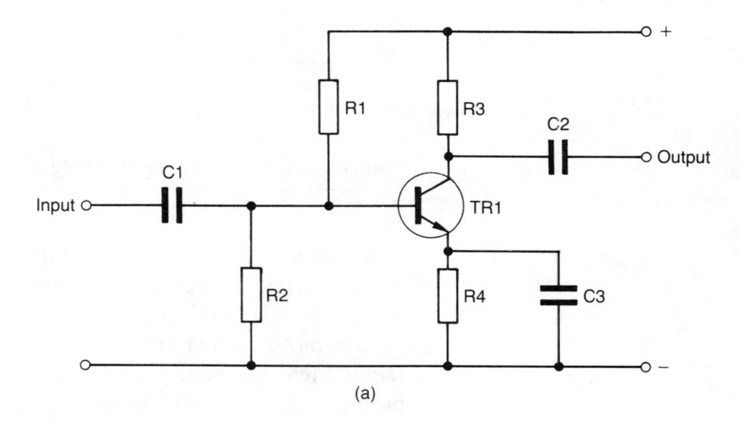

Figure 6.47 *Electrical circuit diagram*

The following are some points relevant to drawing circuit diagrams:

1 Where there is a clear sequence from input to output, the diagrams should be drawn so that the sequence is from left to right or top to bottom, as in Figure 6.47.
2 Conductors are represented by continuous lines. Such lines should, in general, be horizontal or vertical, as shown in Figure 6.47.
3 Care needs to be taken with junctions between conductors to avoid misunderstandings. Figure 6.48 shows the recommended practice. Note that the point at which conductors join is usually indicated by a •. Figure 6.49 shows the recommended practice for where conductors cross without joining.
4 All the switches in a circuit should be shown in the state they would be in when there is no current in the circuit.
5 Component references, or component values, may be given alongside each component.

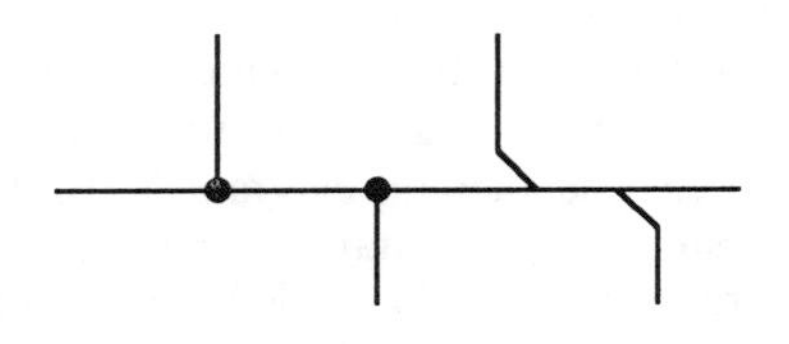

Figure 6.48 *Conductor junctions*

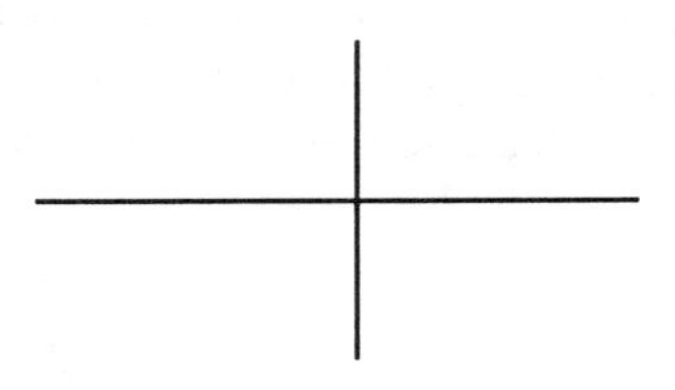

Figure 6.49 *Conductors crossing*

As an illustration, consider Figure 6.50 and the circuit diagram for a motor starter. This circuit is designed to ensure that in the event of a drop in voltage or loss of voltage, the circuit will prevent the automatic restarting of the motor when the full voltage is restored. This is often a safety requirement for motors and equipment driven by motors. There could be problems if the motor suddenly, unexpectedly, started up again. The line of dashes on the diagram is to indicate that the switches A, B and C are linked with the starter coil. When the start button is pushed, a current flows

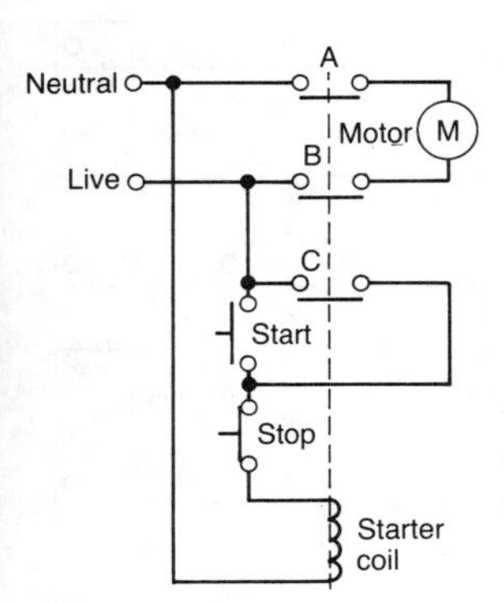

Figure 6.50 *Motor starter*

through the coil and creates a magnetic field which causes the normally open switches A, B and C to close. When the start button is released, there is still a current through the coil because A, B and C are closed. The switches A, B and C are only released by opening the stop switch or as a result of a drop in voltage of the supply giving insufficient current through the coil to hold the switches A, B and C closed. The motor cannot start again, even if the voltage is raised, until the start button is pressed.

Figure 6.51 is an example of a lighting circuit which might be found in the home. The circuit is designed so that the light can be switched on or off by the operation of either switch A or B. With the switches in the positions shown in the figure, the light is off. There is no connection of the live wire to the lamp. If, however, one of the switches, say A, is switched to the other position, then the light will come on. If, however switch B is then moved to the other position, the light is switched off.

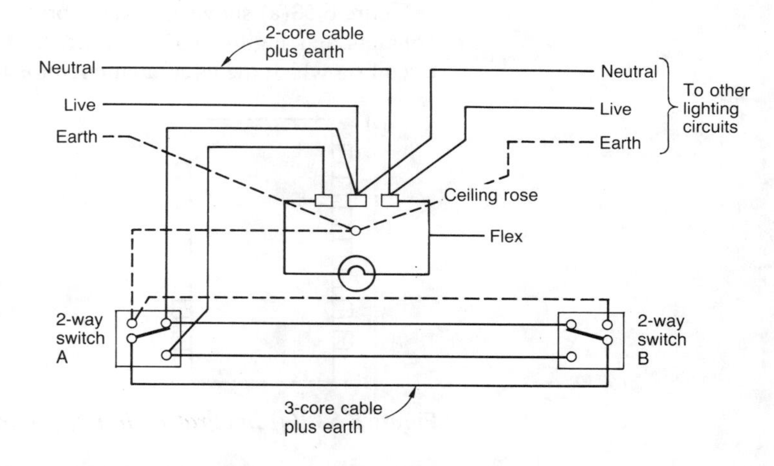

Figure 6.51 *Two-way switching circuit for a lamp*

Electrical circuit diagrams do not show the layout of components in a circuit, merely how the components are interconnected. To show the layout a layout diagram needs to be drawn. Figure 5.8(b) shows such a diagram for the circuit shown in Figure 6.47. As a further illustration, consider the circuit shown in Figure 6.52(a). Figure 6.52(b) and (c) show layout diagrams for both sides of a circuit board on which the circuit could be assembled. Figure 6.52(b) shows the side on which the components are mounted and Figure 6.52(c) the side which has the copper strip. The small circles show where the connecting wires from the components come through the board and are to be soldered to the strip. The larger holes are where the strip has to be drilled away to break contact.

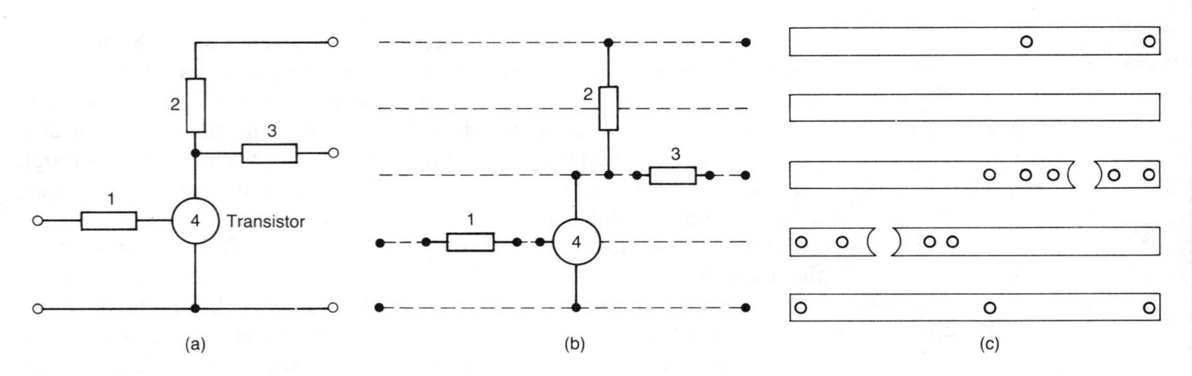

Figure 6.52 *(a) Circuit diagram, (b) component side of board, (c) strip side of board*

Figure 6.53(a) shows the basic form of an integrated circuit which has 14 contacts and Figure 6.53(b) the form that might be taken by the printed circuit on which the integrated circuit is to be mounted.

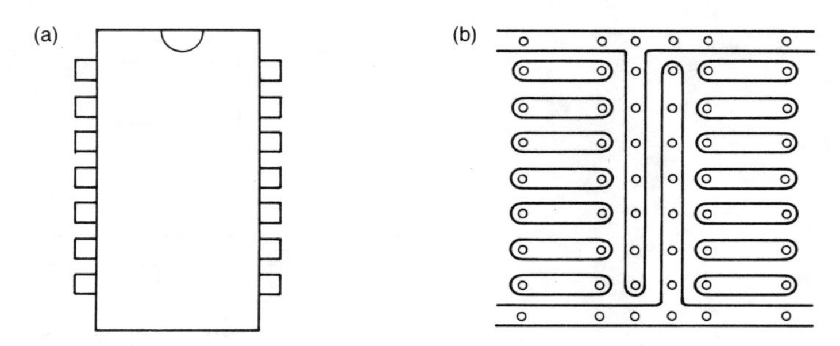

Figure 6.53 *(a) Integrated circuit, (b) printed circuit pattern*

6.4.1 Hydraulic/pneumatic circuit diagrams

Circuit diagrams can be used to show the detailed piping connections between components or items of equipment. They show connections, flow paths and, by the symbols used, the functions of the components. They do not include constructional details or the physical location of elements; to do that a layout diagram is needed. The standards used are given in British Standard 2917. The following are a few points, more details being given in chapter 7.

1. Where there is a clear sequence from input to output, the diagrams should be drawn so that the sequence is from left to right or top to bottom.
2. Pipelines are represented by continuous lines.
3. Junctions between pipelines and crossovers between non-connecting pipelines are shown in the same way as conductors in electric circuit diagrams, i.e. as in Figures 6.48 and 6.49.

4 Valves should be shown in the unoperated state.
5 Component references may be given alongside each component.

An example of a hydraulic circuit diagram for the braking system of a car is given in Figure 5.9. This is semi-pictorial in that double lines have been used for the pipes and the fluid in the pipes indicated by shading. A circuit diagram drawn to the above conventions would only have single lines for the pipes. Another example is shown in Figure 6.54. This is a hydraulic hoist. When the control valve is in the raise position, the hydraulic fluid is directed to the lower part of the cylinder and the pressure pushes the piston up, the fluid above the piston being connected via the control valve to the return line. When the control valve is in the lower position, the hydraulic fluid is directed to the upper part of the cylinder and the pressure pushes the piston down, the fluid below the piston being connected via the control valve to the return line.

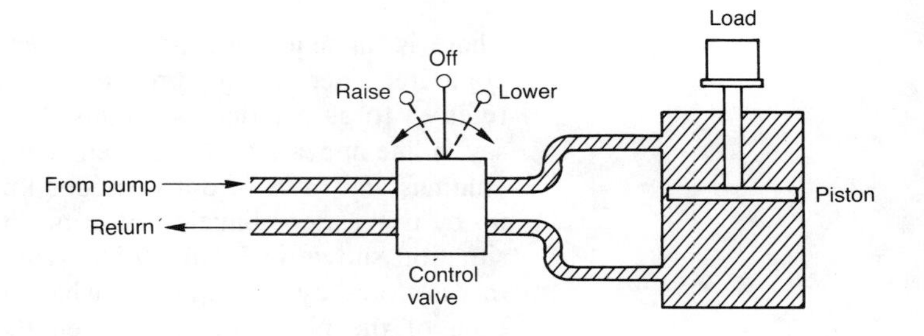

Figure 6.54 *Hydraulic hoist*

6.5 Computer-aided drawing

The discussion so far in this chapter has been on the basis that the drawing was to be carried out manually. However, computer-aided drawing systems for producing engineering drawings are now extremely common. The term *computer-aided draughting* is used for the use of a computer-based system for producing drawings suitable for use in the manufacturing of a component. *Computer-aided design* (CAD) is the use of a computer-based system to assist in translating a requirements into an engineering design and uses a databank of design principles and information on such matters as properties of materials and characteristics of production processes, etc.

Basic features of carrying out computer-aided draughting or design are:

1 Load the software. A very common system for use with IBM compatible microcomputers is AUTODESK's AutoCAD. Another, for use with a minicomputer or a workstation, is DOGS (Drawing Office Graphics System).
2 Use menus which are on the screen or on the surface of a graphics tablet to select commands and with the aid of input devices, such as a mouse, puck, digitizer or light pen, create the required shapes on the screen. A menu is a list of operations from which one can be selected.

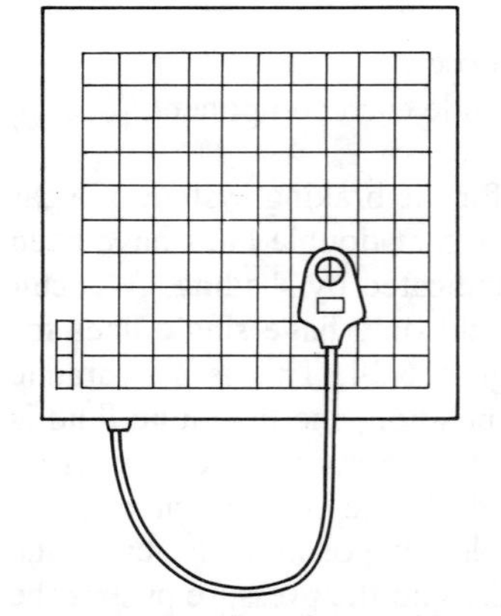

Figure 6.55 *Graphics tablet*

For example, the draw menu might have such operations as line, point, rectangle, regular polygon, circle, etc. A graphics tablet is a flat plate over which a puck can be moved (Figure 6.55), the movement of the puck driving the computer screen cursor to mimic its movement or the puck position can be used to pick options from a menu card. Dimensioning is considerably easier than with the manual method. All that has to be done is to indicate what features have to be dimensioned, the system then works out the dimension from the scaled drawing on the screen and places it at the required position.

3 Manipulate the components within the screen display, e.g. moving parts relative to other parts, changing their scale, rotating them, etc.
4 Store the resulting drawing/design on disk.
5 Send an output from the computer to a printer or plotter to print or plot the drawing/design.

6.5.1 Pixel- and vector-based programs

There is an important distinction between computer-aided draughting or computer-aided design programs and drawing programs that are often referred to as painting programs. It is first necessary to understand how, say, a line appears on the screen. The screen is organised into a number of columns and rows of dot positions known as *pixels* and the image is built up by using a combination of dots. Thus, for example, we might have the situation shown in Figure 6.56. With a painting program the information that is stored by the computer when you draw such a line is the location of each of the pixels on the screen that represent the line. With a CAD program, the information about the line is not stored as the location of each pixel in the line but in terms of the locations of the two end points. Each time the line is to be displayed on a screen, the program derives the information regarding which pixels should be turned on in order to display the line. This method is termed *vector based*, the drawing program being said to be *pixel-based*. With a vector-based program a circle would be defined by the location of its centre and the value of its radius, with a pixel-based program by the location of each pixel round the circumference of the circle.

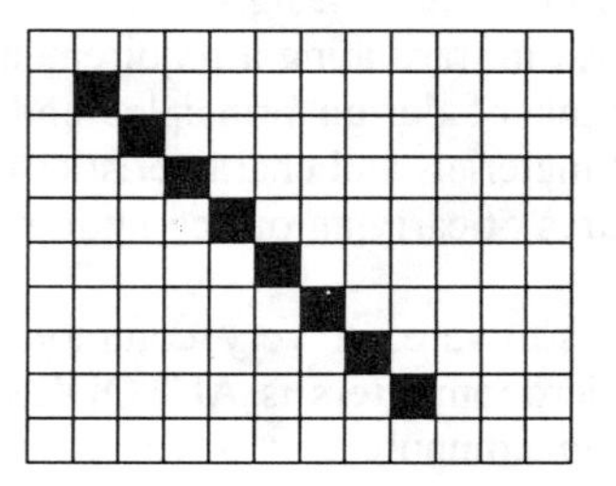

Figure 6.56 *Pixels forming a line*

Vector-based programs have the advantage that the line can more readily be manipulated by the program. With a pixel-based system, to modify the shape or size of a line or curve requires each pixel to be modified in turn, a slow and tedious process. With a vector-based program, all that needs to be done is to change a term or terms in the instructions stored for the line or curve.

6.5.2 Common commands in CAD programs

Many of the commands used in CAD programs are the same in many of the available software packages. The following are just a few of the more common commands to illustrate the types of commands possible and procedures that might be used:

1 *LINE* This requires LINE to be selected from a menu or typed in at the keyboard. In response to this a prompt is supplied requesting the coordinates of the point from which the line should start and the coordinates of the point at which it should finish. Thus to specify a rectangle using the keyboard the entries might be:

Command: LINE. Return key pressed
From point: 40, 100. Return key pressed
To point: 140: 100. Return key pressed
To point: 140: 50. Return key pressed
To point: 40, 50. Return key pressed
To point: Close command selected or types in. Return key pressed

Alternatively we might use a pointing device such as a mouse or a puck on a graphics tablet. Then we might have:

Mouse used to move the cursor on the screen over LINE. Mouse button clicked.
Mouse used to move the cursor to the position on the screen for the first point. Mouse button clicked. This is then repeated for each of the other points.

2 *CIRCLE* This requires CIRCLE to be selected from a menu. In response to this a prompt is supplied listing alternatives by which the circle can be specified. These are by specification of two points through which the circle should pass, three points through which the circle should pass, be tangential or the coordinates of the centre point. One of these methods has to be selected. Then the coordinates have to be given. The diameter or radius has then to be selected and its data given. Thus to specify a circle using the keyboard the entries might be:

Command: CIRCLE. Return key pressed.
3P/2P/TTR/Centre point: Centre point selected. Return key pressed
120: 50 Return key pressed
Diameter/Radius: Diameter selected. Return key pressed
20 Return key pressed

Alternatively we could use a pointing device such as a mouse or a puck with a graphics tablet. Then the sequence might be:

Mouse used to move cursor on screen to the command: CIRCLE. Mouse button clicked.
Mouse used to move cursor to select from 3P/2P/TTR/Centre point. Centre point selected. Mouse button clicked
Mouse used to indicate position of centre of circle. Mouse button clocked
Mouse used to move cursor over the command Diameter or Radius. Diameter selected. Mouse button clicked.
Mouse used to move cursor to position corresponding to circle circumference. Mouse button clicked.

3 *ARC* This command it to enable an arc to be drawn. Depending on the sophistication of the software so the number of options available for carrying out such a drawing can change. Thus when the command ARC is entered or selected, a menu may appear offering a choice of three different points for the specification of the arc, e.g. start, centre, end; start, centre, length (of chord); start, end, angle (included); etc. When an item is selected then the information can be entered from the keyboard or by a pointing device in the same way as with the circle.

4 *DELETE (or ERASE)* This command enables errors or changes to be made to a drawing by deletion. Using a mouse the sequence might be:

Use the mouse to move the cursor over the DELETE command. Select it by clicking the mouse button.
Use the mouse to move the cursor over the line to be deleted. Click the mouse button to select that line. The line flashes.
The prompt then appears on the screen: Delete this line? (Y/N). Entering Y by the keyboard deletes the line.
Another prompt then appears: Another (Y/N)? If no other line to be deleted, enter N from the keyboard.

5 *MOVE* This command is used to move an object on the screen. Thus, for example, to move a rectangle using a mouse the sequence might be:

Use the mouse to move the cursor over the WINDOWS command. Select it by clicking the mouse button.
Now move the cursor to one corner of the area in which the objects are to be edited. Press the mouse button and with it kept pressed move it out to trace a rectangle round the object to be moved. Then release the button.
Use the mouse to move the cursor over the MOVE command. Select it by clicking the mouse button.
Now select the window by clicking the mouse button anywhere within it. The window and its contents can then be dragged to a new position by pulling the window with the mouse button

depressed. When this new position is reached, release the mouse button.

This selecting of an object by putting a window round it can be used with other commands, such as DELETE, to carry out editing of objects. Alternatively, if just a line is to be moved the procedure can be similar to that used with the DELETE command.

6 *COPY* This command allows multiple copies of a particular object to be created. For example, after one circle is drawn to represent a hole, copies of it might be made in order to generate, say, five identical holes in other parts of the object.

As with command MOVE above, draw a window round the object to be copied.
Use the mouse to move the cursor over the COPY command. Select it by clicking the mouse button.
Now select the window by clicking the mouse button anywhere within it. The window and its contents can then be copied to a new position by pulling the window with the mouse button depressed. When this new position is reached, release the mouse button.

Alternatively, if just a line is to be moved the procedure can be similar to that used with the DELETE command.

7 *GRID and SNAP* When GRID is selected a pattern of dots or fine lines appears on the screen as an aid to drawing. Grid pattern can be vertical and horizontal lines or vertical and lines at angles to the horizontal for isometric or oblique drawings. When GRID is selected, a prompt appears indicating the default grid size and asking you to specify a different one if you require it. When the command SNAP is selected, then the movement of the cursor is tied to the grid. Each time the cursor is moved it jumps to the next grid point.

An important aspect of drawing with CAD programs is the use of layers. A drawing can be drawn as a series of layers with different aspects of it on different layers. You can think of the drawing as being made on a number of overlapping sheets of tracing paper. Thus there might be the outline on one layer, hidden features on another, dimensions on another, text on another. Each layer fits precisely over the others. Different line thicknesses, types and widths can be selected for the different layers. Layers can be switched off so that they do not show on the screen and so enable details of complicated drawings to be more easily seen as they are drawn. Coloured drawings can be developed by using different colours for different layers.

With CAD software there will be a library of standard elements, e.g. nuts, bolts, electrical symbols, hydraulic and pneumatic symbols, which can be selected and positioned on drawings.

Problems

Questions 1 to 17 have four answer options: A, B, C and D. You should choose the correct answer from the answer options.

Questions 1 to 6 relate to the following types of lines that can be used in engineering drawing:

A Continuous thick line
B Continuous thin line
C Thin chain line
D Thick chain line

Select the most appropriate type of line for the following applications:

1 A centre line
2 Dimension lines
3 Leader lines
4 Hatching
5 Cutting planes
6 Visible outlines

7 In order to emphasise a note on an engineering drawing it should be:

A Underlined
B Printed in bold characters
C Printed in italics
D Printed using larger characters

8 The most prominent lines on an engineering drawing should be:

A Visible outlines
B Hidden outlines
C Outlines of adjacent parts
D Outlines of revolved sections

9 Decide whether each of the following statements is TRUE (T) or FALSE (F).

Engineering drawings include a number of different types of lines. Leader lines are used to indicate where:
(i) Dimensions are intended to apply.
(ii) Notes are intended to apply.

A (i) T (ii) T
B (i) T (ii) F
C (i) F (ii) T
D (i) F (ii) F

Questions 10 to 13 refers to Figure 6.57 which shows an object and a number of views of it, labelled A, B, C and D. Hidden detail is not shown.

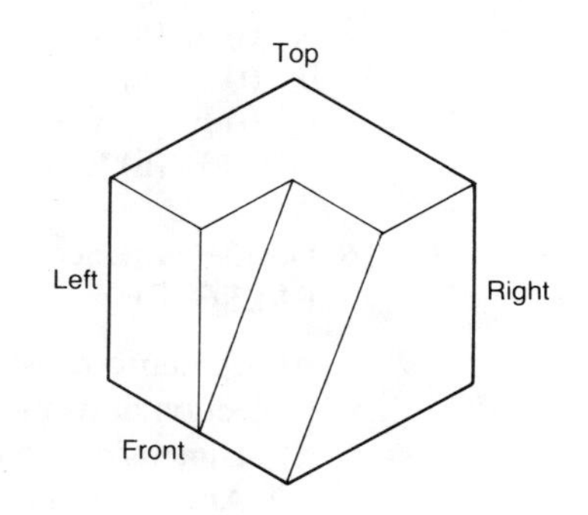

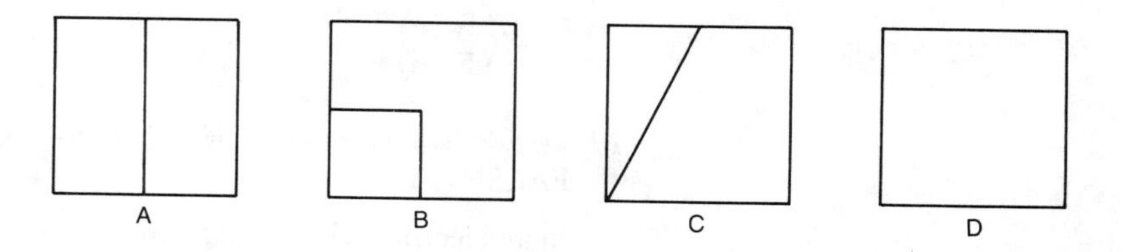

Figure 6.57 *Problems 10 to 13*

State which view represents:

10 The front view.
11 The right side view.
12 The left side view.
13 The plan.

14 Decide whether each of the following statements is TRUE (T) or FALSE (F).

Orthographic projection gives drawings which have:
(i) Dimensions which are scaled versions of those of the actual part.
(ii) The same angles as those of the actual part.

A (i) T (ii) T
B (i) T (ii) F
C (i) F (ii) T
D (i) F (ii) F

15 Decide whether each of the following statements is TRUE (T) or FALSE (F).

Isometric projection gives drawings which have:
(i) The same shape faces as those of the actual part.
(ii) The same angles between faces as those of the actual part.

A (i) T (ii) T
B (i) T (ii) F
C (i) F (ii) T
D (i) F (ii) F

16 Decide whether each of the following statements is TRUE (T) or FALSE (F).

An isometric projection drawing can be obtained from an orthographic projection drawing by transferring:
(i) Dimensions to lines drawn parallel to the isometric axes.
(ii) Angles from the orthographic drawing to the isometric drawing.

A (i) T (ii) T
B (i) T (ii) F
C (i) F (ii) T
D (i) F (ii) F

17 Decide whether each of the following statements is TRUE (T) or FALSE (F).

In an electrical circuit diagram:
(i) Continuous lines are used to represent conductors.
(ii) A junction between conductors is indicated by a • being present at the point concerned.

A (i) T (ii) T
B (i) T (ii) F
C (i) F (ii) T
D (i) F (ii) F

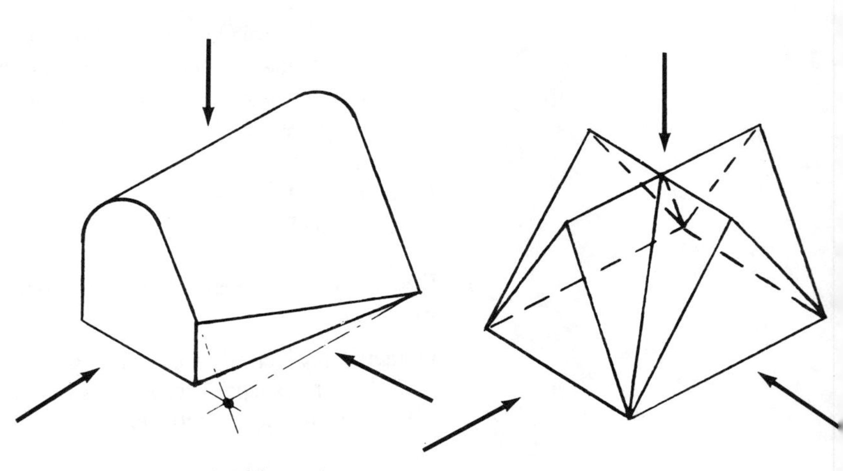

Figure 6.58 *Problem 18*

18 Sketch the views in first angle projection, in the directions indicated, for the items shown in Figure 6.58. Do not show any hidden detail.

19 Sketch the views in third angle projection, in the directions indicated, for the item shown in Figure 6.59. Show all the hidden detail.

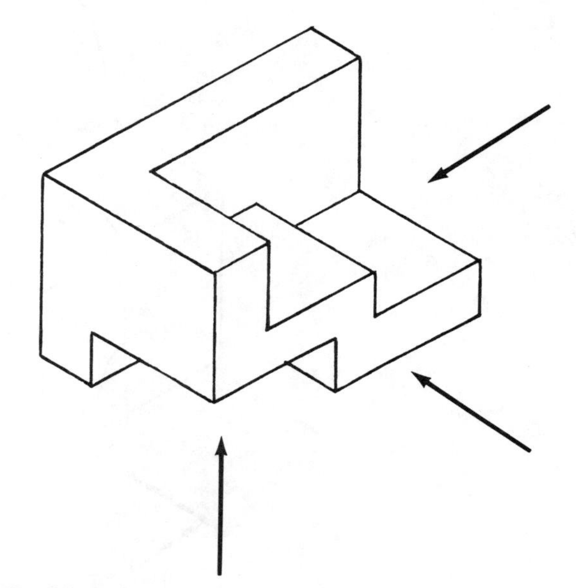

Figure 6.59 *Problem 19*

20 Sketch the views in first angle projection, in the directions indicated, and the indicated sectioned view for the component shown in Figure 6.60.

21 Draw the section view given by the cutting plane AA in Figure 6.61.

22 Sketch the isometric views of the components shown in Figure 6.62.

23 Sketch the oblique projection views of the cheesehead screw and the taper pin shown in Figure 6.63.

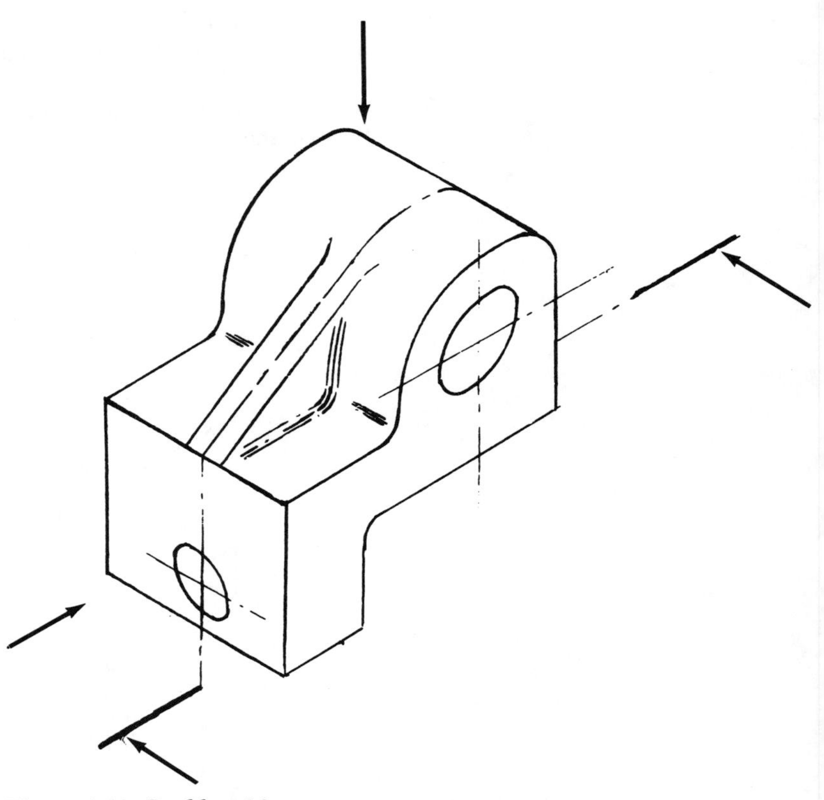

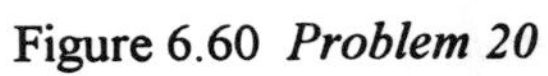
Figure 6.60 *Problem 20*

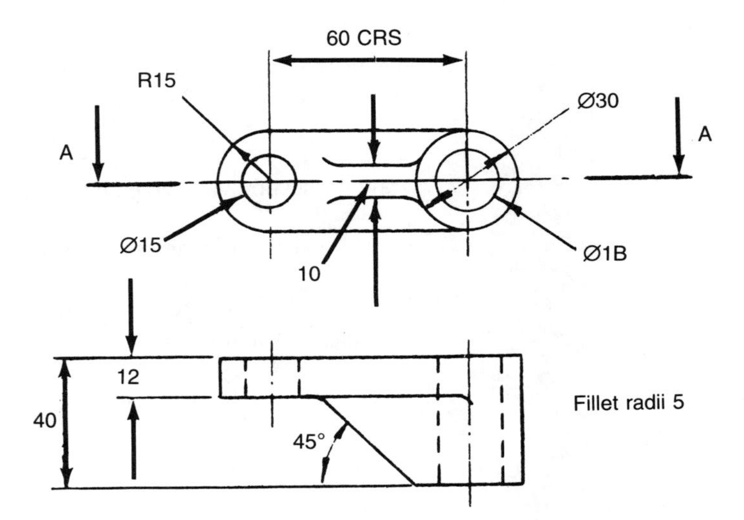

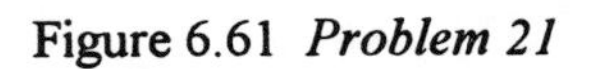
Figure 6.61 *Problem 21*

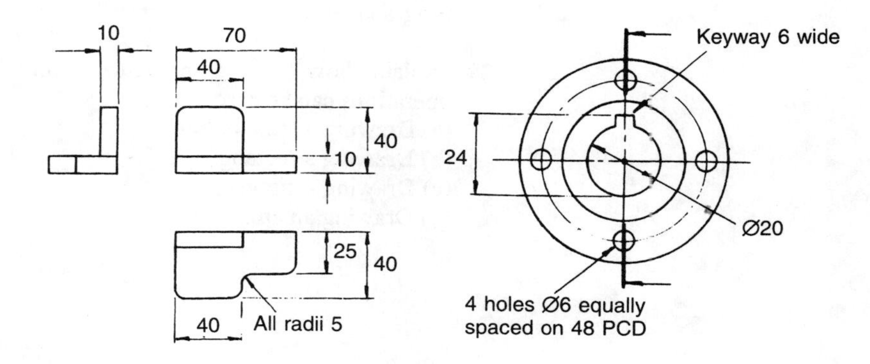

Figure 6.62 *Problem 22*

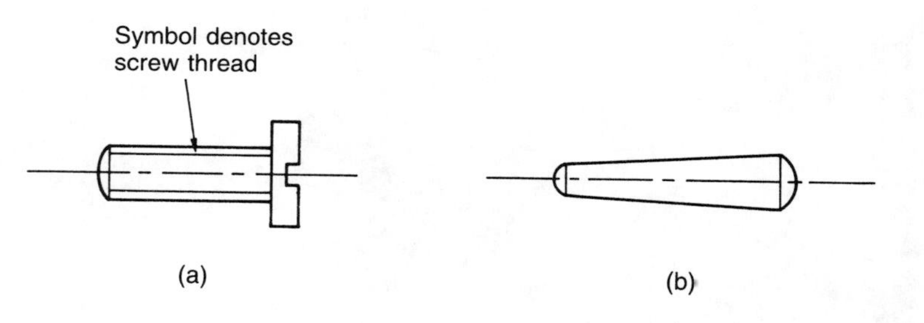

Figure 6.63 *Problem 23*

24 Open up an electrical/mechanical item such as a torch or a mains plug and produce single-part drawings, assembly drawings and an electrical wiring diagram for the item.

25 Examine a circuit mounted on a circuit board and produce a wiring diagram for the component.

THE LIBRARY
GUILDFORD COLLEGE
of Further and Higher Education

26 Open up an electrical toaster and produce single-part drawings, assembly drawings, electrical circuit diagram and an electrical layout diagram for the item.

27 Produce detail and assembly drawings of a car roof rack.

28 Explain the difference between pixel-based and vector-based drawing programs.

29 Explain how, for a particular software program, the following operations can be carried out:
(a) Drawing a straight line.
(b) Drawing a rectangle.
(c) Drawing a circle.
(d) Drawing an arc.

7 Interpreting drawings

7.1 Common features and symbols

Consider the problem of drawing a thread. We might, for example, choose to draw it in the form shown in Figure 7.1(a). This, however, is tedious if there are many screws and takes time. The use of a standard symbol can, however, considerably reduces the time spent on such detail work. The standard symbol is shown in Figure 7.1(b). This consists of the outside diameter of the thread being drawn with a thick line, i.e. 0.7 mm wide, and the root diameter of the thread as a thin line, i.e. 0.3 mm wide.

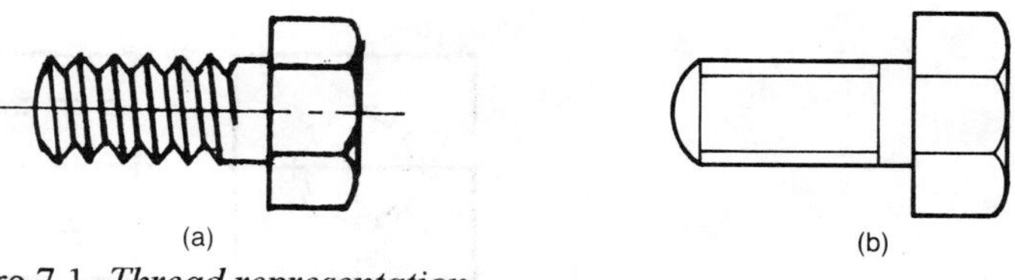

Figure 7.1 *Thread representation*

Likewise in electrical circuits or pneumatic/hydraulic diagrams. Considerable time is saved by representing components by standard symbols. For example, using the standard symbol for an ammeter (Figure 7.2) can considerably simplify a drawing of an electrical circuit.

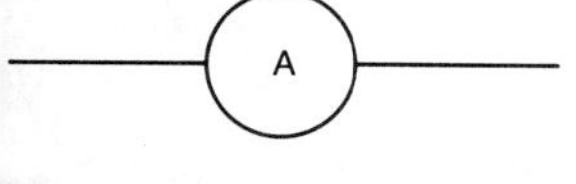

Figure 7.2 *An ammeter*

This chapter is about the symbols used to represent commonly encountered items in engineering drawings and circuit diagrams, and how drawings using such symbols can be related to the physical situation and information extracted.

7.2 Mechanical symbols

Figure 7.3(a) to (k) shows the conventions that are adopted for representing screw threads. For visible screw threads a type A line (see Figure 6.5) is used to define the roots of the threads. For hidden screw threads, the crests and roots are defined using type F lines. Note that in the end view the inner circle is broken, not extending to more than three quadrants. This break represents the start of the screw thread on the bolt. Figure 7.3(b) shows an example of the use of this convention for a threaded fastener, a hexagon head screw.

Figure 7.4 shows how interrupted views can be represented, (a) being a tube and (c) a solid circular shaft. Figure 7.5(a) and (b) shows how circular and square shafts can be represented; (c) rolling bearings; (d) knurling, the left-hand figure being straight knurling and the right-hand one diamond knurling; and (e) transverse details of splines and serrations, the serrations being shown in the left-hand figure.

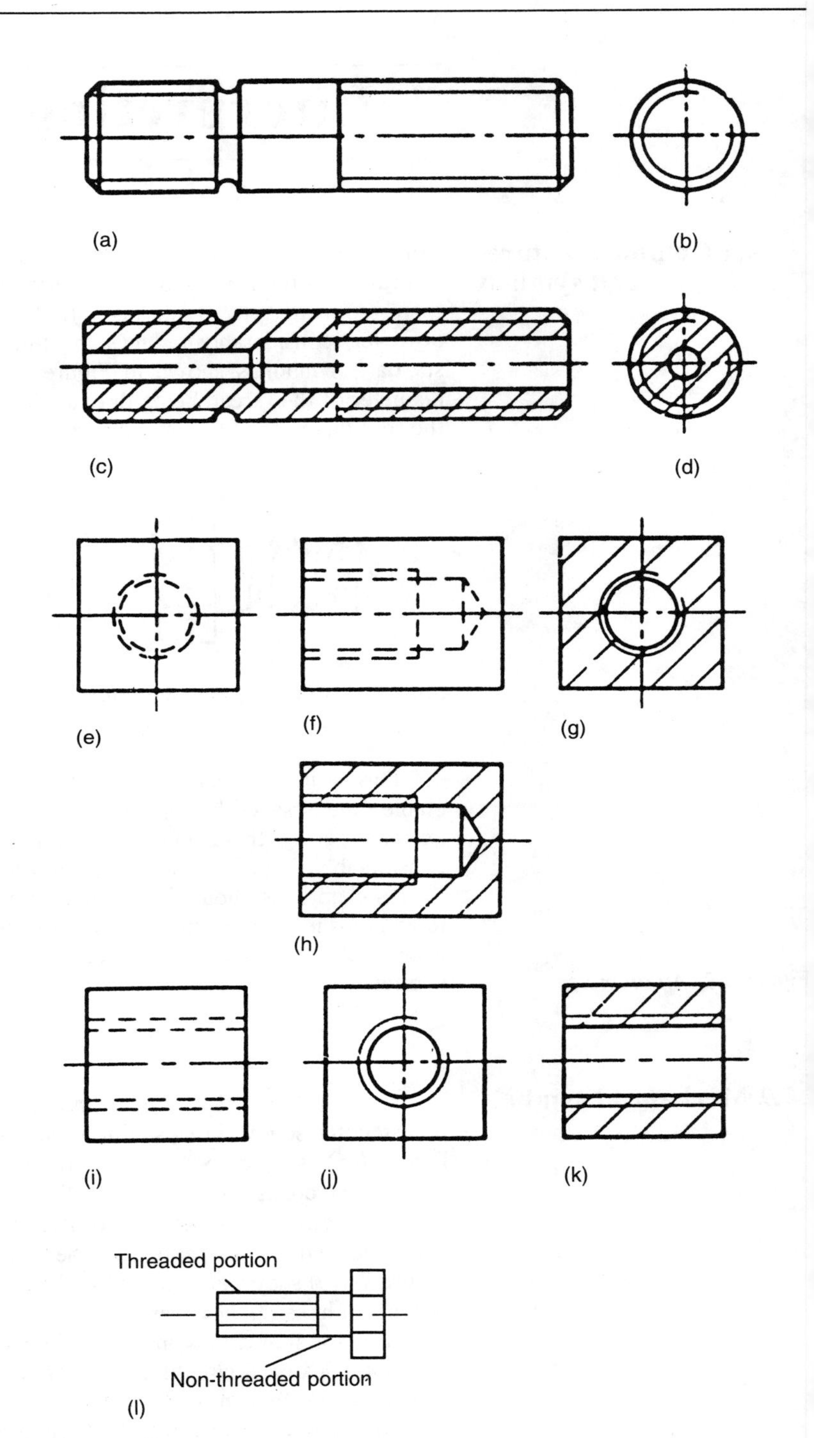

Figure 7.3 *(a)–(k) Conventions for screw thread (Figure 47: BS 308: Part 1: 1993), (l) a hexagon head screw*

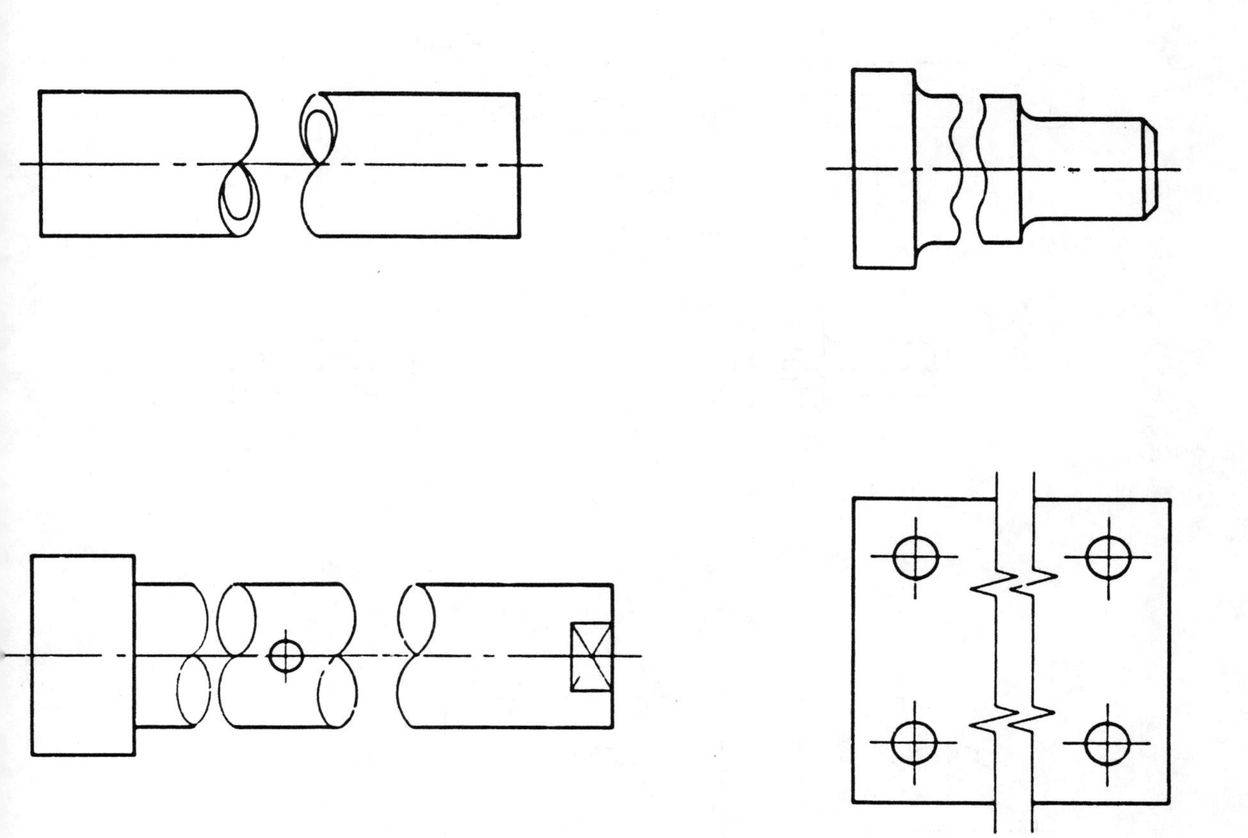

Figure 7.4 *Symbols for indicating an interrupted view (Figure 14: BS 308: Part 1: 1993)*

Example

Select from the symbols given in Figures 7.3 to 7.7 the appropriate symbols for (a) a threaded portion of rod, (b) diamond knurling, (c) a rolling bearing, (d) a cylindrical tension spring, (e) a spur gear.

(a) See part of Figure 7.3(a).
(b) See the right-hand figure in Figure 7.5(d).
(c) See Figure 7.5(c).
(d) See Figure 7.6(a).
(e) See Figure 7.7(a).

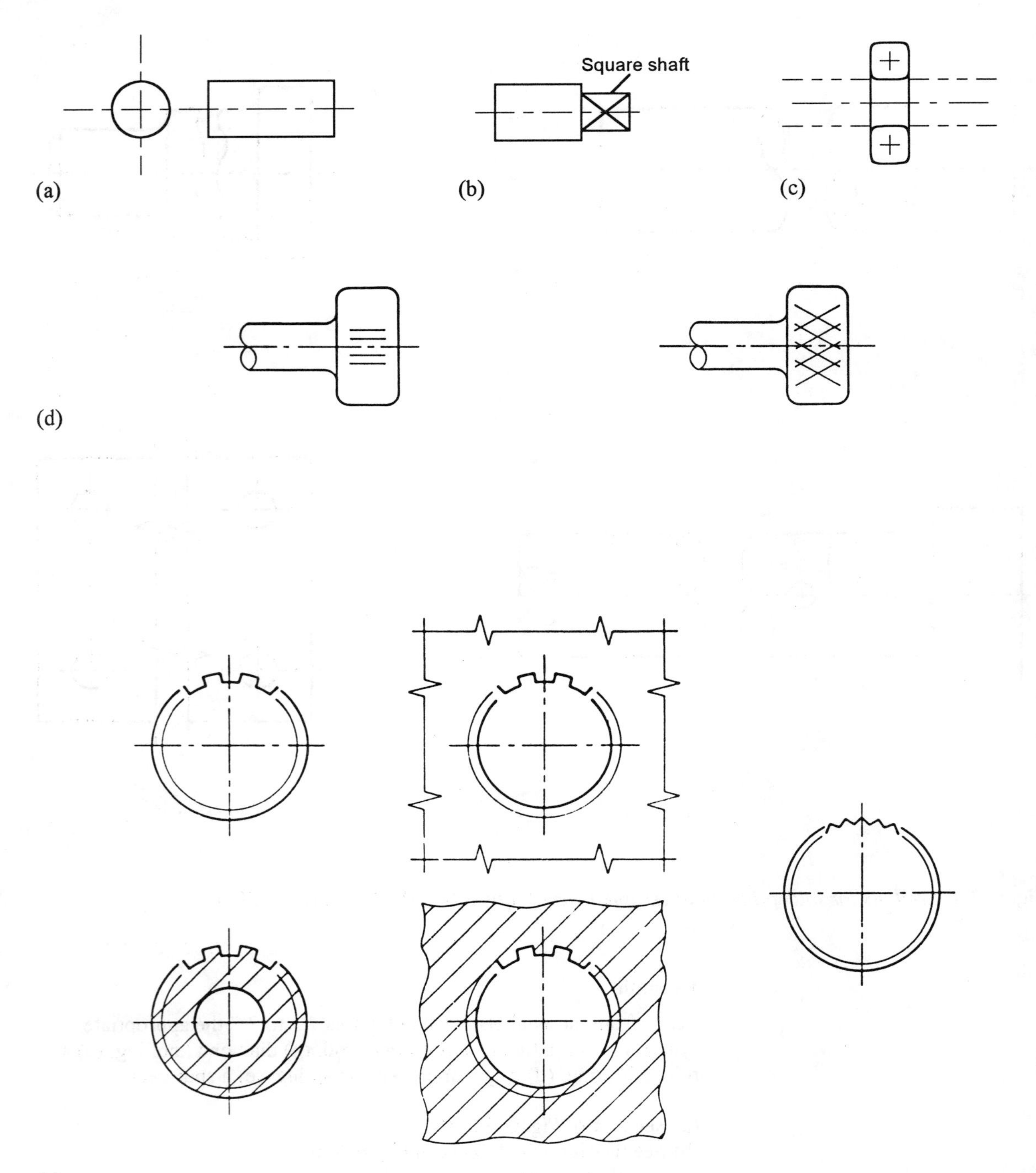

Figure 7.5 *(a) Circular shaft, (b) square shaft, (c) rolling bearing, (d) knurling (Figure 44: BS 308: Part 1: 1993), (e) splines and serrations (Figure 46: BS 308: Part 1: 1993)*

Figure 7.6 shows symbols for tension springs and Figure 7.7 symbols for gears. The above represents just a selection of common symbols, the reader is referred to BS 308 Part 1 for more.

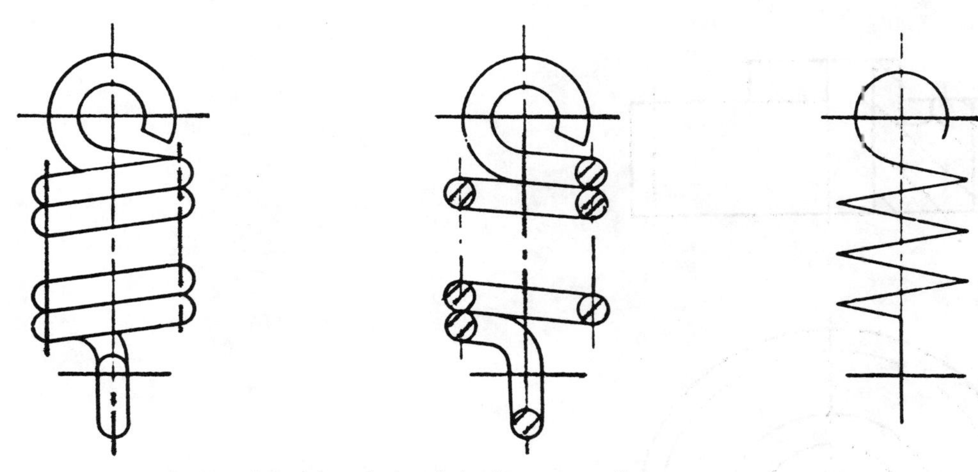

(a) Cylindrical helical tension spring of wire of circular cross section

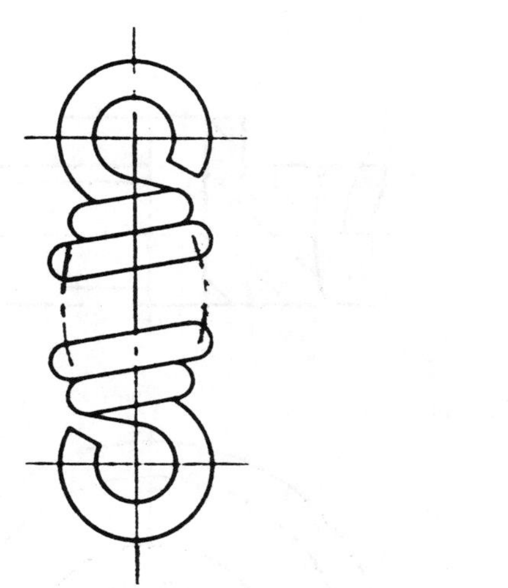

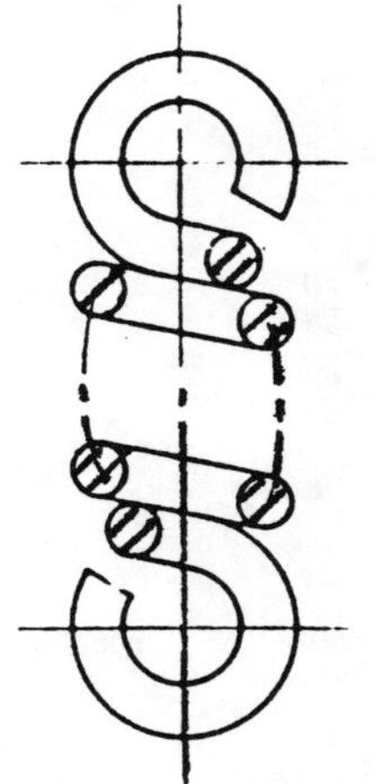

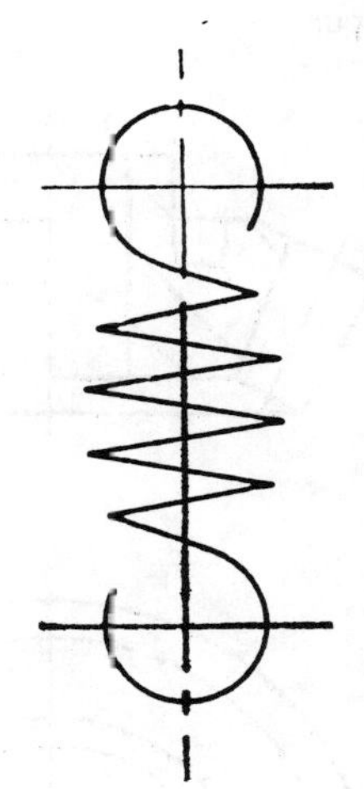

(b) Double-conical helical tension spring of circular cross-section

Figure 7.6 *Symbols for tension springs (Figure 65: BS 308: Part 1: 1993)*

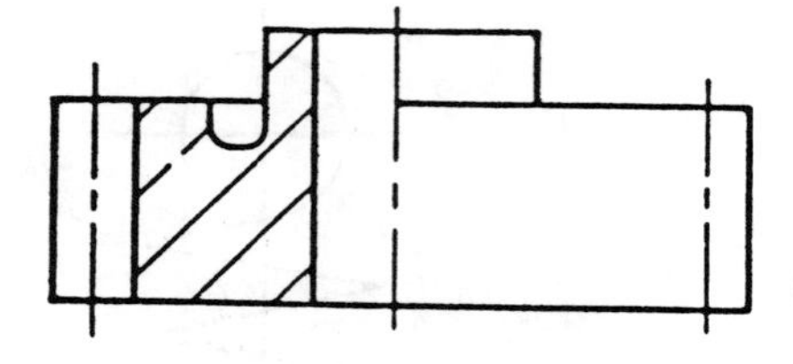

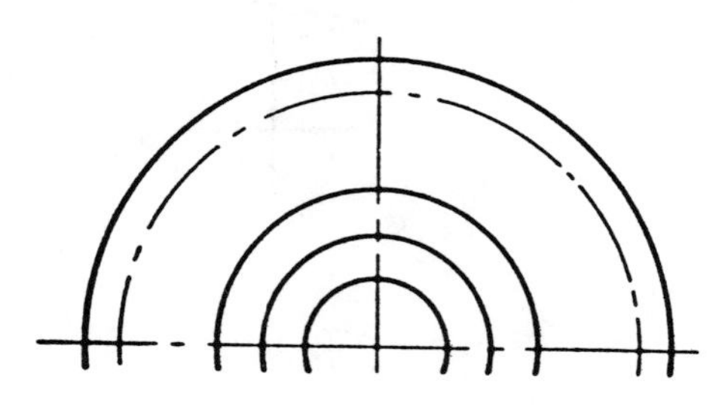

(a) Spur

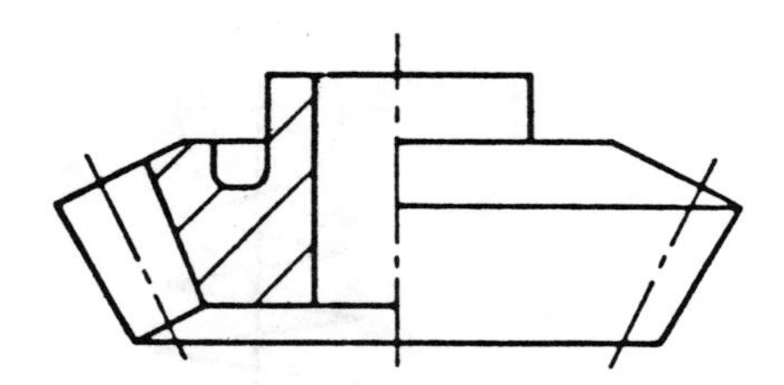

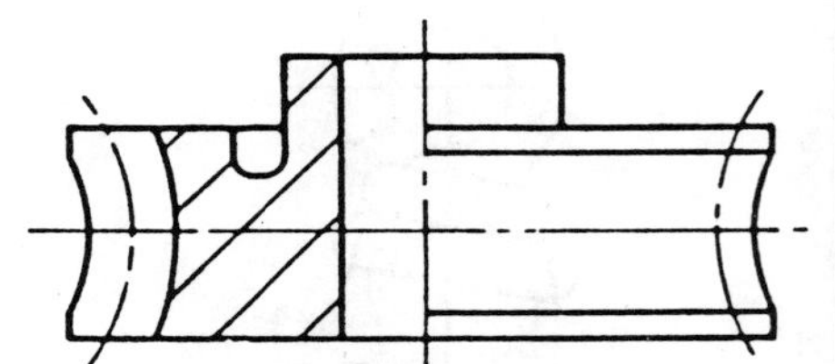

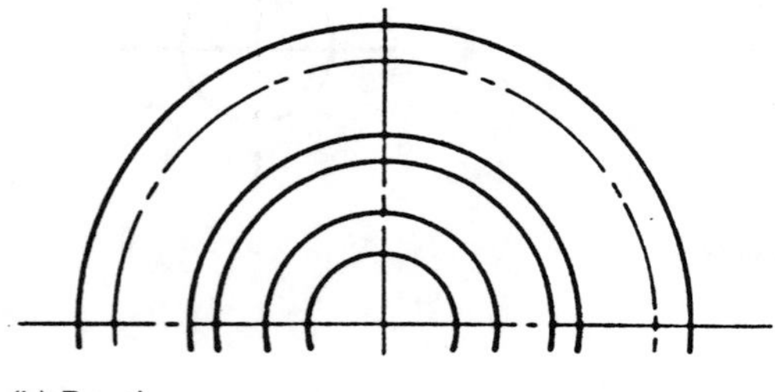

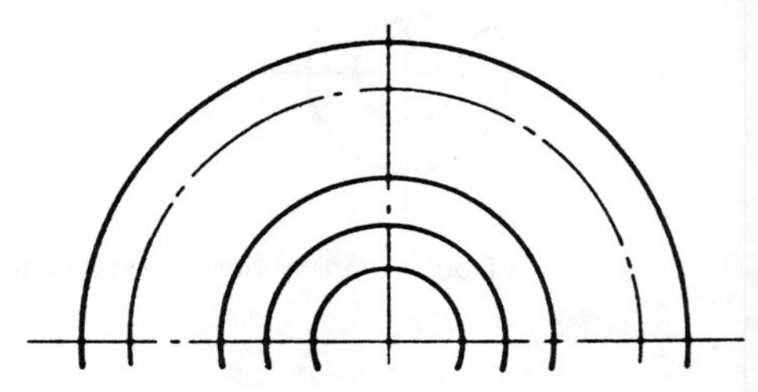

(b) Bevel

(c) Worm wheel

Figure 7.7 *Symbols for gears (Figure 54: BS 308: Part 1: 1993)*

7.2.1 Abbreviations

Table 7.1 shows abbreviations used on engineering drawings for common terms.

Table 7.1 Abbreviations

Term	Abbreviation
Across flats	A/F
Assembly	ASSY
Centres	CRS
Centre line	CL or ℄
Chamfered	CHAM
Cheese head	CH HD
Counterbore, i.e. an additional bore on the same axis as another	C'BORE
Countersunk	CSK
Countersunk head	CSK HD
Diameter, in a note	DIA
Diameter, preceding a dimension	∅
Drawing	DRG
External	EXT
Figure	FIG.
Hexagon	HEX
Hexagon head	HEX HD
Hydraulic	HYD
Insulated or insulation	INSUL
Internal	INT
Left hand	LH
Long	LG
Material	MATL
Millimetre	mm
Maximum	MAX
Minimum	MIN
Number	NO.
Pattern number	PATT NO.
Pitch circle diameter	PCD
Radius, preceding a dimension	R
Required	REQD

Right hand	RH
Round head	RD HD
Screwed	SCR
Sheet	SH
Sketch	SK
Specification	SPEC
Spherical diameter, preceding a dimension	SPHERE ∅
Spherical radius, preceding a dimension	SPHERE R
Spotface, i.e. providing a flat surface just sufficient for washer or bolt head	S'FACE
Square, in a note	SQ
Square, preceding a dimension	□
Standard	STD
Undercut	U'CUT
Volume	VOL
Weight	WT
Maximum material condition	MMC or Ⓜ
Full indicated movement	FIM
Taper, on diameter or width	—▷—

Capital letters are shown in the above list, although lower case letters may be used where appropriate. Full stops are not normally used, except where the abbreviation makes a word which may prove confusing. Thus, for example number is abbreviated to NO. and not NO without a full stop.

Figure 7.8 shows a drawing involving abbreviations. These are:

S'FACE ∅ 8 to indicate spotfacing with a diameter of 8 mm.

C'BORE ∅ 15 × 4 DEEP to indicate counterbores, i.e. additional bores on the same axis, of diameter 15 mm and depth 4 mm.

4 HOLES ∅ 5 EQUALLY SPACED ON 38 PCD to indicate 4 holes of diameter 5 mm equally spaced round a circle of diameter 38 mm.

U'CUT ∅ 15 × 4 DEEP to indicate an undercut, i.e. a groove machined in the component, with a diameter of 15 mm and 4 mm deep.

CHAM ALL SHARP CORNERS 1 AT 45° to indicate all the sharp corners are to be chamfered at 45°.

MATL - MILD STEEL indicates the material to be used is mild steel.

DRG NO. 26379 indicates the drawing number.

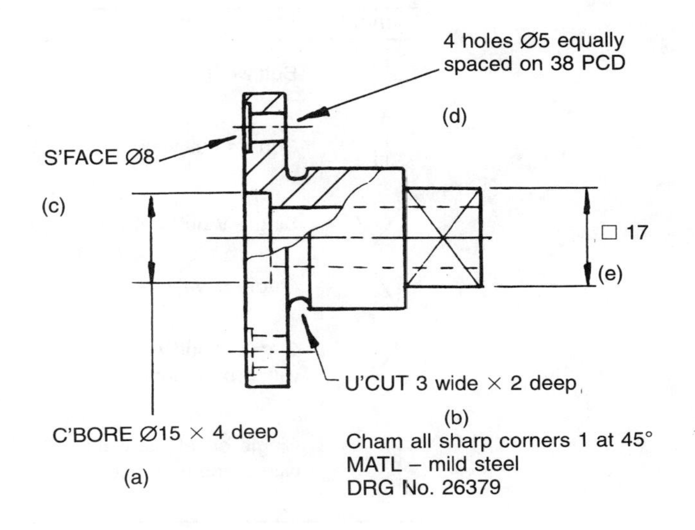

Figure 7.8 *Examples of abbreviations*

7.2.2 Machining symbols

It is often necessary to indicate that a surface has to be machined. This can be done using a symbol of the form shown in Figure 7.9. The angle of the lines in the symbol should be about 60°. Where the part is to be machined all over the symbol may be just followed by the note ALL OVER.

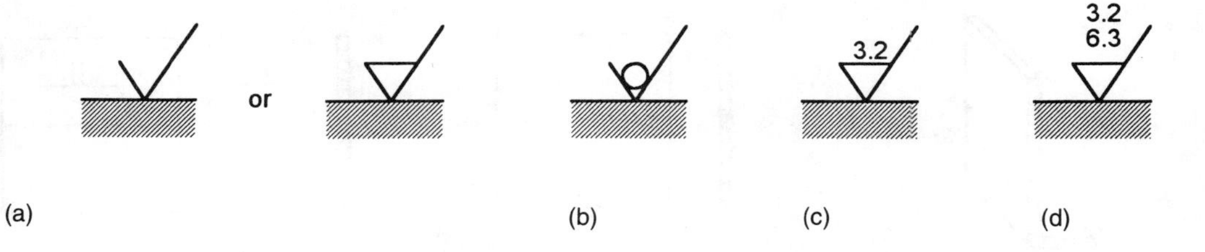

Figure 7.9 *Machining symbols, (a)basic symbols, (b) no machining, (c) machine to a surface texture to the specified roughness, (d) machine to a roughness between 3.2 and 6.3. Note that the roughness number gives the mean deviation in micrometres of surface irregularities.*

7.3 Welding symbols

Figure 7.10 shows the symbols used for the various types of welds and Figure 7.11 some examples, in third angle projection, of the use of such symbols. Figure 7.12 shows the significance of the positioning of the symbol with reference to the leader line.

Symbol	Description	Symbol	Description
	Butt weld		Singe-U butt weld
	Square butt weld		Single-J butt weld
	Single-V butt weld		Backing run
	Single-bevel butt weld		Fillet weld
	Single-V butt weld with broad root face		Plug weld
			Spot weld
	Single-bevel butt weld with broad root face		Seam weld

Figure 7.10 *Welding symbols*

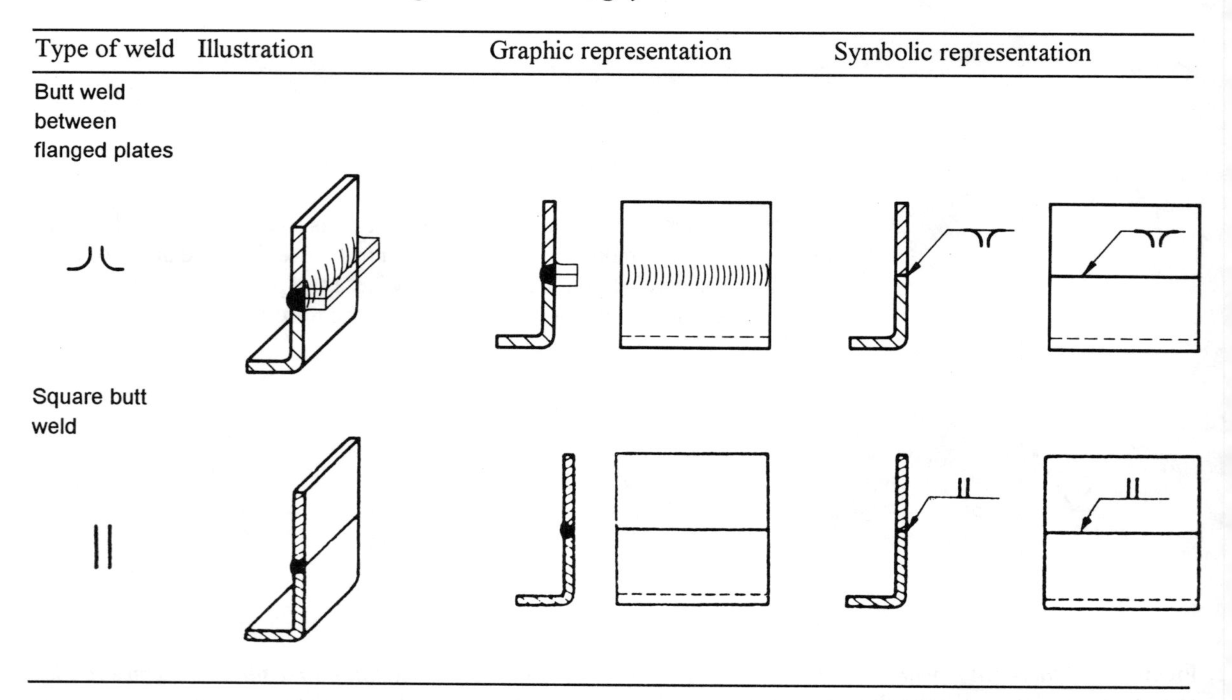

Figure 7.11 *Examples of the use of weld symbols (Table 5: BS 499: Part 2: 1980 (1989))*

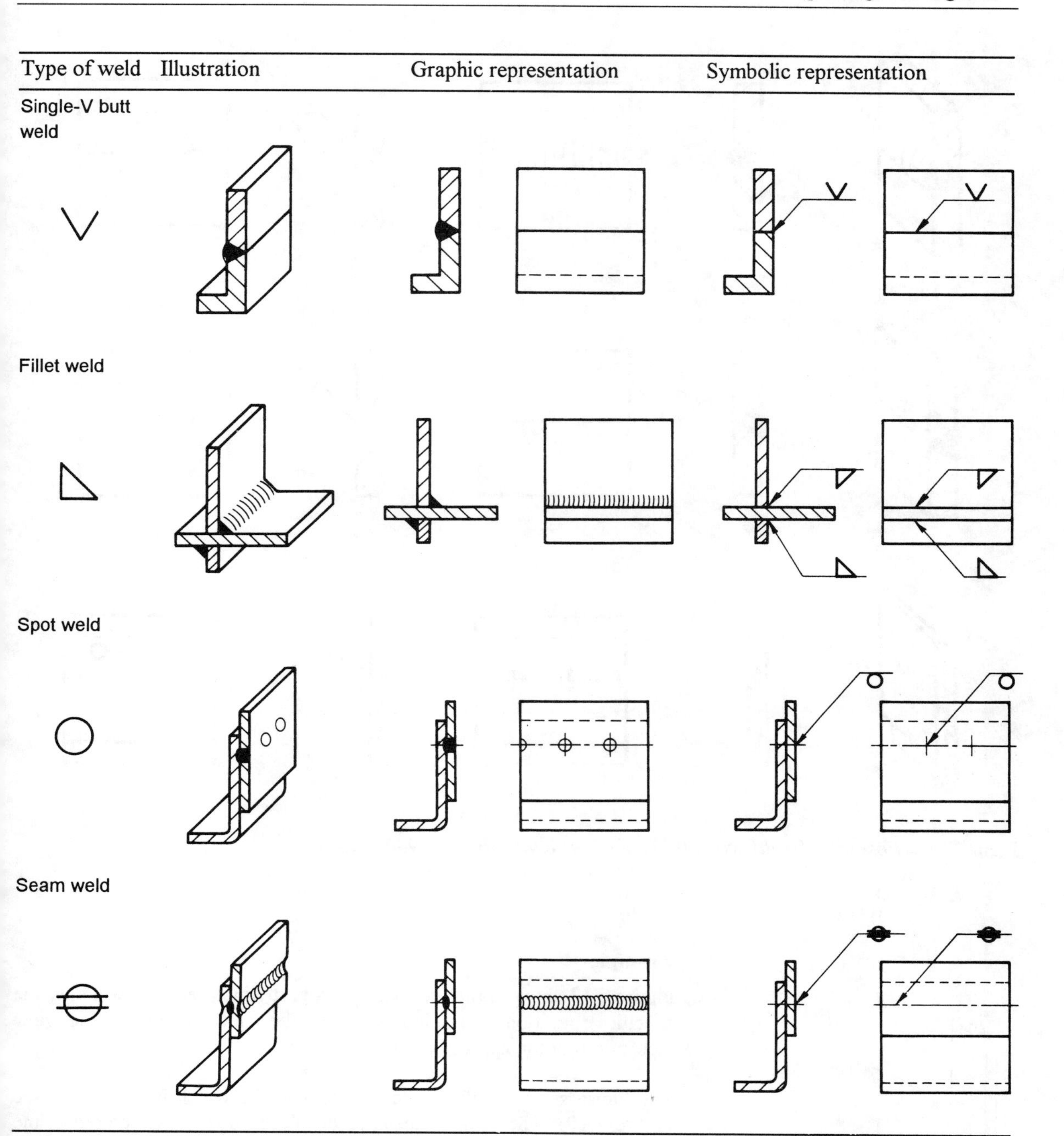

Figure 7.11 *(continued)*

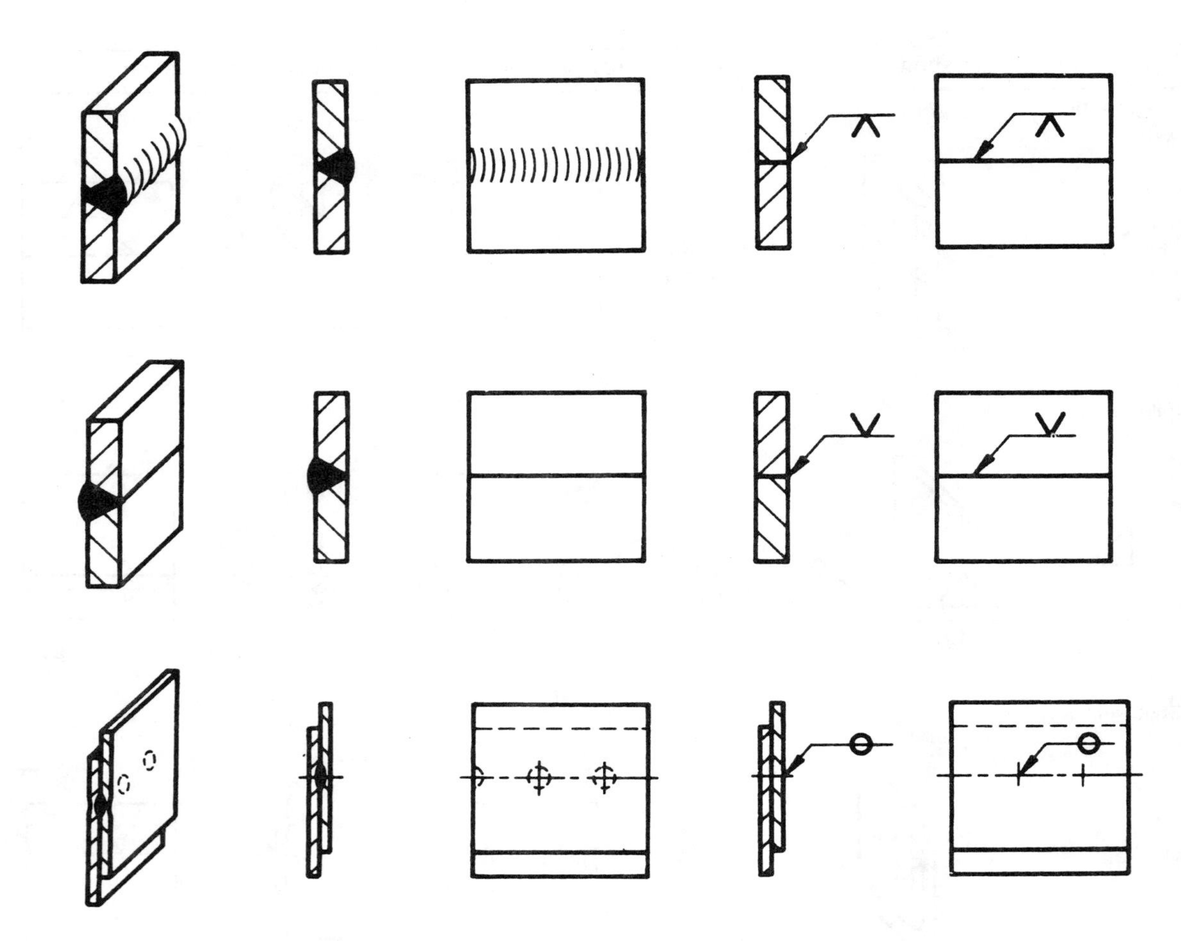

Figure 7.12 *Position of weld symbols (Table 4: BS499: Part 2: 1980 (1982))*

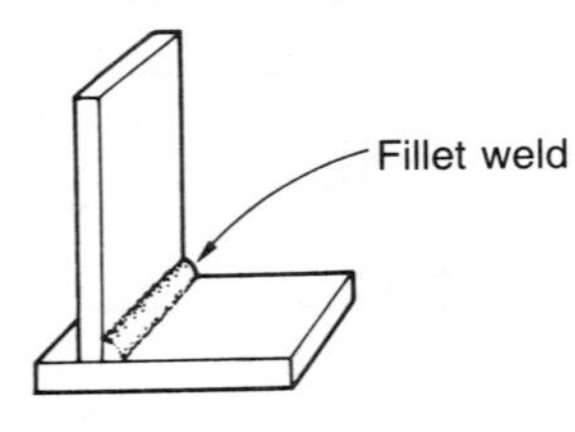

Figure 7.13 *Example*

Example

Figure 7.13 shows a pictorial view of a part which requires welding, the specific operation being indicated on the view. Draw a side view incorporating the appropriate welding symbol.

The figure shows a fillet weld, the symbol for which is a triangle (see Figure 7.10). The symbol needs to be placed below the reference line when the external surface of the weld, i.e. the weld face, is on the side of the joint indicated by the arrow. Thus the symbol needs to be as indicated in Figure 7.14.

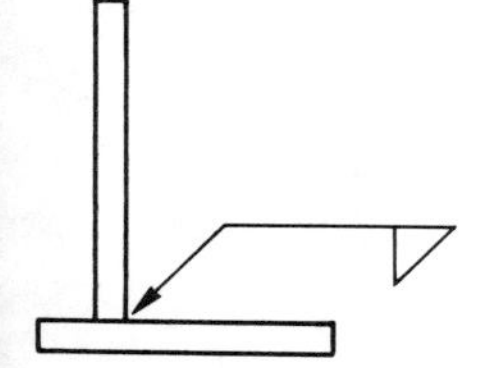

Figure 7.14 *Example*

Example

Figure 7.15 shows a pictorial view of a part which requires welding, the specific operations being indicated on the view. Draw a side view and plan incorporating the appropriate welding symbols.

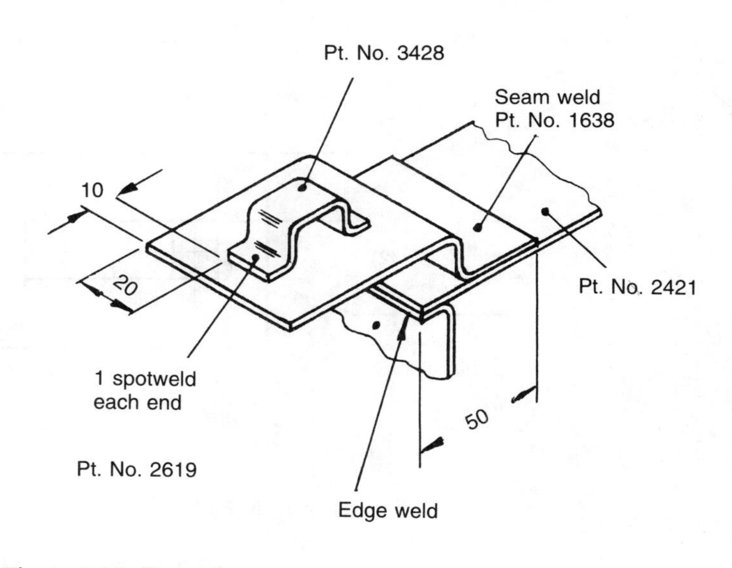

Figure 7.15 *Example*

Figure 7.16 shows the resulting side view and plan. The term edge weld is often used for what the British Standard refers to as a backing or sealing run weld. In the side view, the edge weld symbol is below the reference line since the weld face is on the arrow side of the joint. The spot weld symbol is shown across the reference line as it is a weld made within the plane of the joint. The (2) is used to indicate that there are two such welds. The seam weld symbol indicates the position of that weld.

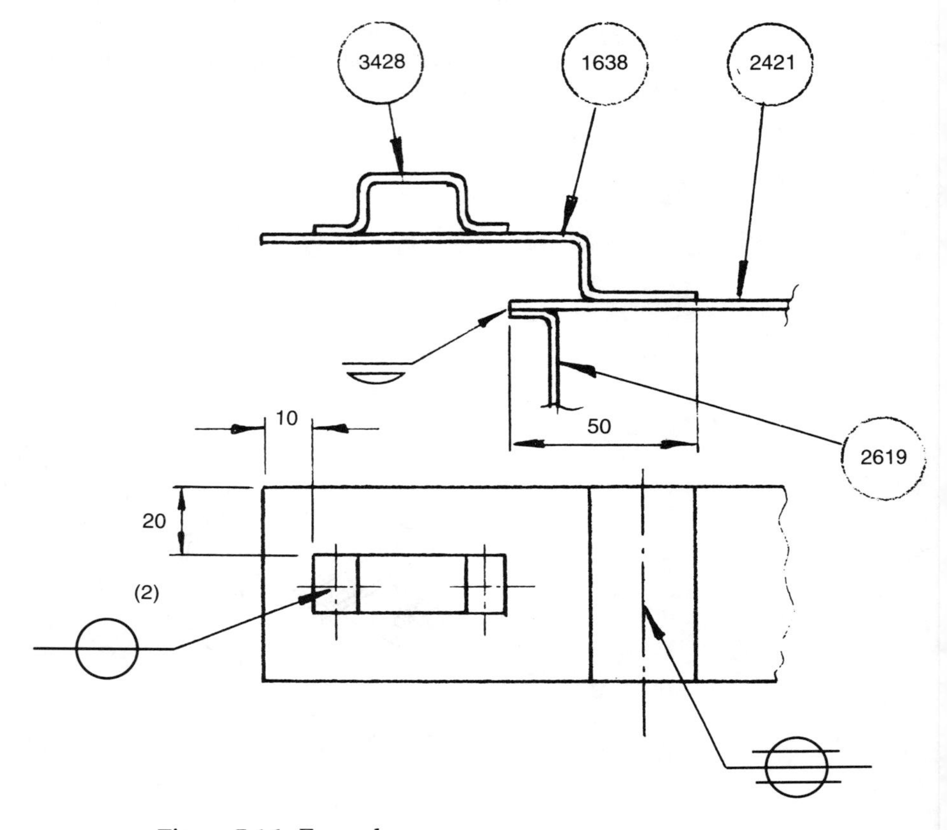

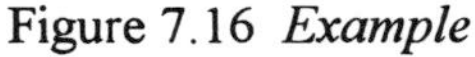
Figure 7.16 *Example*

7.4 Electrical and electronic symbols

Figure 7.17 shows some of the more commonly used electrical and electronic symbols (British Standard BS 3939). Symbols may be built up from symbol elements, e.g. the variability symbol ↗ can be added to that of a resistor to give the symbol for a variable resistor or that of a capacitor to give a variable capacitor. In general, the meaning of a symbol is not altered by changes in its orientation on the page or a mirror-image reversal.

Example

What do the following symbols represent: (a) ○ ,(b) (↑) ,(c) Ⓜ ?

(a) This is the general symbols for an indicating instrument or measuring instrument.
(b) This is the symbol for a galvanometer.
(c) This is the symbol for a motor.

Description	Symbol	Description	Symbol
Direct current or steady voltage	—	Earth	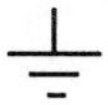
Alternating: general symbol		Frame of chassis not necessarily earthed	
Indicates suitability for use on either direct or alternating supply	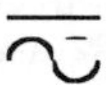	Positive polarity	+
Variability: general symbol		Negative polarity	–
Pre-set adjustment		Signal lamp: general symbol	
Inherent non-linear variability		Filament lamp	
Mechanical coupling: general symbol	- - - -	Conductor or group of conductors: general symbol	—
Example, coupled switches so that when one switch closes the other closes, when one opens the other opens.	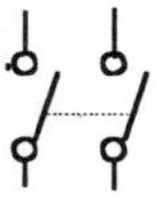	Crossing of conductor symbols on a diagram (no electrical connection)	
Primary or secondary cell The long line represents the positive pole, the short line the negative pole	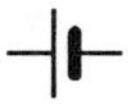	Junction of conductors	
Battery of primary or secondary cells		Double junction of conductors	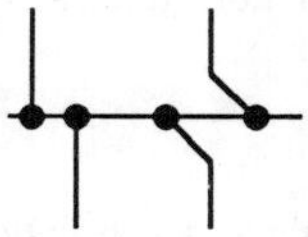
Alternative symbol		Terminal or tag: general symbol	
Ideal voltage source	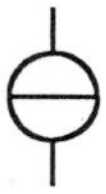	Link normally closed: with two readily separable contacts	
Ideal current source			

Figure 7.17 *Electrical and electronic symbols*

Description	Symbol
Plug (male)	
Socket (female)	
Fuse: general symbol	
Fuse: alternative symbol	
Fixed resistor: general symbol	
Alternative general symbol	
Fixed resistor with fixed tapping (voltage divider)	
Variable resistor: general symbol	
Voltage divider with moving contact (potentiometer)	
Capacitor: general symbol	
Polarised electrolytic capacitor	+
Variable capacitor: general symbol	
Winding (i.e. of an inductor, coil or transformer): preferred general symbol	
Inductor with core (ferromagnetic unless otherwise indicated)	
Transformer with ferromagnetic core	
Transformer with air core	
Generator	G
Motor	M
Indicating instrument, or measuring instrument (galvanometer): general symbol	↑
Ammeter	A
Voltmeter	V

Figure 7.17 *(continued)*

Description	Symbol	Description	Symbol
Contacts: (a) Make contacts (normally open): general symbols	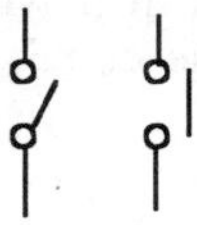	pn diode: (a) general symbol	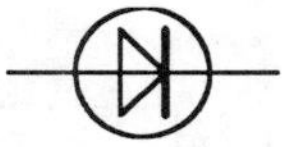
	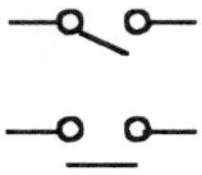	(b) alternative symbol Note: The use of the envelope symbol is optional	
(b) Break contact (normally closed): general symbols	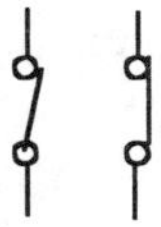	Thyristor: general symbol	
	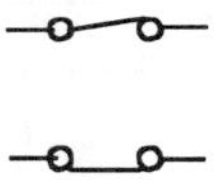	pnp transistor	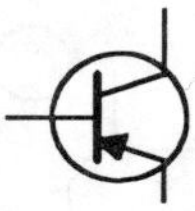
		npn transistor	
Push-button switches, non-locking: (a) make contact	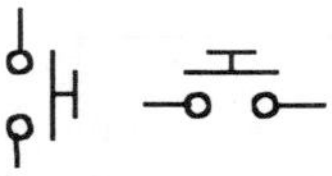	Light-emitting semiconductor diode	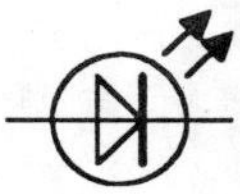
(b) break contact	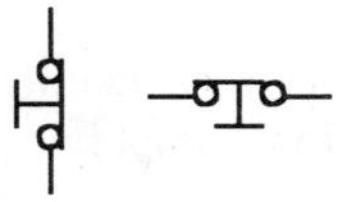	Diac	
Single element relay	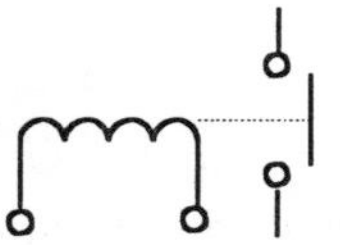	Triac	
Thermistor		Operational amplifier	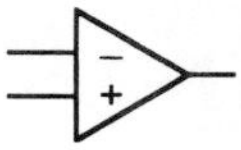 − +
Light dependent resistor (LED)	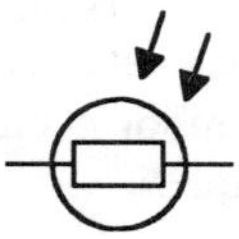	Integrated circuit	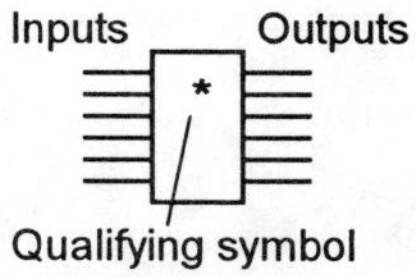 Inputs Outputs * Qualifying symbol

Figure 7.17 *(continued)*

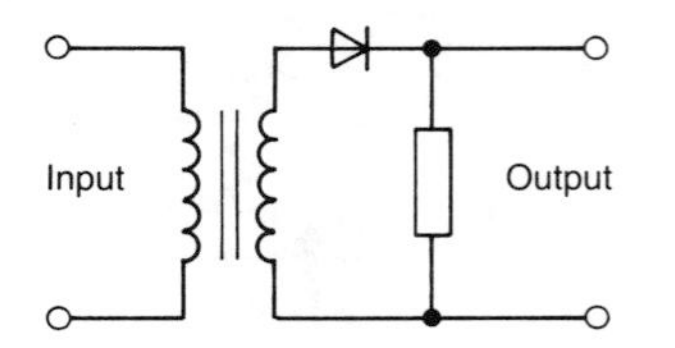

Figure 7.18 *Example*

Example

Figure 7.18 shows a half-wave rectifier circuit with the following components: 1 - a transformer with a ferromagnetic core, 2 - a diode, 3 - a resistor. Redraw the circuit diagram with the symbols for those components.

The circuit diagram, using the symbols given in Figure 7.17 is as shown in Figure 7.19.

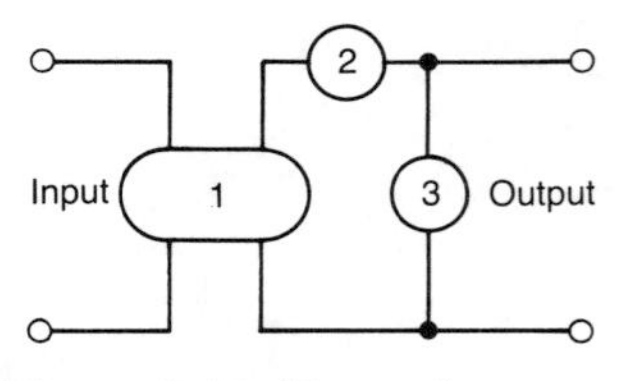

Figure 7.19 *Example*

Example

For the circuit shown in Figure 7.20, a high-pass filter, identify the components 1, 2, 3, 4 and 5.

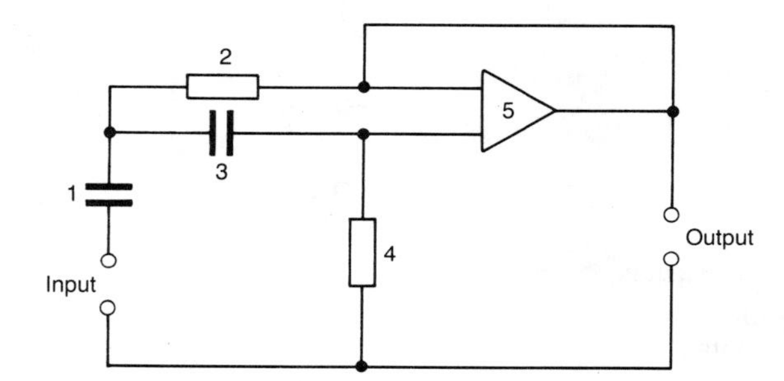

Figure 7.20 *Example*

The components are: 1 - a capacitor, 2 - a resistor, 3 - a capacitor, 4 - a resistor, 5 - an operational amplifier.

Example

Figure 7.21 shows the circuit of an astable multivibrator connected so that alternate lights flash on and off. Identify the components labelled 1, 2, 3 and 4.

The items are: 1 - a signal lamp, 2 - an npn transistor, 3 - a polarised electrolytic capacitor, 4 - a resistor, 5 - a battery consisting of a number of cells. Note that junctions between conductors only occur where there is a •, where no such symbol occurs then the conductor merely cross without a junction being made.

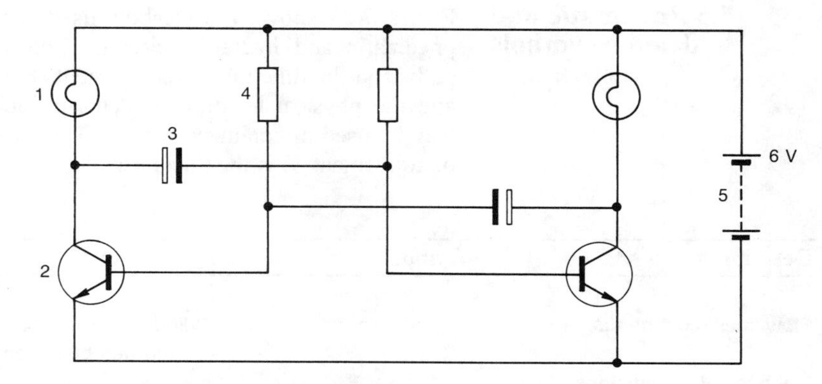

Figure 7.21 *Astable multivibrator*

Example

Figure 7.22 shows a two-stage transistor amplifier. Identify the components to which (a) resistor 4, (b) transistor 8, (c) capacitor 10 are connected?

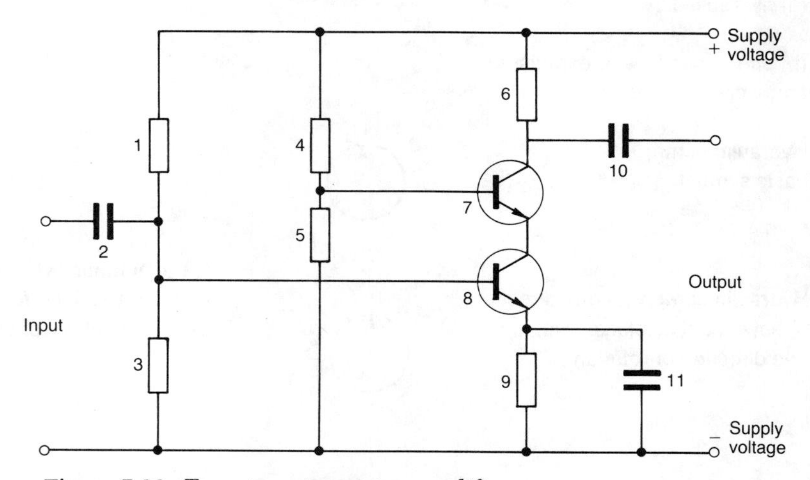

Figure 7.22 *Two-stage transistor amplifier*

(a) Resistor 4 is connected at one end to resistor 5 and transistor 7 and at the other end to the + supply voltage line.
(b) Transistor 8 is connected at one of its terminals to capacitor 2 and resistor 3 (not resistor 8), at another to transistor 7 and at the other to resistor 9 and capacitor 11.
(c) Capacitor 10 is connected to the output terminal and at the other end to transistor 7 and resistor 6.

7.5 Pneumatic and hydraulic symbols

Figure 7.23 shows the symbols used for some basic items that occur in pneumatic and hydraulic circuits. The symbols show connections, flow paths and the functions of components but do not show construction details and the physical location of items on actual components. Some symbols may be used in conjunction with ↗ to indicate they are variable or capable of adjustment. E is the thickness of line.

Description	Symbol	Description	Symbol
Flow lines: continuous		Cylinder: Single acting examples: (a) returned by unspecified force	Detailed
Mechanical connections	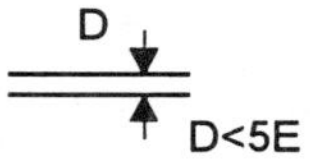 D, D<5E		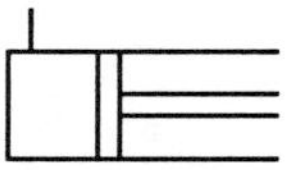
Flow line connections	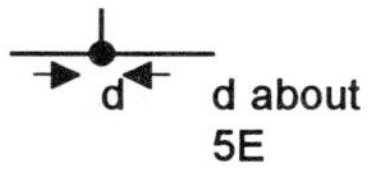 d, d about 5E		Simplified 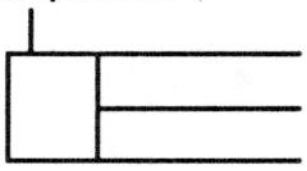
Spring			
The direction of flow and the nature of the fluid: (a) hydraulic flow		(b) returned by spring	Detailed 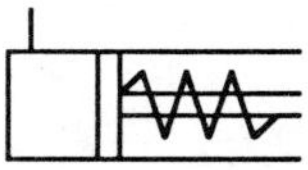
(b) pneumatic flow or exhaust to atmosphere			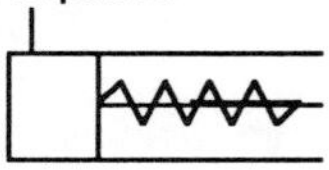
Hydraulic pump: basic symbol			Simplified
Hydraulic pump with one direction of flow, fixed displacement, and one direction of rotation	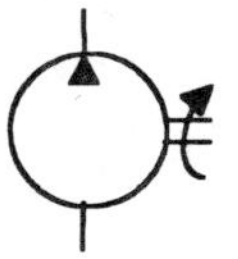	(c) pneumatic cylinder returned by an unspecified force, with exhaust to atmosphere	Detailed 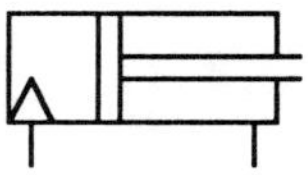
Air compressor: basic symbol	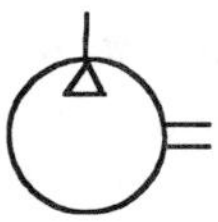		Simplified 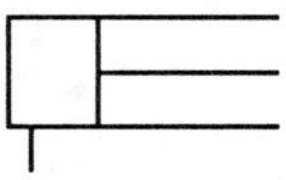

Figure 7.23 *Pneumatic and hydraulic system symbols*

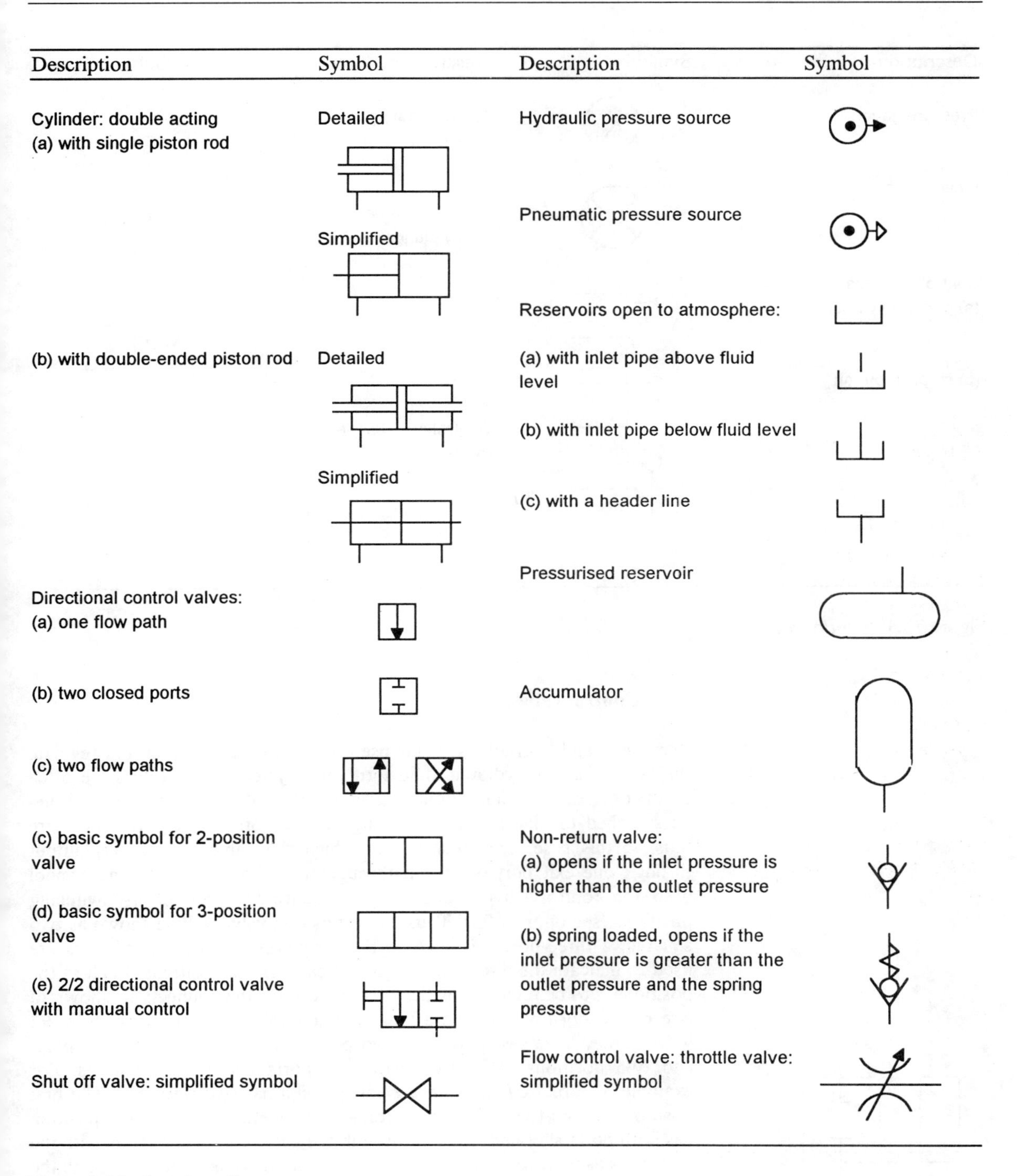

Description	Symbol	Description	Symbol
Cylinder: double acting (a) with single piston rod	Detailed Simplified	Hydraulic pressure source	
		Pneumatic pressure source	
		Reservoirs open to atmosphere:	
(b) with double-ended piston rod	Detailed Simplified	(a) with inlet pipe above fluid level	
		(b) with inlet pipe below fluid level	
		(c) with a header line	
		Pressurised reservoir	
Directional control valves: (a) one flow path			
(b) two closed ports		Accumulator	
(c) two flow paths			
(c) basic symbol for 2-position valve		Non-return valve: (a) opens if the inlet pressure is higher than the outlet pressure	
(d) basic symbol for 3-position valve		(b) spring loaded, opens if the inlet pressure is greater than the outlet pressure and the spring pressure	
(e) 2/2 directional control valve with manual control			
Shut off valve: simplified symbol		Flow control valve: throttle valve: simplified symbol	

Figure 7.23 *(continued)*

Description	Symbol	Description	Symbol
Pressure gauge		(d) by pedal	
Flow meter		Control methods: mechanical (a) by plunger	
Control methods: Muscular (a) general symbol		(b) by spring	
(b) by push button		Control methods: electrical by solenoid	
(c) by lever			

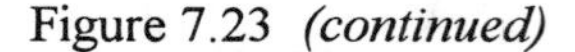

Figure 7.23 *(continued)*

7.5.1 Control valves

Pneumatic and hydraulic systems use control valves to direct and regulate the flow of air or hydraulic fluid through a system. Such valves are often on/off devices. The fluid might thus enters the valve through one port and be switched to give an output through one or other exit port. There are thus, in this case, two positions in which the valve can be set; output through one exit port or output through the other exit port. The symbol used for control valves consists of a square for each of its switching positions (see figure 7.23). Thus a two-position valve will be shown as two squares, a three-position valve as three squares. Arrow-headed lines are used to indicate the directions of flow in each of the positions. Each of the positions is to be regarded as the alternative ways the connections shown to one of the positions can be arranged. Such valves are specified as a 2/2 valve if they have two ports and two positions, 3/2 valve with three ports and two positions, 4/2 valve with four ports and three positions. For example, Figure 7.24 shows a 2/2 valve which has two positions. The first position has the two valve ports connected together. The second position has both ports shut and no flow through the valve can occur. Thus with the flow lines as indicated, i.e. the second position, there is no flow through the valve. If, however, the other position is used as the connection between the lines then flow occurs from the top line to the bottom line.

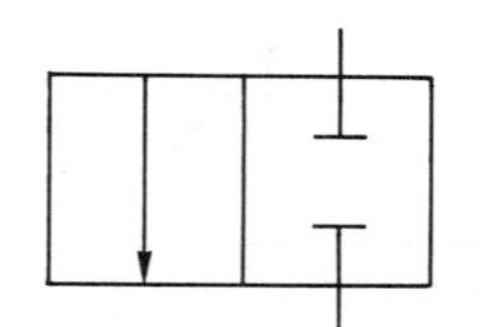

Figure 7.24 *A 2/2 valve*

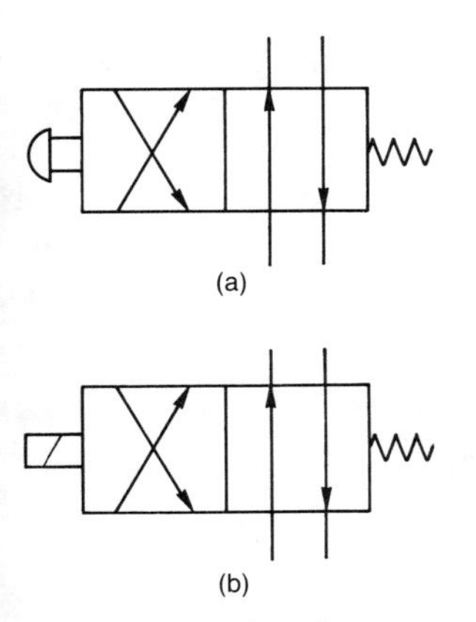

Figure 7.25 *4/2 valve*
(a) push-button operated,
(b) solenoid operated

No indication is given on the symbol as to how the valve can be switched from one position to the other. A number of methods are available to switch a valve from one position to another. For example, there might be a push-button, lever, pedal, spring, solenoid, etc. More than one of these symbols might be used with the valve symbol. Figure 7.25(a) shows how, with a 4/2 valve, a push button can be depressed to cause fluid flow paths to be switched. The push button movement gives the position indicated by the symbols used in the square to which it is attached. When the push button is released, the spring pushes the valve back to its original position and the system reverts back to its initial position. The spring movement gives the position indicated by the symbols used in the square to which it is attached. Instead of a push button 4/2 valve we might use a solenoid-operated valve (Figure 7.25(b)). When the current through the solenoid is switched on, then the position indicated by the square next to the solenoid is given. When the current ceases, the spring pushes the valve back to its original position and the piston retracts.

Example

Figure 7.26 shows a hydraulic circuit. State the functions of the items 1, 2, 3, 4, 5, 6 and 7 and explain how the system can be used to raise or lower the load.

The items are 1 - a reservoir, 2 - an electric motor, 3 - a fixed capacity hydraulic pump with one direction of flow, 4 - a shut-off valve, 5 - a pressure gauge, 6 - a single acting cylinder and 7 - a control valve.

With the raise valve open, fluid flows from the pump to the cylinder and causes the piston to move upwards. With the raise valve closed and the lower valve opened, the fluid flows into the reservoir and the load is lowered. The valve 7 is used as a pressure regulating valve to keep the pressure in the cylinder at a safe level. When the pressure in the cylinder exceeds the safe level then 7 starts to open and bleeds fluid back into the reservoir.

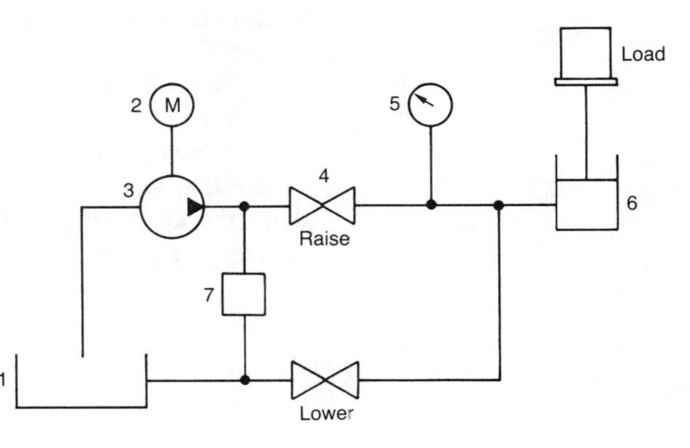

Figure 7.26 *Hydraulic circuit*

Example

Figure 7.27 shows a pneumatic lift system. Explain how the system operates.

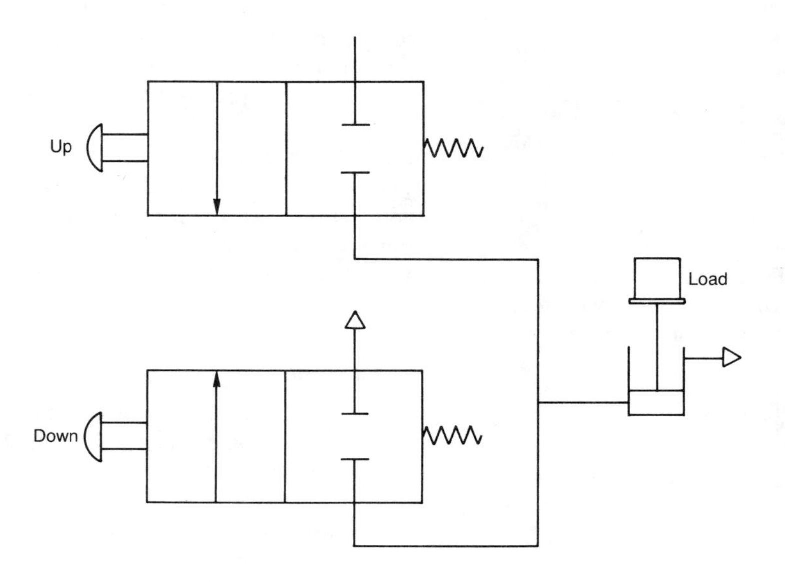

Figure 7.27 *Lift system*

Two push-button 2/2 valves are used with a single-acting cylinder. The up valve is initially in the second position with the air intake from above shut off from being passed through it to the cylinder. When the button on the up valve is pressed then the load is lifted. This is because the first position occurs and air then passes to the cylinder. With the down valve the air from the cylinder is normally shut off from escape to the atmosphere. When the button on the down valve is pressed then the first position occurs and the load is lowered because the air is then allowed to leave the cylinder.

Note that with the valves shown, the open arrow indicates a vent of air to the atmosphere. Thus there is a vent to the atmosphere on the cylinder as a safety feature to stop the piston rising above a safe level.

Problems

Questions 1 to 30 have four answer options: A, B, C and D. You should choose the correct answer from the answer options.

Questions 1 to 3 relate to Figure 7.28 which shows a number of symbols used in engineering drawings of mechanical parts.

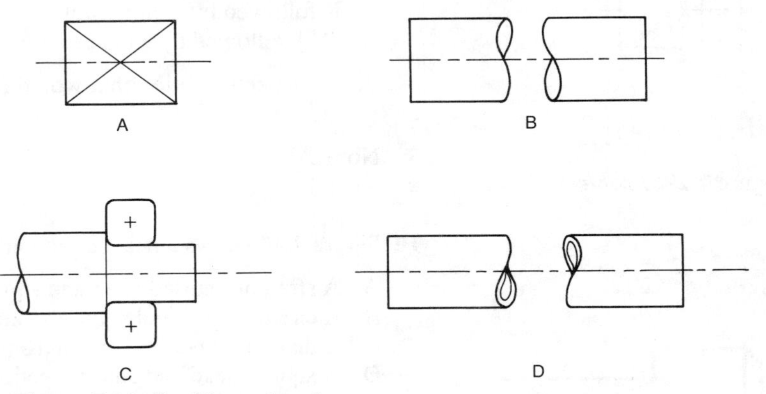

Figure 7.28 *Problems 1 to 3*

Select the symbol which is used to represent:

1 A square section shaft
2 Bearings on a shaft
3 A hollow circular shaft

4 A curve of radius 20 mm is shown in an engineering drawing as:

A CURVE 20
B RAD 20
C R 20
D ∅ 20

5 The term CH HD is used on an engineering drawing for:

A Chamfered head
B Cheese head
C Case hardened head
D Chamfered and hardened

6 The note S'FACE ∅ 6 on an engineering drawing is used for:

A Spherical face of diameter 6 mm
B Spot face area of diameter 6 mm
C Surface with a diameter of 6 mm
D Square face with a diameter of 6 mm

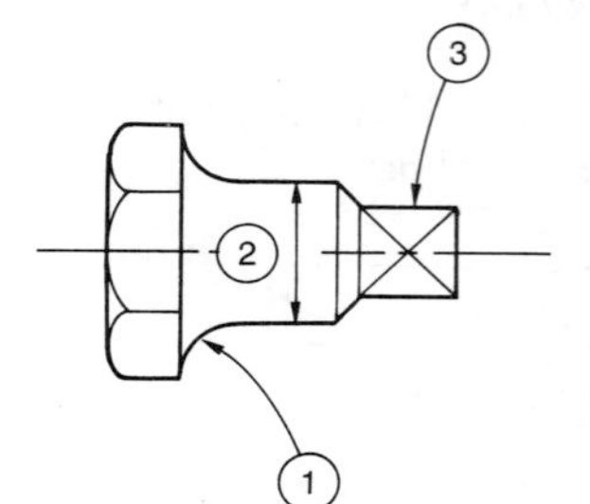

Figure 7.29 *Problems 7 to 9*

Questions 7 to 9 relate to Figure 7.29 which shows a view of a part with features labelled as 1, 2 and 3. Abbreviations which might be used to identify such features are:

A □ followed by a dimension
B ∅ followed by a dimension
C R followed by a dimension
D DIA followed by a dimension

Select the abbreviation that would be used for:

7 Note 1
8 Note 2
9 Note 3

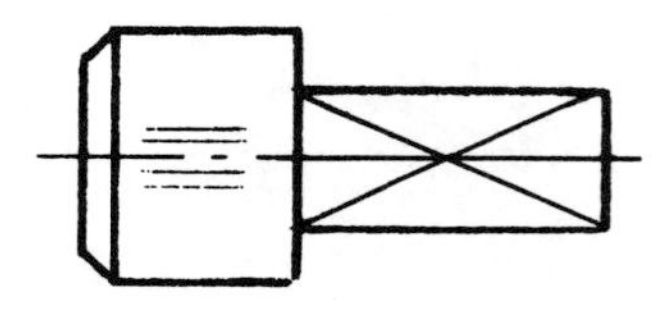
Figure 7.30 *Problems 10*

10 Figure 7.30 shows a headed pin with:

A A straight knurled head and square section shaft.
B A circular head and square section shaft.
C A diamond knurled head and square section shaft.
D A square head and square section shaft.

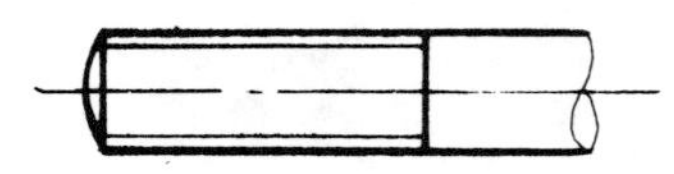
Figure 7.31 *Problem 11*

11 Figure 7.31 shows:

A A hollow circular rod splined at one end
B A solid circular rod grooved at one end
C A hollow circular rod threaded at one end
D A solid circular rod threaded at one end

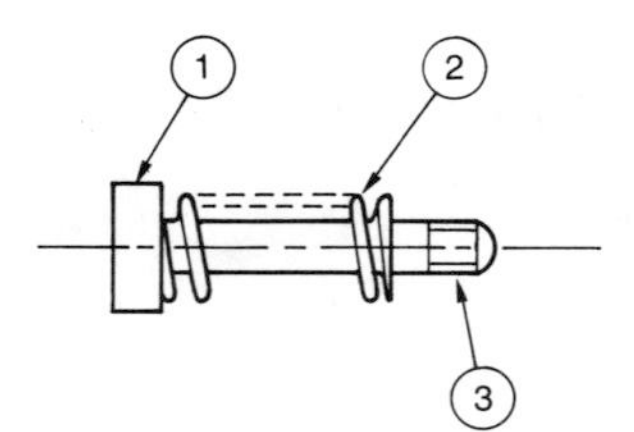

Figure 7.32 *Problems 12 to 14*

Questions 12 to 14 relate to Figure 7.32 which shows a bolt having:

12 For item 1:

A A diamond knurled head
B A straight knurled head
C A plain circular head
D A rectangular head

13 For item 2:

A A cylindrical compression spring
B A cylindrical tension spring
C A cylindrical torsion spring
D A coarse external thread

14 Item 3 shows:

A An external screw thread
B An internal screw thread
C A splined shaft
D A serrated shaft

Questions 15 and 16 relate to Figure 7.33 which shows part of a shaft.

15 Item 1 is:

A An interrupted view of a circular solid shaft
B An interrupted view of a tubular shaft
C An interrupted view of a rectangular shaft
D An interrupted view of a square shaft

16 Item 2 is:

A A rectangular section shaft
B Two rectangular sections on a shaft
C A bearing
D Centres for holes

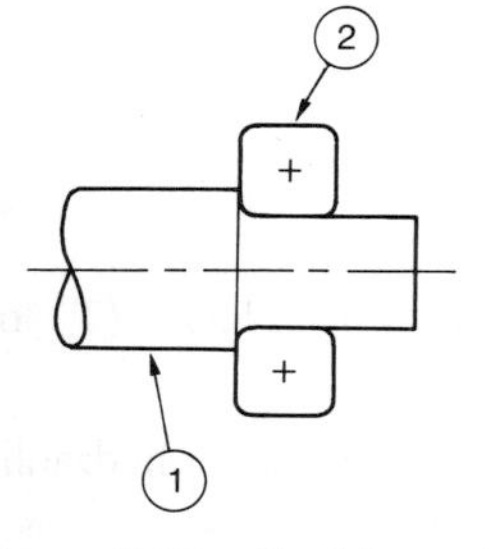

Figure 7.33 *Problems 15 to 16*

17 Decide whether each of the following statements is TRUE (T) or FALSE (F).

Figure 7.34 shows a weld with:
(i) A single V-butt weld.
(ii) The external surface of the weld on the right-hand side.

A (i) T (ii) T
B (i) T (ii) F
C (i) F (ii) T
D (i) F (ii) F

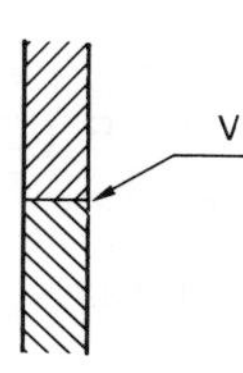

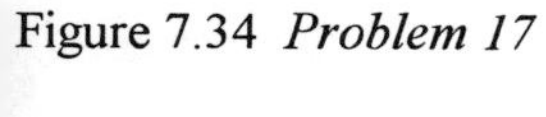

Figure 7.34 *Problem 17*

18 Decide whether each of the following statements is TRUE (T) or FALSE (F).

Figure 7.35 shows a weld with:
(i) A spot weld.
(ii) The weld on the right-hand side of the joint.

A (i) T (ii) T
B (i) T (ii) F
C (i) F (ii) T
D (i) F (ii) F

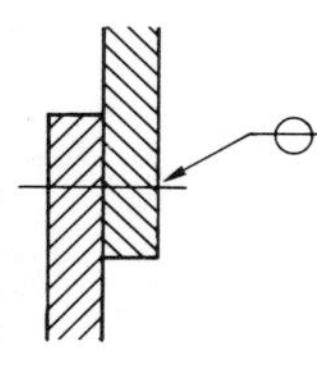

Figure 7.35 *Problem 18*

Questions 19 to 22 refer to Figure 7.36 which shows symbols used in electrical circuit diagrams.

19 Which is the standard symbol for a fixed resistor?
20 Which is the standard symbol for a variable resistor?
21 Which is the standard symbol for a voltage divider?
22 Which is the standard symbol for a fuse?

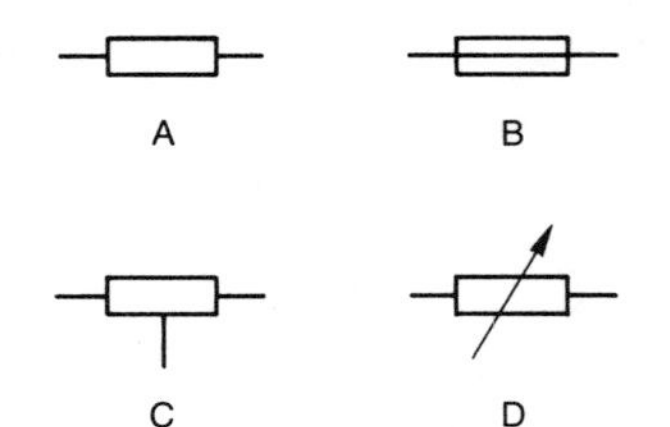

Figure 7.36 *Problems 19 to 22*

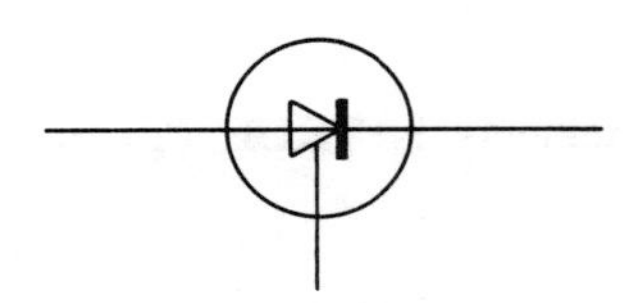

Figure 7.37 *Problem 24*

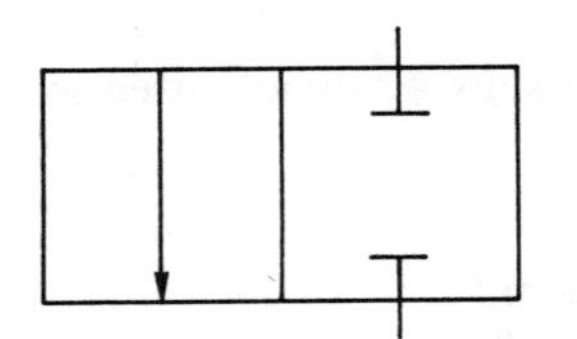

Figure 7.38 *Problem 25*

23 Decide whether each of the following statements is TRUE (T) or FALSE (F).

The standard symbol (M) can be used for a:
(i) Meter
(ii) Motor

A (i) T (ii) T
B (i) T (ii) F
C (i) F (ii) T
D (i) F (ii) F

24 The standard symbol shown in Figure 7.37 is that of a:

A Junction diode
B pnp transistor
C npn transistor
D Thyristor

25 Decide whether each of the following statements is TRUE (T) or FALSE (F).

Figure 7.38 shows a symbol used with pneumatic and hydraulic systems. It shows a directional control valve which has:
(i) Four ports.
(ii) Two positions.

A (i) T (ii) T
B (i) T (ii) F
C (i) F (ii) T
D (i) F (ii) F

26 Decide whether each of the following statements is TRUE (T) or FALSE (F).

With pneumatic and hydraulic systems the symbol + shows:
(i) A junction between pipelines.
(ii) Crossed pipelines.

A (i) T (ii) T
B (i) T (ii) F
C (i) F (ii) T
D (i) F (ii) F

27 Decide whether each of the following statements is TRUE (T) or FALSE (F).

With pneumatic and hydraulic systems the symbol ∇ is used for:
(i) Pneumatic flow.
(ii) Exhaust to atmosphere.

A (i) T (ii) T
B (i) T (ii) F
C (i) F (ii) T
D (i) F (ii) F

Questions 28 to 30 relate to Figure 7.39 which shows a pneumatic circuit involving three components labelled 1, 2 and 3.

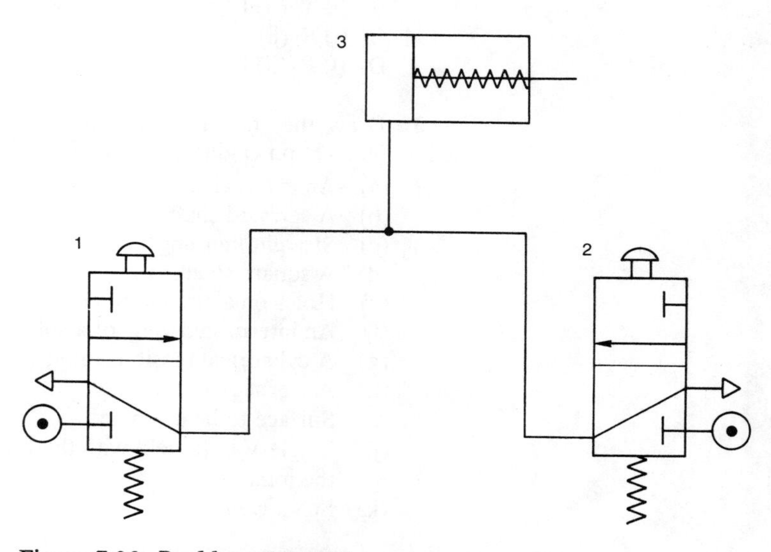

Figure 7.39 *Problems 27 to 30*

Decide whether each of the following statements is TRUE (T) or FALSE (F).

28 Item 1 has:
(i) Three ports.
(ii) Two positions.

A (i) T (ii) T
B (i) T (ii) F
C (i) F (ii) T
D (i) F (ii) F

29 Item 1 is operated by:
(i) A push-button.
(ii) A spring.

A (i) T (ii) T
B (i) T (ii) F
C (i) F (ii) T
D (i) F (ii) F

30 Item 2 is a single acting cylinder returned by a spring which is moved:
(i) To the right when the push-button on 1 is pressed.
(ii) To the left when the push-button in 2 is pressed.

A (i) T (ii) T
B (i) T (ii) F
C (i) F (ii) T
D (i) F (ii) F

31 Draw the standard symbols that are used to represent the following features on engineering drawings:
(a) An external screw thread
(b) A serrated shaft
(c) Straight knurling
(d) A square shaft
(e) Holes on a circular pitch
(f) An interrupted view of a solid circular shaft
(g) A cylindrical tension spring
(h) A worm gear
(i) Surface to be machined
(j) Single V butt weld with the weld face on one particular side of the joint
(k) Seam weld

32 Produce drawings, with sectioned views, which could be used to fully describe the items shown in Figure 7.40.

33 State the standard abbreviations/symbols that can be used to describe the features/dimensions indicated in Figure 7.41.

34 Draw the standard symbols used in circuit diagrams for:
(a) Alternating current
(b) Earth
(c) Fuse
(d) Variable resistor
(e) Capacitor
(f) Air-cored transformer
(g) Motor
(h) Ammeter
(i) pn diode
(j) npn transistor

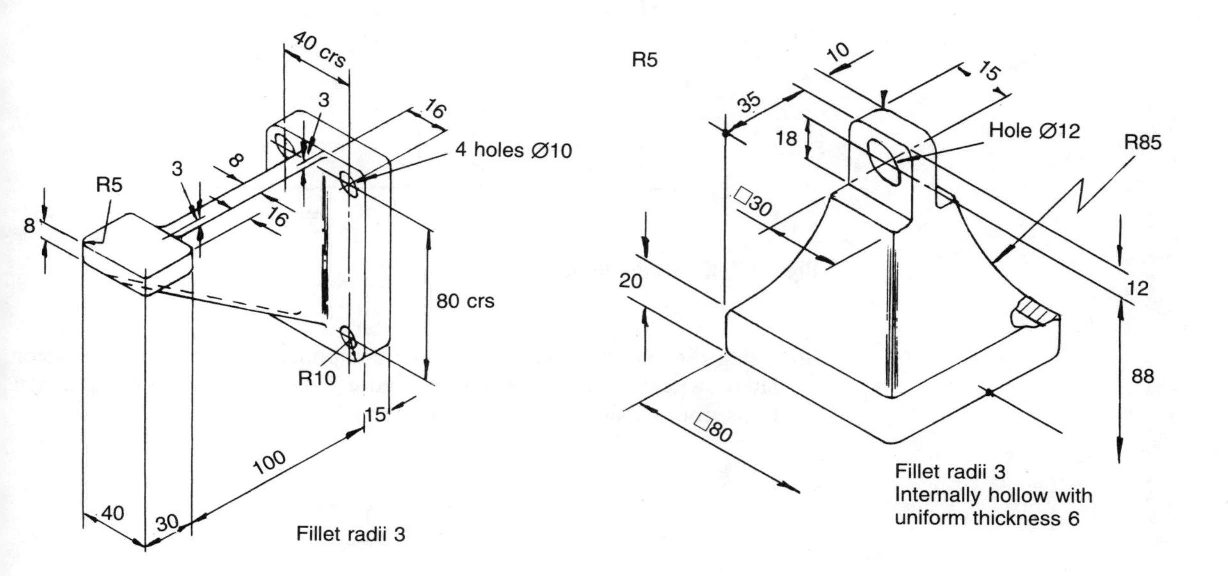

Figure 7.40 *Problem 32*

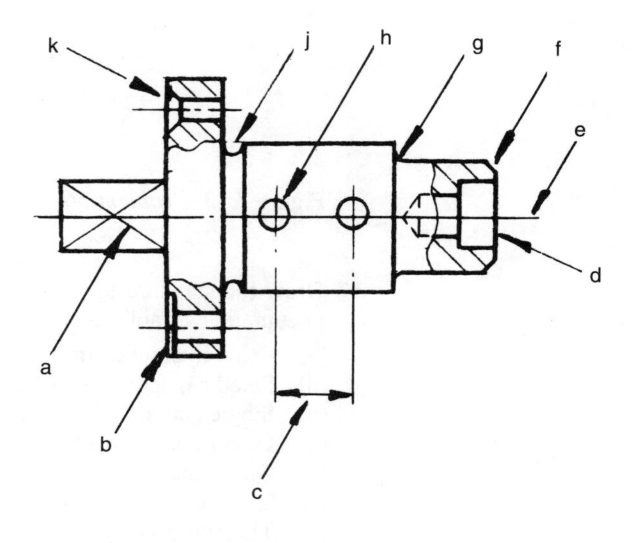

Figure 7.41 *Problem 33*

35 Identify the numbered items in the circuit diagram shown in Figure 7.42.

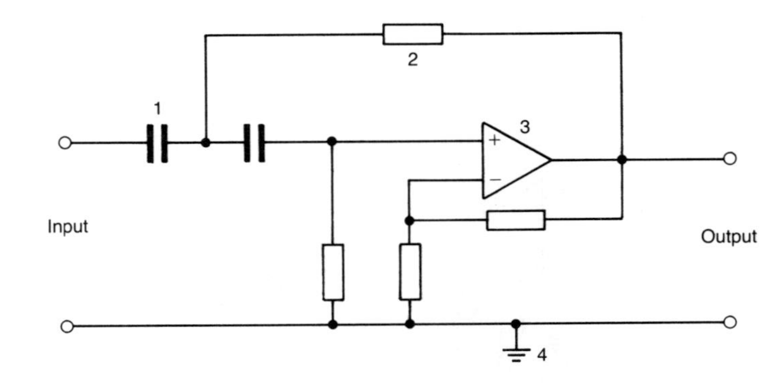

Figure 7.42 *Problem 35*

36 Using the standard symbols for the components, redraw the transistor lamp switching circuit shown in Figure 7.43. 1 is a resistor, 2 a npn transistor, 3 a signal lamp and 4 earth.

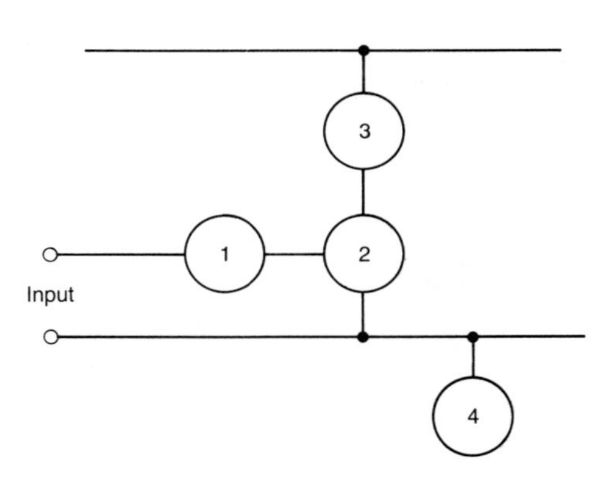

Figure 7.43 *Problem 36*

37 Draw the standard symbols that can be used for the following items in pneumatic/hydraulic systems:

(a) Fixed hydraulic pump with one direction of flow
(b) Fixed capacity pneumatic motor with one direction of flow
(c) Single acting cylinder returned by spring
(d) Directional control valve with on flow path between two ports
(e) Directional control valve with two closed ports
(f) 2/2 directional control valve with two closed ports
(g) Throttle valve
(h) Shut-off valve

(i) Accumulator
(j) Reservoir open to the atmosphere

38 Figure 7.44 shows a two-switch operated pneumatic system. Explain what happens to the piston in the cylinder when (a) A only is pressed, (b) B only is pressed, (c) A and B are both pressed.

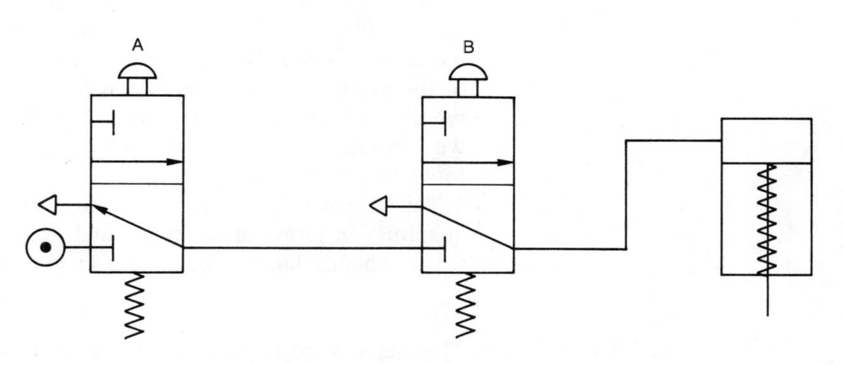

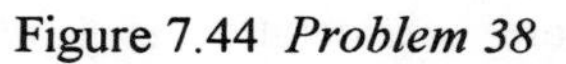

Figure 7.44 *Problem 38*

8 Engineering mathematics

8.1 Relationships

In the study of electrical circuits a relationship can be found between the voltage across a resistor and the current through it. For example we might have a voltage of 2 V when the current is 1 A and a voltage of 4 V when the current is 2 A. The voltage is related to the current. Likewise, in a study of the extension of a spring when subject to forces stretching it, there is a relationship between the extension and the force. With motion of an object we can find a relationship between the acceleration of the object and the force acting on the object.

This chapter is about how we can describe relationships between quantities in terms of equations and graphically, also the basic techniques that can be used in manipulating such equations.

8.2 Fractions

This section deals with the basic core skills of handling fractions and decimals before consideration is given in later sections to dealing with relationships and algebraic equations.

The term on the top of a fraction is called the *numerator* and the term on the bottom the *denominator*. If the top and bottom of a fraction are multiplied by the same number then the value of the fraction is unchanged. This is because we are effectively multiplying the fraction by 1. Thus

$$\frac{1}{2} \text{ has the same value as } \frac{1}{2} \times \frac{4}{4}$$

Suppose we want to add two fractions, e.g. $\frac{1}{2} + \frac{1}{4}$. Note that the answer is *not* $\frac{1}{6}$. To carry out such an addition both the fractions must have the same denominator. This we can achieve by multiplying the denominator of the $\frac{1}{2}$ by 4 and the denominator of the $\frac{1}{4}$ by 2, both then becoming 8. In order not to alter the value of the fraction we need to multiply the top and bottom of each by the same number, hence

$$\frac{1}{2} + \frac{1}{4} \text{ has the same value as } \frac{1}{2} \times \frac{4}{4} + \frac{1}{4} \times \frac{2}{2} \text{ or } \frac{1 \times 4 + 1 \times 2}{2 \times 4} = \frac{6}{8}$$

and for subtraction

$$\frac{1}{2} - \frac{1}{4} \text{ has the same value as } \frac{1}{2} \times \frac{4}{4} - \frac{1}{4} \times \frac{2}{2} \text{ or } \frac{1 \times 4 - 1 \times 2}{2 \times 4} = \frac{2}{8}$$

Example

Simplify the fractions (a) $\frac{1}{2} + \frac{2}{3}$, (b) $\frac{5}{7} - \frac{1}{3}$.

(a) Simplification, in this case, involves bringing both the fractions to the same denominator of 2×3. We thus have

$$\frac{1}{2} \times \frac{3}{3} + \frac{2}{3} \times \frac{2}{2} \text{ and so } \frac{1 \times 3 + 2 \times 2}{2 \times 3} \text{ or } \frac{7}{6}$$

(b) Bringing both fractions to the denominator of 7×3, we have

$$\frac{5}{7} \times \frac{3}{3} - \frac{1}{3} \times \frac{7}{7} \text{ and so } \frac{5 \times 3 - 1 \times 7}{7 \times 3} \text{ or } \frac{8}{21}$$

8.2.1 Multiplication and division of fractions

Suppose we multiply two fractions, e.g. $\frac{1}{2}$ and $\frac{1}{4}$. What we are asking is what is half of a quarter. This can be obtained by multiplying the two numerators to obtain the new numerator and the two denominators to obtain the new denominator, i.e.

$$\frac{1}{2} \times \frac{1}{4} \text{ has the same value as } \frac{1 \times 1}{2 \times 4} = \frac{1}{8}$$

Suppose we divide $\frac{1}{2}$ by $\frac{1}{4}$. What we are asking is how many quarters are in a half, the answer being 2. This is obtained as follows:

$$\frac{1}{2} \div \frac{1}{4} \text{ is } \frac{\frac{1}{2}}{\frac{1}{4}}$$

We can simplify this by multiplying the top and bottom of the fraction by 4 to give

$$\frac{\frac{1}{2} \times 4}{\frac{1}{4} \times 4} = \frac{2}{1} = 2$$

Effectively we take the fraction that is the denominator, invert it and then multiply the numerator fraction by it. The above fraction could have been written as

$$\frac{\frac{1}{2}}{\frac{1}{4}} = \frac{1}{2} \times \frac{4}{1} = \frac{4}{2} = 2$$

Example

Simplify the following fractions:

(a) $\frac{1}{2} \times \frac{4}{3}$, (b) $\frac{1}{4} \times \frac{1}{3}$, (c) $\frac{2}{3} \div \frac{1}{2}$

(a) For this multiplication of two fractions we have

$$\frac{1 \times 4}{2 \times 3} = \frac{4}{6} \text{ or } \frac{2}{3}$$

(b) For this multiplication of two fractions we have

$$\frac{1 \times 1}{4 \times 3} = \frac{1}{12}$$

(c) Multiplying the numerator fraction by the inverted denominator fraction gives

$$\frac{\frac{2}{3}}{\frac{1}{2}} = \frac{2}{3} \times \frac{2}{1} = \frac{4}{3}$$

8.2.2 Decimals

Decimal fractions are a special way of writing fractions that have denominators of 10, 100, 1000, etc. Thus when we have $\frac{1}{10}$ we write it as 0.1 and when we have $\frac{2}{10}$ we write it as 0.2. The period (full stop) in the sequence of digits is the *decimal point*. The first number after the decimal point gives the number of tenths. When we have $\frac{1}{100}$ we write it as 0.01 and when we have $\frac{2}{100}$ we write it as 0.02. The second number after the decimal point gives the hundredths. When we have $\frac{1}{1000}$ we write is as 0.001 and when we have $\frac{2}{1000}$ we write it as 0.02. The third number after the decimal point gives the thousandths. The fourth number after the decimal point gives the ten thousandths, the fifth the hundred thousandths, and so on. Thus 0.12 is $\frac{1}{10} + \frac{2}{100}$, which is the same as $\frac{12}{100}$.

Hence, to interpret a decimal number, read the number after the decimal point as a whole number divided by 1 plus a number of noughts equal to the number of digits after the decimal point. Thus 0.12 is $\frac{12}{100}$ and 0.012 is $\frac{012}{1000}$. if we have the number 1.12 then we have 1 + 0.12 and so 1 + $\frac{12}{100}$. This can be written as $\frac{100}{100} + \frac{12}{100} = \frac{112}{100}$.

To add or subtract decimals, place the numbers in a column with the decimal points lined up. This means tenths above tenths, hundredths above hundredths, etc. We then add or subtract the tenths, the hundredths, etc. This means add or subtract as for whole numbers with the decimal point coming in the result in line with those in the numbers above it. The following examples illustrate this. Thus for addition:

$$\begin{array}{l} 1.35 \\ 0.012 \\ 0.101 \\ \underline{0.0017} \\ 1.4647 \end{array}$$

and for subtraction

$$\begin{array}{r} 1.411 \\ -\underline{0.032} \\ 1.379 \end{array}$$

When a decimal is multiplied by 10 we end with a result in which the decimal point has been moved one place to the right. For example, 0.12 when multiplied by 10 is 1.2.

$$\frac{12}{100} \times 10 = \frac{12}{10} = 0.12$$

When a decimal is multiplied by 100 we end up with the result in which the decimal point has been moved two places to the right. Thus 0.12 when multiplied by 100 is 12.

$$\frac{12}{100} \times 100 = 12$$

When a decimal is multiplied by $\frac{1}{10}$, i.e. 0.1, we end up with the result that the decimal place is moved one place to the left. Thus 0.12 multiplied by 0.1 is 0.012.

$$\frac{12}{100} \times \frac{1}{10} = \frac{12}{1000} = 0.012$$

When a decimal is multiplied by $\frac{1}{100}$, i.e. 0.01, we end up with the result that the decimal place is moved two places to the left. Thus 0.12 when multiplied by 0.01 is 0.0012.

$$\frac{12}{100} \times \frac{1}{100} = \frac{12}{10000} = 0.0012$$

Consider the division of two fractions, say 0.12 divided by 0.03. We can write the 0.12 as a fraction $\frac{12}{100}$ and the 0.03 as a fraction $\frac{3}{100}$. Then, using the rule developed for the division of fractions in the previous section of this chapter,

$$\frac{0.12}{0.03} = \frac{12}{100} \div \frac{3}{100} = \frac{12}{100} \times \frac{100}{3} = 4$$

We could have considered this problem in another way. If we take the fraction $\frac{0.12}{0.03}$ and multiply the numerator and denominator by the same number then the value of the fraction is not changed. Thus, in this case we choose from 10, 100, 1000, etc. the one which makes the denominator a whole number. Thus

$$\frac{0.12}{0.03} \times \frac{100}{100} = \frac{12}{3} = 4$$

Another example of this is to consider the problem of dividing 0.15 by 0.0005. Then we have

$$\frac{0.15}{0.0005} = \frac{0.15}{0.0005} \times \frac{10000}{10000} = \frac{1500}{5} = 300$$

What we have effectively done is to move the decimal point in the divisor to the right until it becomes a whole number and similarly moved the decimal point of the number being divided by the same number of places. Then division is carried out.

As a further example, consider the problem of dividing 10.23 by 3.3. We have thus to evaluate the fraction $\frac{10.23}{3.3}$. Multiplying the numerator and denominator by 10, to make the denominator a whole number, gives 102.3 to be divided by 33. We can do this in the same way as with whole numbers, but placing the decimal point in the result directly over that in the 102.3, i.e.

$$\begin{array}{r|l} & 3.1 \\ \cline{2-2} 33 & 102.3 \\ & \underline{99} \\ & 33 \\ & \underline{33} \\ & 0 \end{array}$$

To change a fraction into a decimal, divide the numerator by the denominator, as in the above illustration, and write the result in decimal form. Some commonly encountered fractions in engineering and their decimal equivalents are:

$\frac{1}{8}$	0.125
$\frac{1}{4}$	0.25
$\frac{3}{8}$	0.375
$\frac{1}{2}$	0.5
$\frac{5}{8}$	0.625
$\frac{3}{4}$	0.75
$\frac{7}{8}$	0.875

A calculator can be used to carry out operations with decimals. Thus if we need to divide 45.6 by 0.32, the procedure is:

1 Press the keys 4, 5, the decimal point, and then 6.
2 Press the ÷ key.
3 Press the keys decimal point, 3 and then 2.
4 Press the = key. The displayed result is 142.5.

Example

Determine to two decimal places the value of 0.456 divided by 1.23 (a) without using a calculator, (b) using a calculator.

(a) Without using a calculator we have to determine the value of the fraction $\frac{0.456}{1.23}$. To make the denominator a whole number we multiply it by 100. Similarly multiplying the numerator by 100 gives 45.6 divided by 123. Thus

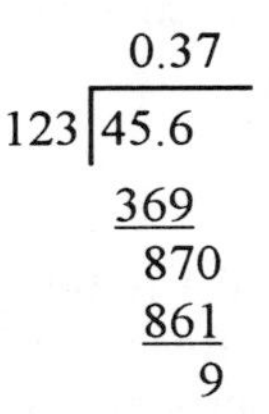

(b) Using a calculator, the sequence is:

1 Press the keys decimal point, 4, 5 and then 6.
2 Press the key ÷.
3 Press the keys 1, decimal point, 2 and then 3.
4 press the = key. The result is, to two decimal places 0.37.

8.3 Proportionality

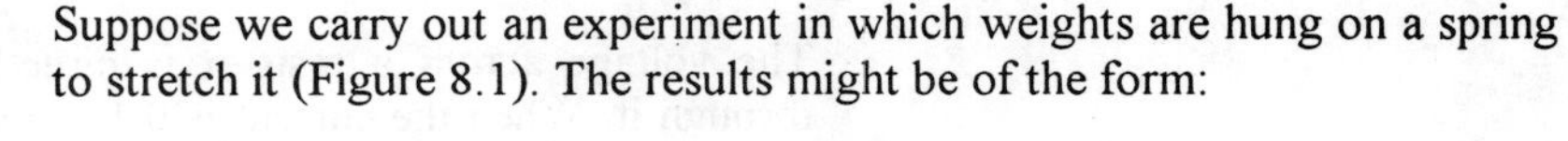

Suppose we carry out an experiment in which weights are hung on a spring to stretch it (Figure 8.1). The results might be of the form:

Force in newtons	0	1	2	3
Extension in mm	0	12	24	36

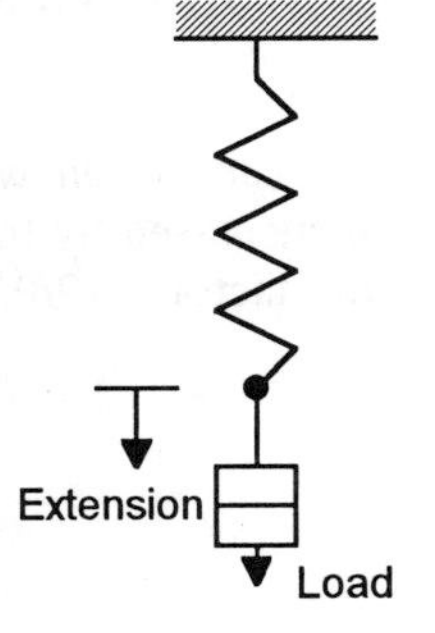

Figure 8.1 *Stretching a spring*

From this we can draw the conclusion that the extension depends on the force applied and that doubling the force doubles the extension, trebling the force trebles the extension. When we have such a relationship we say that the extension is *directly proportional* to the force. This can be written as

extension $\propto$ force

The symbol $\propto$ stands for the words 'proportional to'.

Consider another experiment in which the voltage is measured across a resistor for different currents through it (figure 8.2). We might obtain the following results:

Current in amps	0	1	2	3
Voltage in volts	0	2	4	6

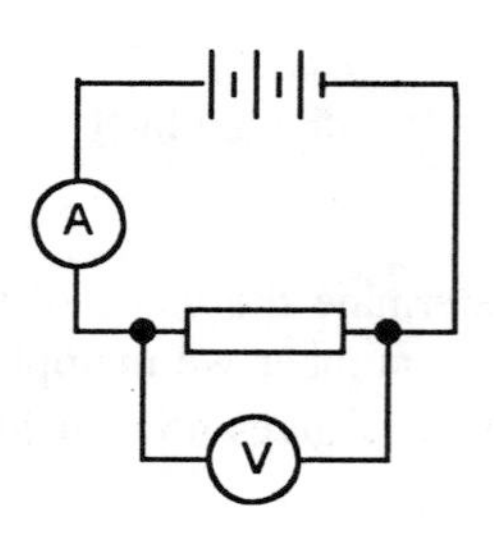

Figure 8.2 *Voltage–current relationship*

From this we can draw the conclusion that the voltage depends on the current and that doubling the current doubles the voltage, trebling the current trebles the voltage. The voltage is *directly proportional* to the current.

$$\text{voltage} \propto \text{current}$$

There are many situations in engineering where we have one physical quantity directly proportional to another. More examples are discussed later in this chapter in the discussion of graphs and chapter 9 will give further examples.

Example

The following are the results obtained for a freely falling object of the distance fallen by a freely falling object at different times. Is the distance fallen directly proportional to the time taken?

Time in tenths of seconds	0	1	2	3
Distance in millimetres	0	50	200	450

When the time is doubled the distance increases by a factor of four. When the time is trebled the distance increases by a factor of nine. Thus the distance is *not* directly proportional to the time.

Example

The voltage across a resistor is directly proportional to the current through it. When the current is 0.1 A the voltage is 6 V. What will be the voltage when the current is 0.12 A?

Since the voltage is proportional to the current this means when we increase the current by some factor that the voltage is increased by the same factor. Thus when the current is increased by the factor 0.12/0.1 then the voltage becomes

$$\text{voltage} = \frac{0.12}{0.1} \times 6 = 7.2 \text{ V}$$

8.3.1 The use of letters

In algebra *letters* are used to represent quantities. Thus if we have the relationship

$$\text{voltage} \propto \text{current}$$

then if we let V represent the value of the voltage across a resistor and I the value of the current through it, we can write it as

$V \propto I$

We can replace the proportional sign by an equals sign and a constant. Thus we can write

$V = RI$

where R is a constant. The term *equation* is used when there is an exact balance between what is on one side of the equals sign (=) and what is on the other side. Thus with the equation $V = RI$, the numerical value on the left-hand side of the equals sign must be the same as the numerical value on the right-hand side. If we consider the data we had earlier, i.e.

Current in amps	0	1	2	3
Voltage in volts	0	2	4	6

then for the equation to be true we must have when V has the value 2 then the value of RI must have the value 2. Since I is 1 at this voltage then we must have R with the value 2. When V has the value 4 then RI must have the value 4 and, since I is 2 at this voltage, we have R with the same value of 2. R is a constant and has the same value as long as the voltage is proportional to the current and the equation holds.

Example

Determine the equation, and the constant of proportionality, relating the force and extension for the spring, described by figure 8.1, giving the following data for the relationship between force and extension.

Force in newtons	0	1	2	3
Extension in mm	0	12	24	36

The extension is proportional to the force and so we can write

extension $\propto$ force

Writing this as an equation, with e being used to represent the extension and F the force, gives

$e = kF$

where k is a constant. If we put the data values into the equation we can determine k. Thus

$12 = k \times 1$

$24 = k \times 2$

$$36 = k \times 3$$

Each of these equations gives $k = 12$. Thus the equation, when the quantities have the units given in the table, is

$$e = 12F$$

8.3.2 Powers

If we have 2 we can represent this as 2^1, i.e. 2 to the power 1. If we have 2×2 then we can represent this as 2^2, this being referred to as 2 to the power 2 or 2 squared. Likewise $2 \times 2 \times 2$ is 2^3, this being referred to as 2 to the power 3 or 2 cubed. If we have 10 we can represent this as 10^1, i.e. 2 to the power 1. If we have 10×10 we can represent this as 10^2, this being referred to as 10 to the power 2 or 10 squared. Likewise $10 \times 10 \times 10$ is 10^3, this being referred to as 10 to the power 3 or 10 cubed. The power is the number of times we must multiply a number by itself.

The equation for the area A of a circle of radius r is $A = \pi r^2$. This means that the area is given by $A = \pi \times r \times r$. With this equation, the area of a circle is not directly proportional to the radius. We can, however, say that the area is proportional to the square of the radius.

Example

The relationship between the volume V of a sphere of radius r and its radius is that the volume is proportional to the cube of the r. Write an equation describing this relationship.

We have

$$V \propto r^3$$

and so

$$V = Kr^3$$

Note that the volume is not directly proportional to the radius.

8.3.3 Units

Letters are also used to represent units. Thus we might have a voltage of two volts and write this as 2 V. The V here does not represent a quantity but is just a shorthand way of writing the unit of voltage. In textbooks, units are written as V but when the letter is used to represent a quantity it is written in italics as V.

It is not only numerical values that must balance when there is an equation. The units of the quantities must also balance. For example, the area A of a rectangle is the product of the lengths b and w of the sides, i.e.

$A = bw$. The units of the area must therefore be the product of the units of b and w if the units on both sides of the equation are to balance. Thus if b and w are both in metres then the unit of area is metre × metre, or square metres (m^2).

Velocity is distance/time and so the unit of velocity can be written as the unit of distance divided by the unit of time. For the distance in metres and the time in seconds, then the unit of velocity is metres per second, i.e. m/s. For the equation describing straight line motion of $v = at$, if the unit of v is metres/second then the unit of the at must be metres per second. Thus, if the unit of t is seconds, we must have

$$\text{unit of } at = \frac{\text{metre}}{\text{second}} = \frac{\text{metre}}{\text{second}^2} \times \text{second}$$

or, in symbols

$$\frac{\text{m}}{\text{s}} = \frac{\text{m}}{\text{s}^2} \times \text{s}$$

Hence the unit of the acceleration a must be m/s^2 for the units on both sides of the equation to balance.

Example

When a force F acts on a body of mass m it accelerates with an acceleration a given by the equation $F = ma$. If m has the unit of kg and a the unit m/s^2, what is the unit of F?

For equality of units to occur on both sides of the equation

$$\text{unit of } F = \text{kg} \times \frac{\text{m}}{\text{s}^2} = \text{kg m/s}^2$$

This unit is given a special name of the newton (N).

8.4 Manipulating equations

If we have the equation $x + 12 = 20$, then the numerical value of the left-hand side of the equation, i.e. $x + 12$, must be the same as the numerical value on the right-hand side of the equation, i.e. 20. It does not matter what x represents, the expressions on the left and right of the equals sign have the same numerical value. We can then consider what value x must have if this equality is to be true. This is termed solving an equation. *Solving an equation* means finding the particular numerical value of the unknown which makes the equation balance. Thus, for the above equation, x must have the value 8 for the value of the expression on the left-hand side of the equals sign to equal the value on the right. A basic rule which would have enabled us to determine that result is: *adding the same quantity to, or subtracting the same quantity from, both sides of an equation does not change the equality*. Thus if we subtract 12 from both sides of the above equation, then

THE LIBRARY
GUILDFORD COLLEGE
of Further and Higher Education

$$x + 12 - 12 = 20 - 12$$

and so

$$x = 20 - 12 = 8$$

As a check of the answer obtained, we can replace x by the value found. The equality must still hold. Thus $8 + 12 = 20$. The equality holds.

Another rule which can be used is: *multiplying, or dividing, both sides of an equation by the same non-zero quantity does not change the equality.* Thus if we have the equation $3x = 18$, then if we divide both sides of the equation by 3 we obtain

$$\frac{3x}{3} = \frac{18}{3}$$

and thus we have $x = 6$. As a check of the answer obtained, we can replace x by the value found. Thus $3 \times 6 = 18$, the equality still holds.

If we have the equation

$$\frac{2x}{3} = \frac{1}{4}$$

then we can multiply both sides of the equation by 3 to give

$$3 \times \frac{2x}{3} = 3 \times \frac{1}{4}$$

and so

$$2x = \frac{3}{4}$$

If we now divide both sides of the equation by 2 we then obtain

$$\frac{2x}{2} = \frac{\frac{3}{4}}{2} = \frac{3}{8}$$

and so $x = \frac{3}{8}$.

In general, whatever mathematical operation we do to one side of an equation, provided we do the same to the other side of the equation then the balance is not affected. Thus we can manipulate equations without affecting their balance by, for example:

1 Adding the same quantity to both sides of the equation.
2 Subtracting the same quantity from both sides of the equation.
3 Multiplying both sides of the equation by the same quantity.
4 Dividing both sides of the equation by the same quantity.
5 Squaring both sides of the equation.

The following examples illustrate some of these manipulations.

Example

Solve the equation $2x - 4 = x + 1$.

If we add 4 to each side of the equation then

$$2x - 4 + 4 = x + 1 + 4$$

and so $2x = x + 5$. If we now subtract x from both sides of the equation

$$2x - x = x + 5 - x$$

and so we have $x = 5$ as the solution. We can check this by replacing x in the original equation by this value. Then we have $2 \times 5 - 4 = 5 + 1$.

Example

Solve the equation $\frac{2}{3}x = 8$.

Multiplying both sides of the equation by 3 gives

$$3 \times \tfrac{2}{3}x = 3 \times 8$$

and so $2x = 24$. Dividing both sides of the equation by 2 gives

$$\frac{2x}{2} = \frac{24}{2}$$

and so the solution is $x = 12$. We can check this by replacing x in the original equation by this value. Then we have $\frac{2}{3} \times 12 = 8$.

Example

The voltage V across a resistance R is related to the current I through it by the equation $V = IR$. When $V = 2$ V then $I = 0.1$ A. What is the value of the resistance?

Writing the equation with the numbers substituted gives

$$2 = 0.1R$$

Multiplying both sides of the equation by 10 gives

$$2 \times 10 = 10 \times 0.1R$$

Hence $20 = 1R$ and so the resistance R is 20 Ω. We can check this by replacing R in the original equation by this value to give $2 = 0.1 \times 20$.

Example

The mass of a piece of wood is directly proportional to its volume. If a block of the wood of volume 200 cm^3 has a mass of 160 g, write an equation relating the mass and volume.

The relationship is

$$\text{mass} \propto \text{volume}$$

Thus if we represent the mass by m and the volume by V we can write an equation

$$m = kV$$

where k is some constant. Since m = 160 g when V = 200 cm^3, then

$$160 = k \times 200$$

Hence, dividing both sides of the equation by 200, we have

$$\frac{160}{200} = k$$

and k = 0.80 g/cm^3. The equation is thus, with these units,

$$m = 0.80V$$

8.4.1 Transposition

There is another way of considering the manipulation operations discussed above. Consider again the examples discussed there.

For the equation $x + 12 = 20$, subtracting 12 from each side gives

$$x + 12 - 12 = 20 - 12$$

and so $x = 20 - 12$. Effectively what we have done with this operation is to move the +12 from the left-hand side of the equation to the right-hand side when it becomes −12, as illustrated below This is termed *transposition*.

$$x + 12 = 20$$

$$x \qquad = 20 - 12$$

With the equation

$$\frac{2x}{3} = \frac{1}{4}$$

We can solve this by multiplying both sides of the equation by 3 and dividing both sides by 2 to give

$$x = \frac{3 \times 1}{2 \times 4}$$

We can consider that the operations we have carried out are equivalent to having the 2 in the numerator of the fraction on the left moved across the equals sign and downwards to become the denominator on the right and the 3 moved from the denominator on the left moved across the equals sign to become the numerator on the right, the term *cross-multiplying* often being used to describe this operation. Thus the number that is being used to multiply a quantity on the left-hand side of the equation becomes when transposed to the right-hand side a division. A number that is being used to divide a quantity on the left-hand side of the equation becomes when transposed to the right-hand side a multiplication.

The term *transposition* is used when a quantity is moved from one side of an equation to the other side. The following are basic rules for use with transposition:

1 A quantity which is added on the left-hand side of an equation side becomes subtracted on the right-hand side.
2 A quantity which is subtracted on the left-hand side of an equation becomes added on the right-hand side.
3 A quantity which is multiplying on the right-hand side of an equation becomes a dividing quantity on the left-hand side.
4 A quantity which is dividing on the left-hand side of an equation becomes a multiplying quantity on the right-hand side.

Example

Solve the equation $2x - 4 = x + 1$.

This is a repeat of the example in the previous section. Now consider the same example with transposition. We have $2x - 4 = x + 1$. If we move the -4 from the left-hand side to the right-hand side of the equation it becomes $+4$ and we obtain $2x = x + 1 + 4$. If we move the $+x$ from the right-hand side to the left-hand side of the equation it becomes $-x$ and so we have $2x - x = 1 + 4$. Hence, as before, the solution is $x = 5$.

Example

Solve the equation $\frac{2}{3}x = 8$.

This is a repeat of the example in the previous section. Now consider the same example with transposition. Using transposition, the multiplication by 2 on the left-hand side becomes a division by 2 on the right-hand side. Hence we can write

$$\frac{x}{3} = \frac{8}{2}$$

The division by 3 on the left-hand side of the equation becomes a multiplication by 3 on the right-hand side. Thus $x = \frac{8}{2} \times 3 = 12$.

Example

The voltage V across a resistance R is related to the current through it by the equation $V = IR$. When $V = 2$ V then $I = 0.1$ A. What is the value of the resistance?

This is the example considered in the previous section. Now by transposition, the equation $2 = 0.1R$ has the multiplication by 0.1 on the right-hand side of the equation becoming a division by 0.1 on the left-hand side. Hence

$$\frac{2}{0.1} = R$$

Thus, as before, $R = 20\ \Omega$.

8.4.2 Manipulation of formulae

Suppose we have the equation $F = kx$ and we want to solve the equation for x in terms of the other quantities. Writing the equation as

$$kx = F$$

then transposing the k from the left-hand side to the right-hand side (or dividing both sides by k) gives the solution

$$x = \frac{F}{k}$$

As another example, consider the following equation for the variation of resistance R of a conductor with temperature θ:

$$R_t = R_0(1 + \alpha\theta)$$

where R_t is the resistance at temperature θ, R_0 the resistance at 0°C and α a constant called the temperature coefficient of resistance. We might need to rearrange the equation so that we express α in terms of the other variables. As a first step we can multiply out the brackets to give

$$R_t = R_0 + R_0\alpha\theta$$

Transposing the R_0 from the left-hand to the right-hand side of the equation gives

$$R_t - R_0 = R_0\alpha\theta$$

Reversing the sides of the equation so that we have the term involving α on the left-hand side, we then have

$$R_0\alpha\theta = R_t - R_0$$

Transposing the $R_0\theta$ from the left-hand to right-hand sides gives

$$\alpha = \frac{R_t - R_0}{R_0\theta}$$

Example

The surface area A of a closed cylinder is given by the equation

$$A = 2\pi r^2 + 2\pi rh$$

where r is the radius of the cylinder and h its height. Rearrange the equation to give h in terms of the other variables.

Transposing the $2\pi r^2$ term gives

$$A - 2\pi r^2 = 2\pi rh$$

Reversing the sides the equation gives

$$2\pi rh = A - 2\pi r^2$$

Transposing the $2\pi r$ from the left-hand to right-hand sides gives

$$h = \frac{A - 2\pi r^2}{2\pi r}$$

Example

The density ρ of a block of material is related to its mass m and volume V by the equation

$$\rho = \frac{m}{V}$$

Rearrange the equation to give the mass in terms of the other quantities.

Transposing the V from the right- to the left-hand sides of the equation gives

$$\rho V = m$$

Thus, reversing the sides of the equation, $m = \rho V$.

8.5 Graphical presentation of data

Graphs are an indispensable element in engineering. They enable relationships and trends in experimental data to be more easily seen than is possible by just looking at the numbers. This section is a consideration of graphs and how they can be used to describe relationships. Other graphical methods of describing relationships, such as bar charts and pie charts, are discussed in section 8.7.

An essential feature of all graphical methods of representing data is the concept of a *scale*. This is the representation of one quantity by some other quantity so that one is proportional to the other. For example, we might represent voltage in volts by distance along in line in millimetres or some other unit of distance such as the number of squares on a sheet of graph paper. Figure 8.3 shows such a scale with the voltage being represented by distance along this sheet of paper from the distance mark corresponding to the zero voltage.

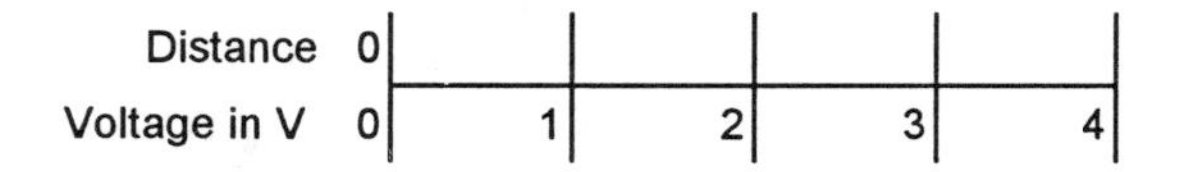

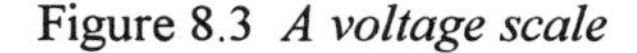
Figure 8.3 *A voltage scale*

If we want a scale for distance along this sheet of paper to represent current, we might have the scale shown in Figure 8.4.

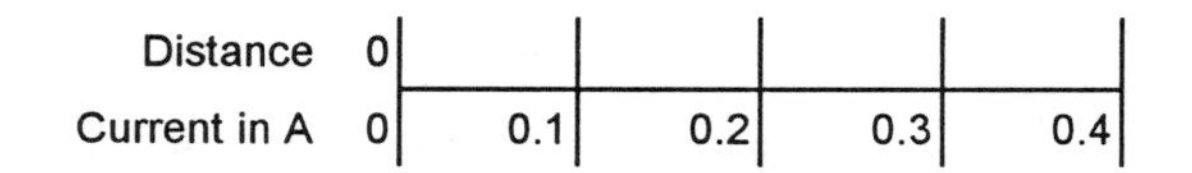

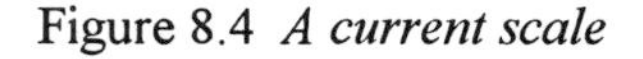
Figure 8.4 *A current scale*

As will be seen with bar charts and pie charts in section 8.7, it is not only distance that we can use to represent a quantity.

8.5.1 Graphs

A graph has two axes at right angles, each of these axes being given a scale. Thus, suppose we want one scale to represent voltages and the other scale currents. If we use the scales given above in Figures 8.3 and 8.4 then we have the graph axes shown in Figure 8.5.

If we now want to represent a voltage of 2 V occurring when the current is 0.2 A, then we locate the point on the graph which is a horizontal distance corresponding to that voltage from the zero voltage line and a vertical distance corresponding to that current from the zero current line. Figure 8.5 shows the lines used to locate such a point. We can repeat this for each of the voltage–current pairs of values. Figure 8.6 shows this for

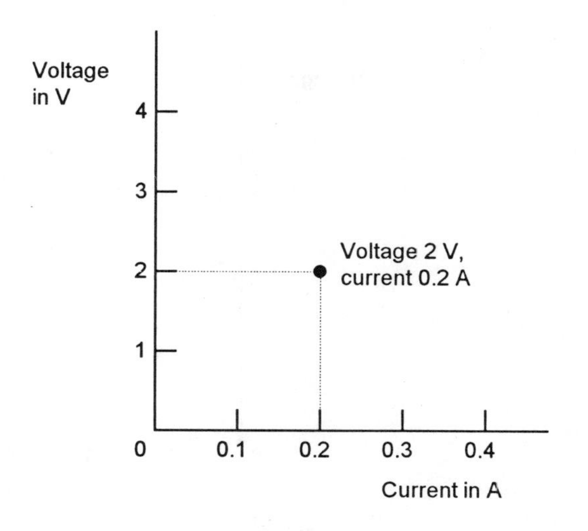

Figure 8.5 *Graph axes and plotting the point: voltage 2 V, current 0.2 A*

when the voltage is 1 V and the current 0.1 A. Hence, when other values are used we obtain a series of points on the graph. A line then drawn through the points enables the relationship between the voltage and current to be seen at a glance, as in Figure 8.7. See the examples at the end of this section for further examples of graph plotting.

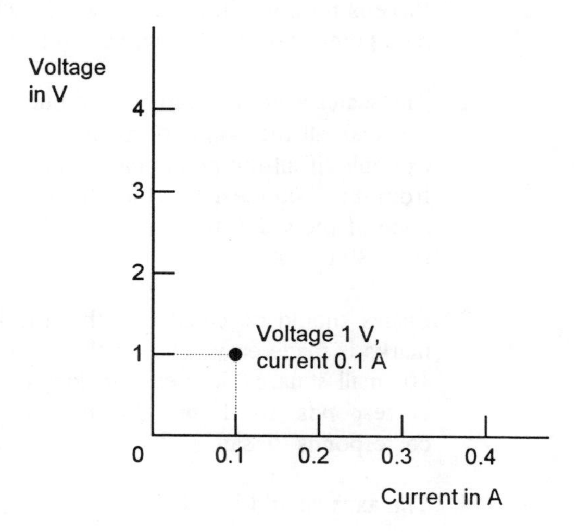

Figure 8.6 *Plotting the point: voltage 1 V, current 0.1 A*

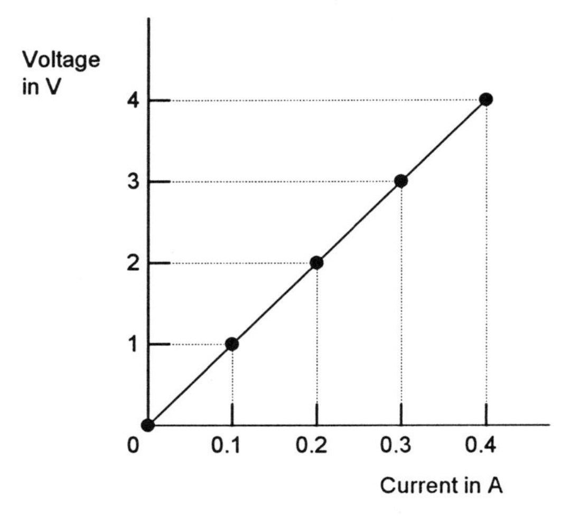

Figure 8.7 *Graph of current plotted against voltage for the values: current 0, voltage 0; current 0.1 A, voltage 1 V, current 0.2 A, voltage 2 V; current 0.3 A, voltage 3 V.*

In selecting the scales to be used for the axes of a graph and in plotting the graph, several points should be taken into account.

1 The scales should be chosen so that the points to be plotted on the graph occupy the full range of the axes used for the graph. For example, there is no point in having a graph with scales from 0 to 100 if all the data points have values between 0 and 10.

2 The scales need not start at zero but can start at some other value if this prevents all the points occurring within a small area of the graph. For example, if all the points have values between 80 and 100 then a scale from 0 to 100 means that all the points are concentrated in just the end zone of the scale. It is better, in this situation, to have a scale running from 80 to 100.

3 Scales should be chosen so that the location of points between scale marks is made easy. Thus with graph paper subdivided into squares of 10 small squares, it is easy to locate a point of 0.2 if one large square corresponds to 1 but much more difficult if one large square corresponds to, say, 3.

4 The axes should be labelled with the quantities they represent and their units.

5 The data points should be clearly marked, e.g. a large dot might be used.

6 With experimental data there will be some errors associated with the values being plotted. This will result in some scatter of the points and so the best line should be drawn. This is a smooth line for which there is the same amount of scatter on one side of the line as the other.

Example

The following data is obtained experimentally for the voltage across a resistor at different electrical currents. Plot the data as a graph.

Voltage in volts	0	0.5	1.0	1.5	2.0	2.5
Current in amps	0	0.12	0.18	0.30	0.42	0.48

If we have the voltages plotted along the vertical scale, then a suitable scale would be from 0 to 2.5. Taking the current to be the horizontal scale, then a suitable scale would be one from 0 to 0.5. Thus we need to plot the point for which the voltage is 0 when the current is 0; the voltage is 0.5 when the current is 0.12; the voltage is 1.0 when the current is 0.18; etc. Figure 8.8 shows the resulting graph. The points are experimental data and so there is some degree of error likely. The best line through the data points is the straight line shown.

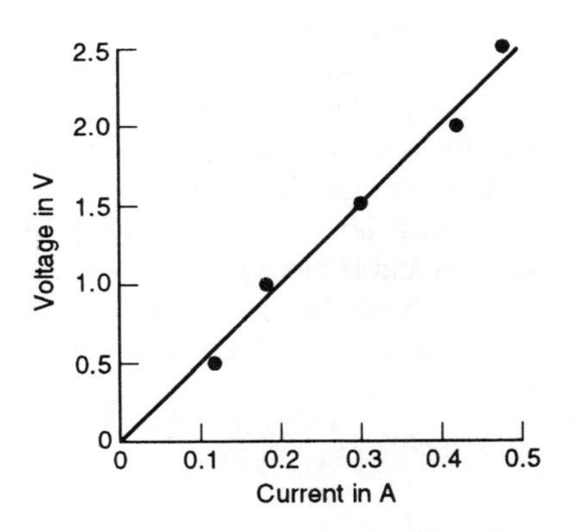

Figure 8.8 *Example*

Example

The following data is for the length of a metal rod at different temperatures. Plot the data as a graph.

Length in m	20.000	20.004	20.008	20.012	20.016
Temperature in °C	20	40	60	80	100

Starting the scale for the length at 0 would squash all the data into the region of the scale devoted to 20. Thus it is better to start this scale at 20 and have the scale going from 20 to 20.016. For the temperature scale, we can have the scale starting at 20 and going to 100. Figure 8.9 shows the resulting graph. All the points lie on a straight line.

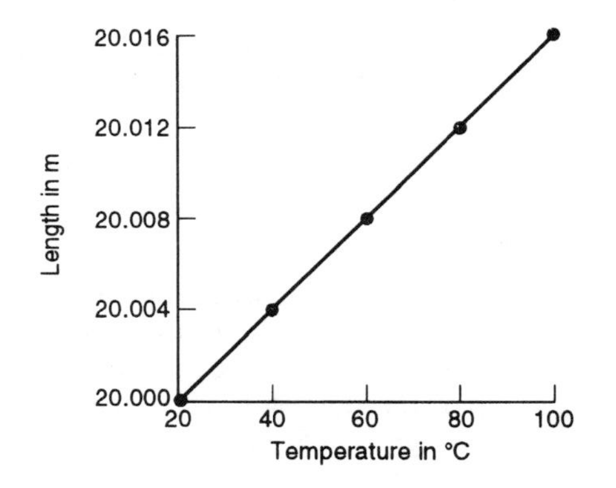

Figure 8.9 *Example*

8.5.2 Gradients of graphs

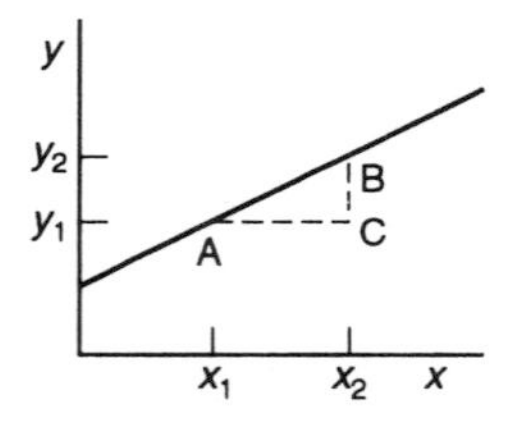

Figure 8.10 *Gradient*

Consider the line in the graph shown in Figure 8.10. This graph is of some quantity y plotted against x. The y, for example, might represent voltage and the x current. From point A to point B on the line there is a vertical rise of BC in a horizontal distance of AC. We say that the line between the points A and B has a *gradient* (slope) of BC/AC. Since BC is the difference in the y values between B and C it is $y_2 - y_1$. Since AC is the difference in the x values between A and C it is $x_2 - x_1$. Thus

$$\text{gradient} = \frac{\text{BC}}{\text{AC}} = \frac{y_2 - y_1}{x_2 - x_1}$$

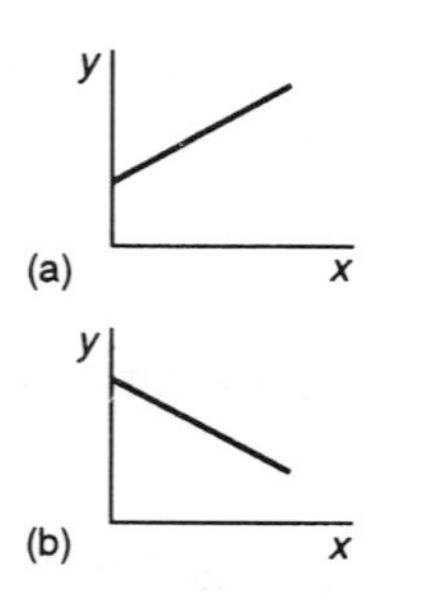

Figure 8.11 *(a) Positive gradient, (b) negative gradient*

With a straight line graph it does not matter what points we consider along the line, the gradient is the same at all positions and so is a constant.

If y increases as x increases then y_2 is greater than y_1 and x_2 greater than x_1. Thus the gradient is a positive quantity. If, however, y decreases as x increases then y_2 is less than y_1 and x_2 greater than x_1. The gradient is then a negative quantity. Figure 8.11 shows these lines with positive and negative gradients. Thus, for example, Figure 8.11(a) might be a graph of the velocity of an object and how it varies with time when the velocity is increasing and the object thus speeding up. Figure 8.11(b) might be a graph of the velocity of an object and how it varies with time when the velocity is decreasing and the object thus slowing down.

The gradient of a straight line is the same for all points on the line. However, the gradient of a curve is not a constant and varies from point to point along the curve. The gradient of a particular point on a curve is defined as being the gradient of the tangent to the curve at that point. Figure 8.12 illustrates this, the gradient at the point P being BC/AC.

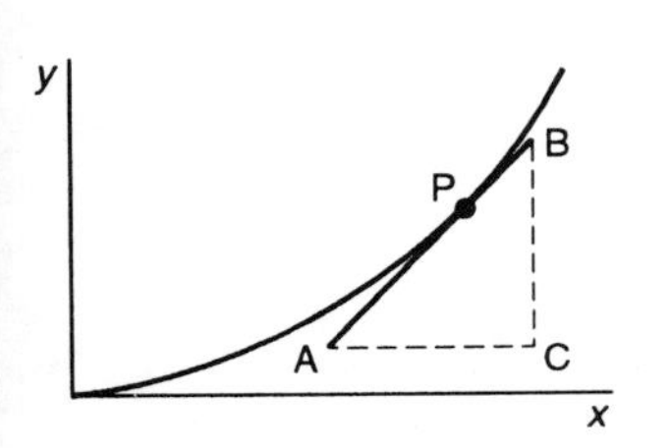

Figure 8.12 *Gradient of a curve*

Example

Determine the gradient of the straight line graph of voltage against current given in Figure 8.13 (note that this is the graph derived earlier in an example as Figure 8.8).

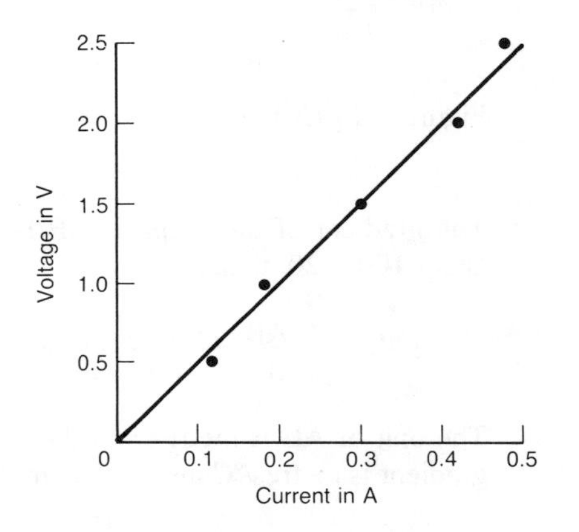

Figure 8.13 *Example*

The gradient of the graph is AB/BC. Since AB = 2.5 – 0 and BC = 0.5 – 0, then

$$\text{gradient} = \frac{\text{AB}}{\text{BC}} = \frac{2.5}{0.5} = 5$$

The unit of AB is volts and the unit of BC is amps. Thus the unit of the gradient is volts/amps. This unit is given a special name, ohms. Thus the gradient is 5 ohms.

Example

Determine the gradient of the straight line graph of length against temperature given in Figure 8.14 (note that this is the graph derived earlier in an example as Figure 8.9).

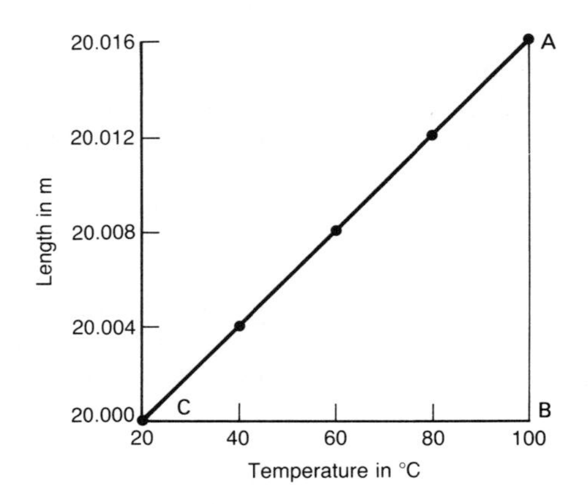

Figure 8.14 *Example*

The gradient of the graph is AB/BC. Since AB = 20.016 – 20.000 and BC = 100 – 20, then

$$\text{gradient} = \frac{\text{AB}}{\text{BC}} = \frac{0.016}{80} = 0.0002$$

The unit of AB is metres and the unit of BC is °C. Thus the unit of the gradient is metres/°C and so the gradient is 0.0002 metres/°C.

8.5.3 Equation for straight line graph

Consider the straight line graph shown in Figure 8.15. This shows a line which has $y = c$ when $x = 0$. c is said to be the *intercept* of the line with the y axis. Point A has thus the values of $y = c$ when $x = 0$. If point B has the value y and x, then if we deonote the gradient by the symbol m,

$$\text{gradient } m = \frac{y - c}{x - 0}$$

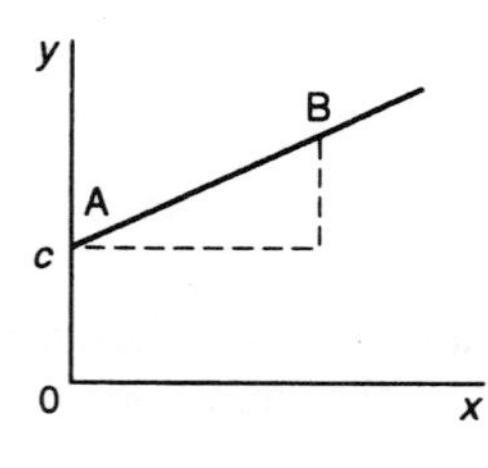

Figure 8.15 *Straight line graph*

We can rearrange this equation, by transposing the x or multiplying both sides of the equals sign by x, to give

$$mx = y - c$$

Further rearrangement gives

$$y = mx + c$$

This is the *general equation for a straight line graph* with slope m and which has an intercept with the y axis of c.

Example

The following is the data obtained from experimental measurements of the load W lifted by a machine and the effort E expended. Plot a graph of E against W and determine the gradient of the best straight line through the points and its intercept with the E axis.

E in newtons	18	27	32	43	51
W in newtons	40	80	120	160	200

Figure 8.16 shows the graph with the best straight line drawn through the data points. The gradient of the graph is

$$\text{gradient} = \frac{\text{AB}}{\text{BC}} = \frac{51 - 10}{200 - 0} \text{ or about } 0.21$$

Since the unit of AB is newtons and that of BC is newtons, the unit of the gradient is newtons/newtons. This means that it is just a number with no unit. The intercept with the E axis is at 10 newtons. We can thus describe the relationship by an equation of the form $y = mx + c$ and so $E = 0.21W + 10$.

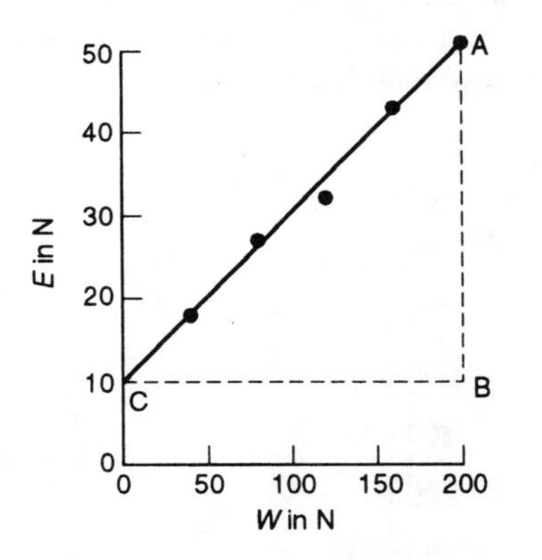

Figure 8.16 *Example*

8.5.4 Proportionality

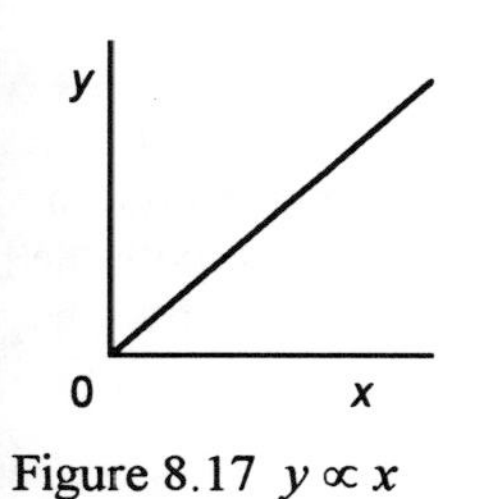

Figure 8.17 $y \propto x$

For a quantity y to be proportional to another quantity x we must have the two related by an equation of the form

$$y = kx$$

where k is a constant. Such a relationship must therefore have a graph for which the gradient is constant and equal to k and the intercepts with the y and x axes must be at 0. The graph must like that in Figure 8.17. Thus for the earlier example and Figure 8.13, the voltage is proportional to the current.

A straight line which has an intercept with the y axis which is not zero does not have y proportional to x. Thus for the example above and Figure 8.16, the effort is not proportional to the load.

8.6 Bar and pie charts

To illustrate how bar and pie charts can represnt data, consider the following data for the production output of three parts over the first three months of the year.

	January	February	March
Part A	100	200	400
Part B	200	100	100
Part C	500	600	300

We can graphically represent this data by the use of a *vertical bar chart*. With such a chart, a vertical bar is used to represent each part in each month, the height of the bar being proportional to the quantity produced. Figure 8.18 shows such a chart. At a glance we can see how the production of each part changes month by month and also how the productions of each part in each month compare.

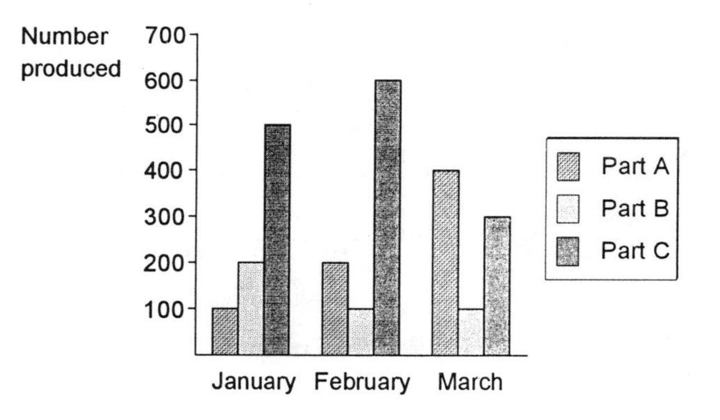

Figure 8.18 *Vertical bar chart*

Alternatively we could show the data as a *horizontal bar chart*, the length of a bar then being proportional to the number of items produced. Figure 8.19 shows the same data. Again, at a glance we can see how the production of each part changes month by month and also how the productions of each part in each month compare.

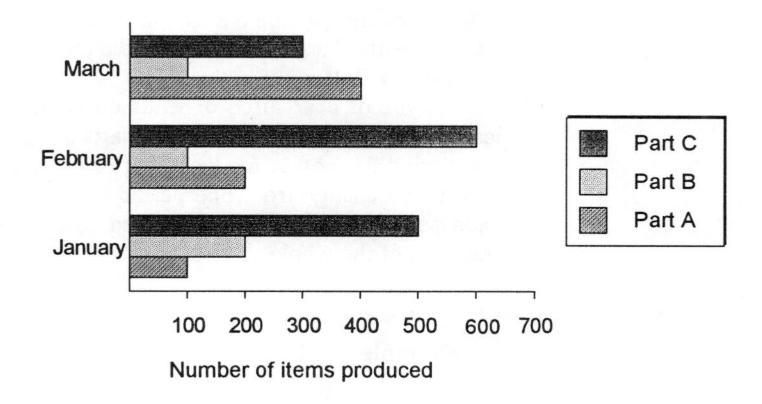

Figure 8.19 *Horizontal bar chart*

If we want to see how the total production varies month by month and the breakdown between the parts in each month, we can use a ***stacked horizontal bar chart*** or a *stacked vertical bar chart*. Figure 8.20 shows such a stacked bar chart using the above data.

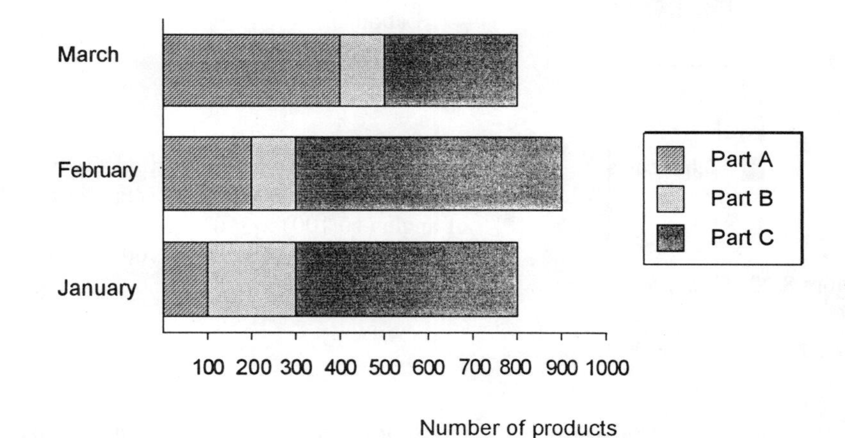

Figure 8.20 *Stacked bar chart*

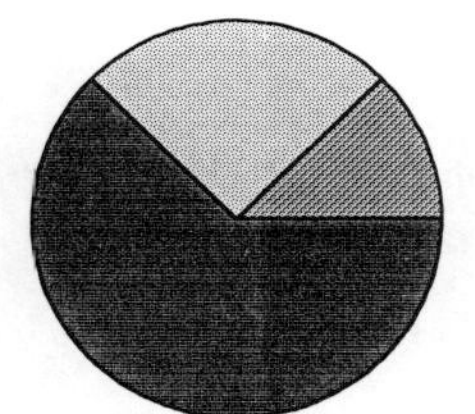

Figure 8.21 *Pie chart*

Suppose we just wanted to show the percentages of the total number of products which are A, B and C in a particular month. For this we can use a *pie chart*. A pie chart is circular and resembles a pie, the angular segments of the pie for each product being proportional to the percentage of the total that is that product. Figure 8.21 illustrates this for the above data for the month of January. Thus part A has the fraction 100/800 of the pie. Since a full pie coresponds to an angle of 360° then this slice for part A corresponds to a segment with an angle of (100/800) × 360 = 45°. Part B has a fraction

200/800 of the pie and a segment with an angle of (200/800) × 360 = 90°. Part C has the fraction 500/800 of the pie and so a segment with an angle of (500/800) × 360 = 225°.

The type of chart used depends on what information is being communicated. Vertical or horizontal bar charts are good for showing changes from one period to another for a number of item, stacked bar charts for showing how totals change from one period to another and giving a breakdown at each period. Pie charts can be used to compare values as percentages or fractions of the whole.

Example

A chart is required to show the breakdown of the costs of making a particular product. Suggest a possible graphical method of presenting this data.

A pie chart can be used. It would have segments corresponding to each of the cost factors. For example, if we have the following data then the chart could look like Figure 8.22.

	Cost in £
Labour	50
Materials	10
Other costs	40

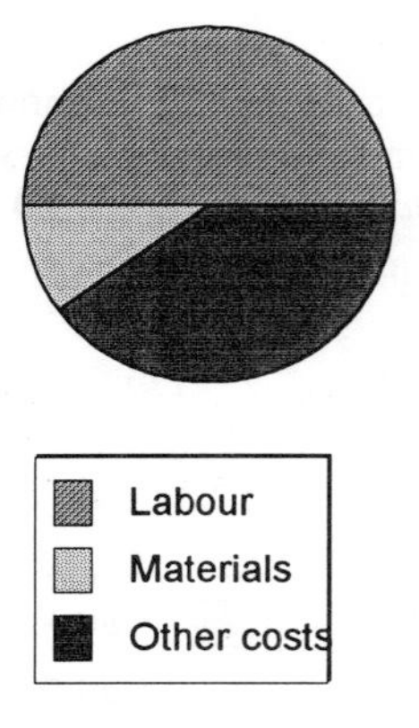

Figure 8.22 *Example*

Labour would have the fraction 50/100 of the pie, i.e. a segment of angle (50/100) × 360° = 180°, materials 10/100 of the pie, i.e. a segment of angle (10/100) × 360° = 36°, and other costs 40/100 of the pie, i.e. a segment of angle (40/100) × 360° = 144°.

Problems

Questions 1 to 16 have four correct answer options: A, B, C and D. You should choose the correct answer from the answer options.

1 The result of adding the fractions $\frac{1}{2}$ and $\frac{1}{5}$ is

A $\frac{1}{3}$
B $\frac{1}{7}$
C $\frac{1}{10}$
D $\frac{7}{10}$

2 The result of subtracting $\frac{1}{3}$ from $\frac{1}{2}$ is

A $\frac{1}{1}$
B $\frac{1}{5}$
C $\frac{1}{6}$
D $\frac{5}{6}$

3 The result of multiplying $\frac{1}{5}$ by $\frac{1}{3}$ is

A $\frac{1}{2}$
B $\frac{1}{8}$
C $\frac{1}{15}$
D $\frac{8}{15}$

4 The result of adding the decimals 0.107 and 0.054 is

A 0.0647
B 0.151
C 0.161
D 0.647

5 The frictional force when a block of metal slides across a horizontal surface is proportional to the weight of the block. When the weight is 12.3 N the frictional force is 5.2 N. Thus, when the weight is 25.4 N the frictional force is:

A 2.5 N
B 10.7 N
C 18.3 N
D 60.0 N

6 The change in resistance of a coil of wire is proportional to the change in temperature. For a temperature change θ of 20°C the resistance changes by 0.1 Ω. The change in resistance is related to θ by:

A Change in resistance = 0.005θ
B Change in resistance = 0.05θ
C Change in resistance = 20θ
D Change in resistance = 200θ

7 The distance travelled by an object moving with a constant velocity is proportional to the time for which it has been moving. When the time is 5 s the distance travelled is 20 m. The equation relating the distance, denoted by s, and the time, denoted by t, is:

A $s = 0.25t$
B $s = 4t$
C $s = 5t$
D $s = 20t$

8 Decide whether each of the following statements is TRUE (T) or FALSE (F).

The equation relating the quantity of heat Q needed to melt a mass m of a solid at its melting point is $Q = mL$. This means that:
(i) The quantity of heat is proportional to the mass.
(ii) If the unit of heat is the joule and that of mass the kilogram, then the unit of L is joule/kilogram.

A (i) T (ii) T
B (i) T (ii) F
C (i) F (ii) T
D (i) F (ii) F

9 Decide whether each of the following statements is TRUE (T) or FALSE (F).

The equation relating the acceleration a of an object and the force F acting on it is $F = ma$. This means that:
(i) The acceleration is proportional to the force.
(ii) If the unit of m is kg and that of a is m/s^2, then the unit of F is kg m/s^2.

A (i) T (ii) T
B (i) T (ii) F
C (i) F (ii) T
D (i) F (ii) F

10 Decide whether each of the following statements is TRUE (T) or FALSE (F).

The equation relating the kinetic energy E of a moving object and its velocity v is $E = \frac{1}{2}mv^2$. This means that:
(i) The kinetic energy is proportional to the velocity.
(ii) If the unit of kinetic energy is the joule J and that of velocity m/s, then the unit of m is Js^2/m^2.

A (i) T (ii) T
B (i) T (ii) F
C (i) F (ii) T
D (i) F (ii) F

11 Decide whether each of the following statements is TRUE (T) or FALSE (F).

The equation relating the quantity of heat Q needed to melt a mass m of a solid at its melting point is $Q = mL$. We can rearrange the equation to write it in the form:

(i) $L = \dfrac{Q}{m}$

(ii) $m = \dfrac{Q}{L}$

A (i) T (ii) T
B (i) T (ii) F
C (i) F (ii) T
D (i) F (ii) F

12 Decide whether each of the following statements is TRUE (T) or FALSE (F).

The equation relating the voltage V across a resistor of resistance R when the current is I is $V = IR$. We can rearrange the equation to write it in the form:
(i) $R = VI$
(ii) $I = \dfrac{V}{R}$

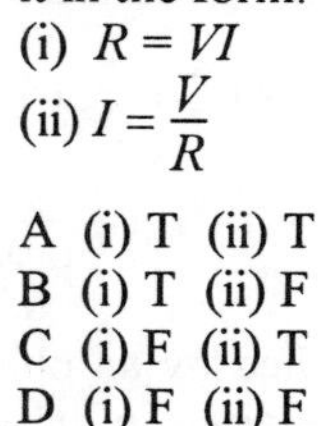

A (i) T (ii) T
B (i) T (ii) F
C (i) F (ii) T
D (i) F (ii) F

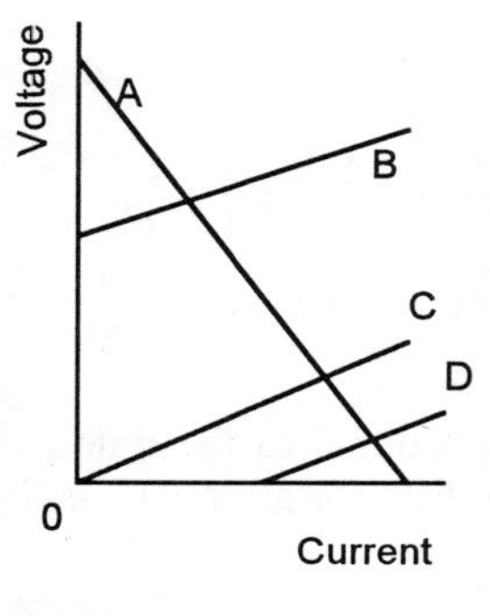

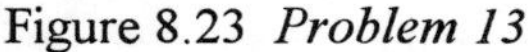
Figure 8.23 *Problem 13*

13 For which graph in Figure 8.23 is the voltage proportional to the current?

14 Decide whether each of the following statements is TRUE (T) or FALSE (F).

The gradient of a graph of y plotted against x must be positive if:
(i) The graph passes through the 0 point.
(ii) y increases as x increases

A (i) T (ii) T
B (i) T (ii) F
C (i) F (ii) T
D (i) F (ii) F

15 Decide whether each of the following statements is TRUE (T) or FALSE (F).

The equation describing the relationship between the resistance R of a wire and its temperature θ is $R = 10\theta$. This means that:
(i) The resistance is proportional to the temperature.
(ii) The resistance increases when the temperature increases.

A (i) T (ii) T
B (i) T (ii) F
C (i) F (ii) T
D (i) F (ii) F

16 Decide whether each of the following statements is TRUE (T) or FALSE (F).

A pie diagram is a good way of showing:
(i) How costs are broken down between the various constituent parts of a product.
(ii) How the labour costs involved in a product change with time.

A (i) T (ii) T
B (i) T (ii) F
C (i) F (ii) T
D (i) F (ii) F

17 Simplify the following fractions and decimals:

(a) $\frac{3}{5}+\frac{1}{3}$, (b) $\frac{1}{5}-\frac{1}{6}$, (c) $\frac{4}{7}+\frac{1}{3}$, (d) $\frac{2}{3}\times\frac{4}{7}$, (e) $\frac{1}{5}\times\frac{2}{3}$, (f) $\frac{3}{4}\div\frac{1}{3}$,

(g) $\frac{4}{5}\div\frac{1}{7}$, (h) 0.150 + 1.01, (i) $\frac{1.3}{0.021}$, (j) 0.451 – 0.045, (k) $\frac{0.1}{0.05}$

18 The distance s travelled by an object is proportional to the time t that has elapsed since it started in motion. If the distance after 10 s is 2 m, what will be the distance after 12 s?

19 Rearrange to the following equation to obtain the values of x:

(a) $2x + 4 = 5 - x$, (b) $x - 4 = 12$, (c) $2x = \frac{3}{4}$, (d) $\frac{3}{2}x = \frac{2}{5}$, (e) $\frac{x+1}{4} = 3$,

(f) $x - 3 = \frac{x+1}{2}$, (g) $\frac{x-1}{2} = \frac{x+2}{3}$, (h) $\frac{1}{2}(x+2) = 5$, (i) $4(x + 2) = 10$

20 The area A of a circle is given by the equation

$$A = \frac{\pi d^2}{4}$$

Solve the equation for d given that the area is 5000 mm^2.

21 The velocity v of an object is given by the equation

$$v = 10 + 2t$$

Solve the equation for the time t when v = 20 m/s.

22 The length L of a bar of metal at t°C is related to the length L_0 at 0°C by the equation

$$L = L_0(1 + 0.0002t)$$

What is the length at 100°C for a bar of length 1 m at 0°C?

23 The specific latent heat L of a material is defined by the equation

$$L = \frac{Q}{m}$$

where Q is the heat transfer needed to change the state of a mass m of the material. If Q has the unit of joule (J) and m the unit of kg, what is the unit of L?

24 The density ρ of a material is defined by the equation

$$\rho = \frac{m}{V}$$

What is the unit of density if m has the unit kg and V the unit m^3?

25 Rearrange the following equations to give the indicated quantity in terms of the remaining quantities:

(a) $p_1 - p_2 = h\rho g$, for h, (b) $\rho = \dfrac{RA}{L}$, for R, (c) $E = V - Ir$, for r,

(d) $I\alpha = T - mr(a + g)$, for T, (e) $s = ut + \dfrac{Ft^2}{2m}$, for F,

(f) $E = \dfrac{mgL}{Ax}$, for x, (g) $\dfrac{M}{I} = \dfrac{E}{R}$, for R, (h) $E = \frac{1}{2}mv^2$, for m.

26 The following data is for the force applied to stretch a spring and the extension that results. Plot the data as a graph.

Force in N	0	0.2	0.4	0.6	0.8	1.0
Extension in mm	0	5	10	15	20	25

27 The following data is for the distance travelled and the time at which the distance is measured. Plot the data as a graph.

Distance in m	10	15	20	25	30	35
Time in s	10	12	14	16	18	20

28 The following data is for the volume V and the temperature θ of a gas. Determine whether V is proportional to θ and the equation relating V and θ.

V in cubic metres	19.7	20.5	21.6	22.7	23.1	23.5
θ in °C	20	30	40	50	60	70

29 The following data is for the resistance R of a wire for different lengths L of that wire. Determine whether R is proportional to L and the equation relating R and L.

R in ohms	2.1	4.3	6.3	8.3	10.5
L in metres	0.5	1.0	1.5	2.0	2.5

30 The following data is for the velocity v of an object at different times t. Determine whether v is proportional to t and the equation relating v and t.

v in m/s	4.5	5.1	5.5	5.9	6.5
t in s	5	10	15	20	25

9 Engineering science

9.1 Engineering systems

In this chapter a number of basic engineering systems and the physical quantities necessary to explain their operation are considered. The term *engineering system* is used for some item around which we can draw an imaginary line and consider its operation in terms of inputs and output from it. One way of representing a system is as box into which inputs feed and outputs emerge. The following are some examples.

For example, consider the motion of some object. The object could be as simple as a trolley on a horizontal surface or as complex as a car on a road. The input to the system to start the object in motion is a force and the output is motion (Figure 9.1). To describe the motion we need to define some scientific terms such as velocity and acceleration. To begin to offer an explanation for how the system operates we need to establish the relationship between force and motion.

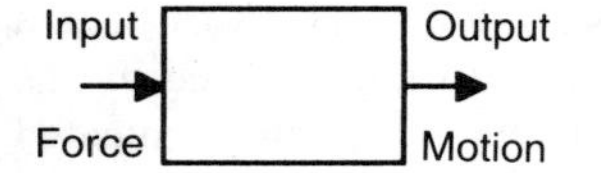

Figure 9.1 *An engineering system*

As another example, consider the boiling of water in an electric kettle. The input to this system is electricity from the mains supply and the output is boiling water and steam (Figure 9.2). To begin to offer an explanation for how the system operates we need to consider the heating of the water in the kettle by the electricity, the rise in temperature of the water when heated and the boiling phenomena. We need to have an idea of what alternating current is, this being the form of current that is supplied by the mains, and define terms such as electrical power, heat capacity, latent heat, etc.

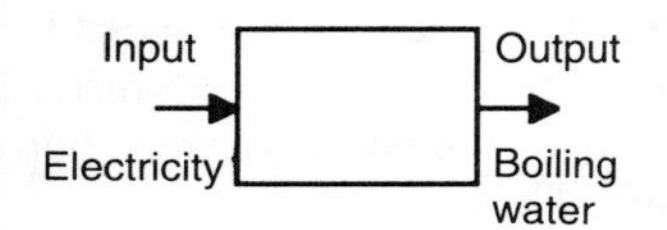

Figure 9.2 *Electric kettle system*

As another example, consider a spring balance. This has an input of a force and an output of an extension, this resulting in a pointer moving across a scale. Figure 9.3 illustrates such a system. To gain an understanding of how the scale can be calibrated in terms of force we have to know the relationship between the force and the spring extension. Only then can the position of the pointer be given a force value.

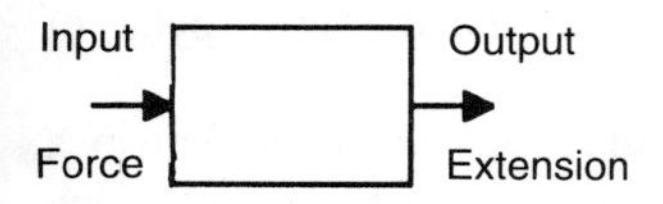

Figure 9.3 *Spring balance system*

9.1.1 The language of science

Explanations involve the language and methods of science. The following are some general terms commonly used in science:

1 A scientific *fact* results from experimental observation and can be reproduced if the experiment is repeated. Thus if the extension of a particular spring is measured for a particular force, then if the experiment is repeated the same value, within the limits set by the accuracy of the experiment, is obtained.

2 A scientific *law* or *principle* is a statement that summarises a range of observations and presents a general idea of importance. Thus Ohm's law is used to describe the relationship between voltage and current for

a resistor. This law states that the voltage is proportional to the current. It arises from observations made of voltages at a number of currents.

3 A scientific *theory* is an imaginative picture used to relate a number of observations by a simple explanation. Thus we have the theory that atoms consist of a positively charged nucleus around which orbit negatively charged electrons.

4 A scientific *model* is an imaginative picture used to aid discussions of some phenomena. Thus as a model for an electric current in a wire we might consider it to be rather like water moving through a pipe.

9.2 Motion in a straight line

To describe motion we need to define some scientific terms. Consider a motor car moving along a road. We can talk of the distance moved by the car along the road. We can talk of the speed of the car, this being indicated by the speedometer, as how much distance the car will cover in one hour if it keeps going at that same speed. Thus a speed of 60 km/h (kilometres per hour) means that if the car continued at that speed for one hour that it would cover 60 km. *Speed* is the rate at which the car covers distance.

If a car travels a distance of, say, 120 km in 2 hours, its average speed is 60 km/h for the journey. The term average is used because the speedometer is extremely unlikely to be at a constant 60 km/h for the entire journey, but might vary between zero and, perhaps, 100 km/h. Thus

$$\text{average speed} = \frac{\text{total distance covered}}{\text{total time taken}}$$

If we make the time taken to cover a distance to be very small, then it is reasonable to talk of the *instantaneous speed*, or just use the term speed, of the car.

A speed of 60 km/h does not tell us that the car will be 60 km in a straight line away in one hour, it may go round in a circular path or some other form of path and be at where it started or indeed almost anywhere. To make life easier, in science we use the term *speed* when we do not know the path being followed by the car and so cannot tell where its speed will get it to after some time, and the term *velocity* when we mean speed in a straight line in some particular direction. Thus a velocity of 60 km/h in a northerly direction means that the car will be, if it keeps going at that speed for an hour, 60 km distant along a straight line in a northerly direction. If we have two cars travelling north at 60 km/h, then they have the same speed of 60 km/h and the same velocity of 60 km/h due north. However, if one car travels due north at 60 km/h and the other due south at 60 km/h, they will have the same speed of 60 km/h but their velocities will be different, one being 60 km/h due north and the other 60 km/h due south.

In everyday language, to accelerate means to go faster, i.e. to increase speed. Thus with a car we might note the speedometer reading in successive seconds and have:

Time in seconds	0	1	2
Speed in km/h	0	5	10

This means that the speed increases by 5 km/h for each second. The car is then said, in everyday language, to have an acceleration of 5 km/h per second. In scientific language, *acceleration* is defined as the change of velocity (note that to define the scientific term we use velocity and not speed) in unit time. In the above data the acceleration is constant or uniform for the times given since the velocity increases by the same amount in each second. However, suppose we just had:

Time in seconds	0	5
Speed in km/h	0	25

We do know that the speed of the car changed by the same amount in each second, i.e. accelerated at the same rate for each second, and so we define an average acceleration as:

$$\text{average acceleration} = \frac{\text{change in velocity}}{\text{time taken for the change}}$$

Thus with the above data for the five seconds, the velocity change is from 0 to 25 km/h in 5 s, and so

$$\text{average acceleration} = \frac{25 - 0}{5} = 5 \text{ km/h per second}$$

When the velocity decreases we talk about *deceleration* or *retardation* or *negative acceleration*. The term negative acceleration is used because the velocity after the time is less than the start value. Thus if we take the average acceleration as (final velocity minus the initial velocity)/time, the result comes out with a negative sign.

Note that an acceleration of m/s per second is written as m/s^2.

Example

A car has an average speed of 50 km/h for a journey of distance 200 km. How long does it take to cover that distance?

$$\text{average speed} = \frac{\text{total distance covered}}{\text{total time taken}}$$

Multiplying both sides of the equation by the total time taken gives

$$\text{average speed} \times \text{total time taken} = \text{total distance covered}$$

Dividing both sides of the equation by the average speed, gives

$$\text{total time taken} = \frac{\text{total distance covered}}{\text{average speed}}$$

Thus

$$\text{total time taken} = \frac{200}{50} = 4\ \text{h}$$

The units of the time are hours since we have

$$\frac{\text{km}}{\text{km/h}} = \frac{\text{km} \times \text{h}}{\text{km}} = \text{h}$$

Example

A car is moving in a straight line along a road with a speed of 40 km/h. When the accelerator is pressed the speed increases to 60 km/h in 5 s. What is the average acceleration?

$$\text{average acceleration} = \frac{\text{change in velocity}}{\text{time taken for the change}}$$

Hence

$$\text{average acceleration} = \frac{60 - 40}{5} = 4\ \text{km/h per second}$$

The unit is km/h per second since when we put the units in the above equation we have

$$\frac{\text{km/h}}{\text{s}}$$

9.2.1 Investigating motion

In studying the motion of a car we can use the speedometer. However, to study in the laboratory the motion of an object we usually measure the distance it covers in various time intervals. An ordinary stopwatch is unsuitable for the accurate measurement of time intervals of less than a second and so special timing devices have to be used. One such device is the *ticker tape timer* (Figure 9.4). This has a small hammer on the under surface of a vibrating steel strip. The hammer strikes a disc of carbon paper under which a paper tape is pulled. The strip is made to vibrate at a constant frequency and so makes dots on the tape at equal intervals of time. Thus if the tape is attached to some moving object, it is pulled through the ticker tape vibrator and a record obtained of how the distance moved by the object varies with time.

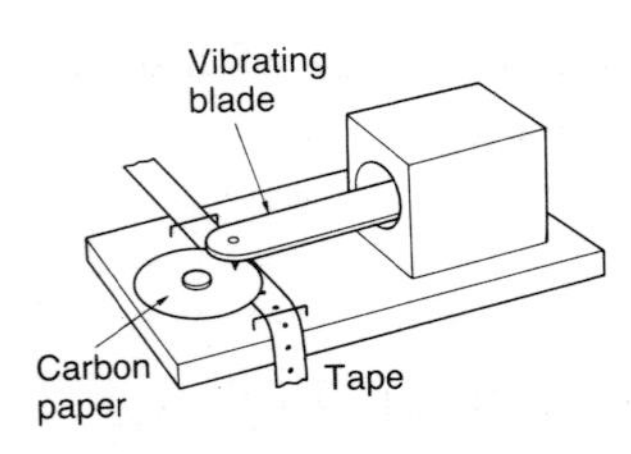

Figure 9.4 *The principle of the ticker tape vibrator*

Suppose we have the object moving with a constant velocity. The ticker tape might then look like that shown in Figure 9.5. The distance between successive dots is the same. If we call the time between dots a tick, then equal distances are covered in equal ticks of time.

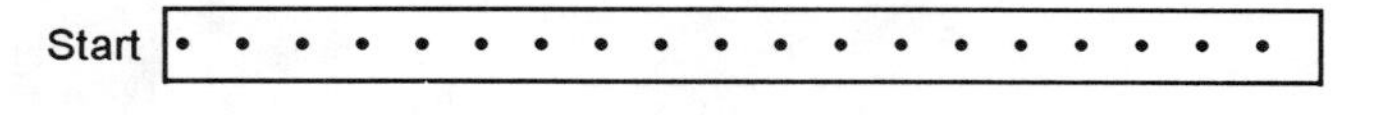

Figure 9.5 *Ticker tape with constant velocity*

If the object giving Figure 9.5 moved with a higher, but still constant, speed then the tape might look like that shown in Figure 9.6. The distances covered in successive ticks are the same but greater than those given by the tape in Figure 9.5.

Figure 9.6 *Ticker tape with constant velocity but faster than that giving Figure 9.5.*

Now consider what the tape would look like if we attached it to a trolley which was allowed to roll down an incline. Figure 9.7 shows such a tape. The distance covered in successive ticks is increasing. Thus the velocity is not constant but increasing, i.e. the object is accelerating.

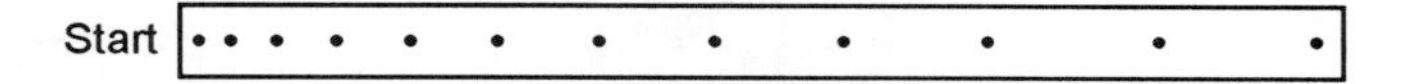

Figure 9.7 *Ticker tape with freely falling object*

We can see how the velocity of an object varies with time if we construct a form of graph called a *tape chart*. Suppose we cut the tape into lengths so that each length was the distance covered in 1 tick. The length of the tape would then be a measure of the average velocity over the time of 1 tick. In practice, the tape is generally cut into lengths containing an equal number of ticks, perhaps ten ticks. Thus the length of the tape gives the average velocity over that ten tick interval. The tape chart is then obtained by pasting successive ten-tick lengths of tape side by side on a sheet of paper. The resulting chart then shows how the average speed in each ten-tick interval changes with time.

Figure 9.8(a) shows the type of chart that would be obtained by cutting up a tape obtained with constant velocity motion, e.g. that obtained in Figure 9.5. Figure 9.8(b) shows the type of chart that would be obtained by cutting up the tape obtained with a higher velocity constant motion, e.g. that obtained in Figure 9.6. Figure 9.8(c) shows the type of tape that would be obtained for an object moving with constant acceleration. The velocity increases by the same amount for each successive ten-tick interval. Figure 9.8(d) shows the type of tape that would be produced for an object moving with a non-constant acceleration. Figure 9.8(e) shows the type of tape that would be produced by an object slowing down and so having a negative, but uniform, acceleration.

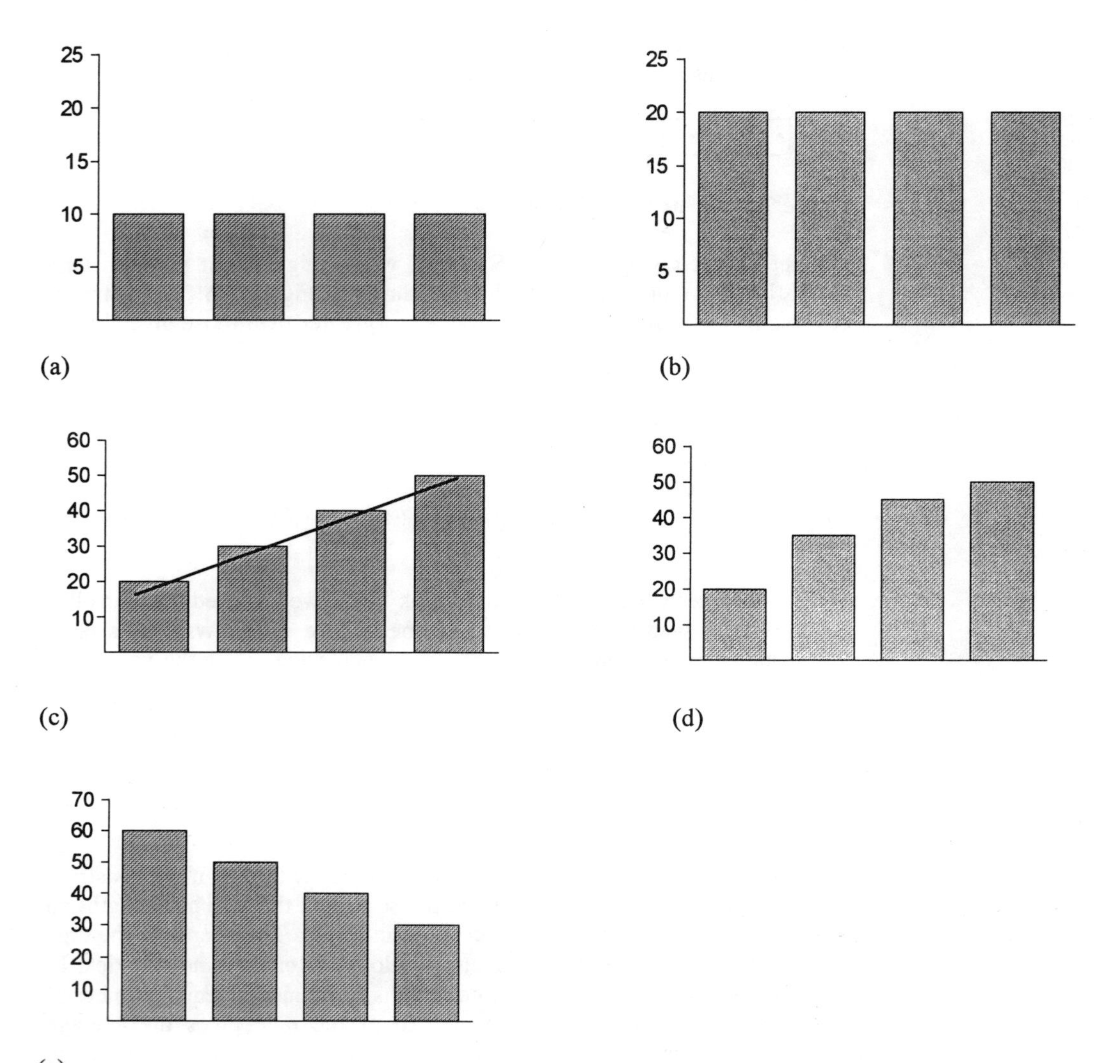

Figure 9.8 *Tape charts: (a) uniform velocity, (b) higher uniform velocity than (a), (c) uniform acceleration, (d) non-uniform acceleration, (e) negative acceleration. The vertical scale is the length of a ten-tick tape in centimetres. The horizontal scale represents time, since the space occupied by each tape corresponds to a tiome interval of ten ticks.*

Consider a special type of motion, *free fall*. If a mass is attached to the ticker tape and allowed to freely fall then the resulting tape chart with ten-tick strips is like that shown in Figure 9.9. As the chart indicates, the motion is uniform acceleration. The average velocity increase from one tape to the next one is about 37 cm per ten-tick strip. With the timer making 50 dots per second, then ten ticks corresponds to $\frac{1}{5}$ second. Thus the change in velocity is about 185 cm/s every ten ticks. This is thus an acceleration of about 925 cm/s per second or 9.25 m/s per second, i.e. 9.25 m/s^2. The value generally taken for the acceleration of free fall is about 9.8 m/s^2.

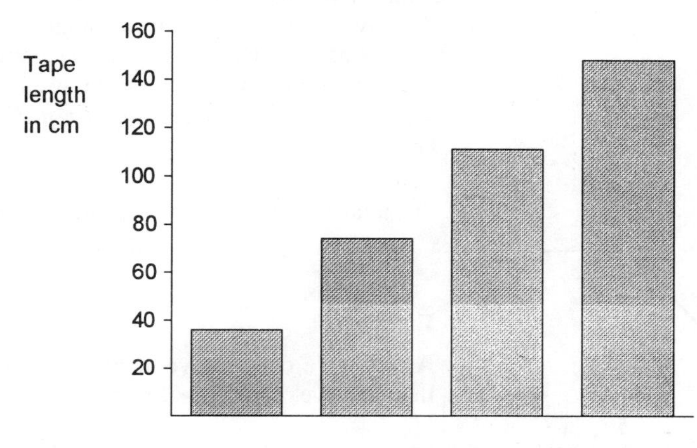

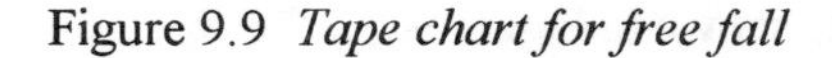

Figure 9.9 *Tape chart for free fall*

9.2.2 Distance–time graphs

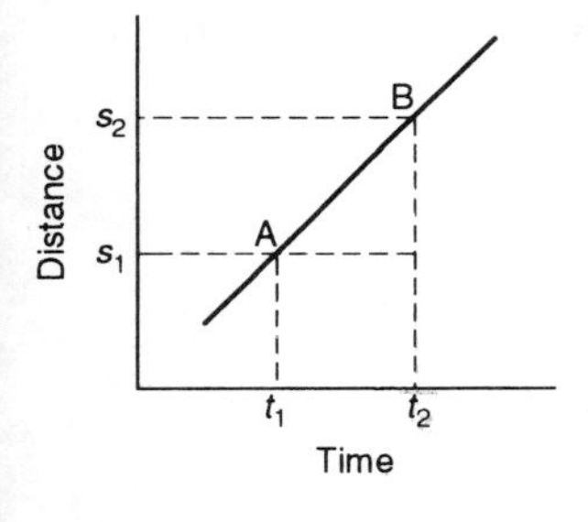

Figure 9.10 *Distance–time graph*

The tape charts in Figures 9.8 and 9.9 are really graphs of velocity plotted against time (in ticks), because each successive strip happens ten ticks later than the one before it and so each tape width represents one ten-tick. In addition to velocity–time graphs we can also have distance–time graphs.

If the distance moved by an object in a straight line, from some reference point on the line, is measured for different times then a distance–time graph can be plotted. Since velocity is the rate at which distance along a straight line changes with time then for the distance–time graph shown in Figure 9.10, where the distance changes from s_1 to s_2 when the time changes from t_1 to t_2, the velocity is

$$\text{velocity} = \frac{s_2 - s_1}{t_2 - t_1}$$

But this is just the gradient of the graph. Thus

$$\text{velocity} = \text{gradient of distance–time graph}$$

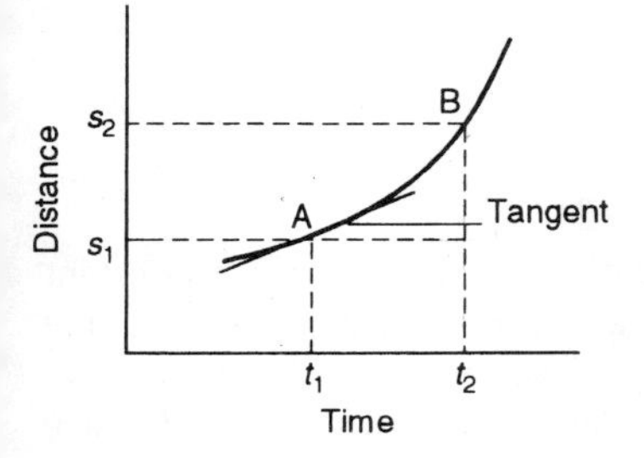

Figure 9.11 *Distance–time graph*

In Figure 9.10 the graph is a straight line graph. The distance changes by equal amounts in equal intervals of time. There is thus a uniform velocity.

Consider the situation when the graph is not a straight line, as in Figure 9.11. Equal distances are not now covered in equal intervals of time and the velocity is no longer uniform. $(s_2 - s_1)$ is the distance travelled in a time of $(t_2 - t_1)$ and thus the average velocity over that time is

$$\text{average velocity} = \frac{s_2 - s_1}{t_2 - t_1}$$

The smaller we make the times between A and B then the more the average is taken over a smaller time interval and so the more it approximates to the instantaneous velocity. An infinitesimal small time interval means we have the tangent to the curve. Thus if we want the velocity at an instant of time, say A, then we have to determine the gradient of the tangent to the graph at A. Thus

instantaneous velocity = gradient of tangent to the distance–time graph at that instant

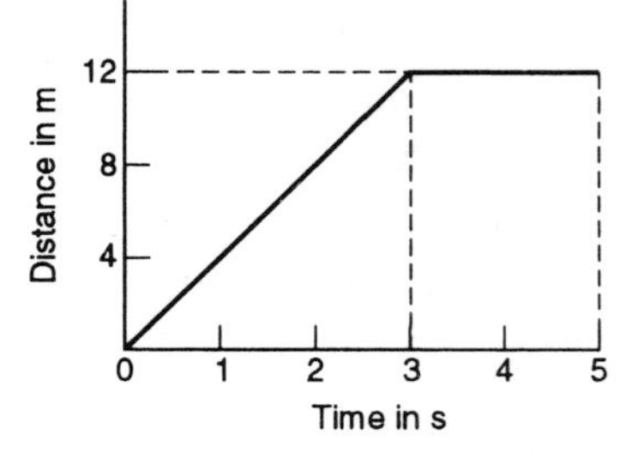

Figure 9.12 *Example*

Example

A car has a constant velocity of 4 m/s for the time from 0 to 3 s and then zero velocity from 3 s to 5 s. Sketch the distance–time graph.

From 0 to 3 s the gradient of the distance–time graph is 4 m/s. From 3 s to 5 s the velocity is zero and so the gradient is zero. The graph is thus as shown in Figure 9.12.

Example

Figure 9.13 shows an example of a distance–time graph. Describe how the velocity is changing in the motion from A to F.

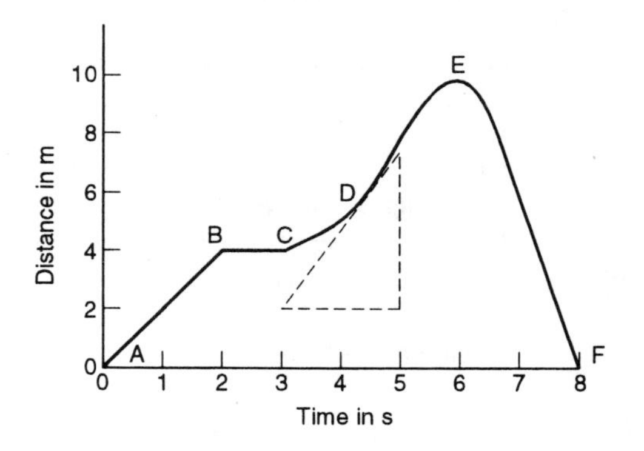

Figure 9.13 *Example*

From the start position at A to B, the distance increases at a constant rate with time. The gradient is constant over that time and so the velocity is constant and equal to (4 – 0)/(2 – 0) = 2 m/s.

From B to C there is no change in the distance from the reference point and so the velocity is zero, i.e. the object has stopped moving. The gradient of the line between B and C is zero.

From C to E the distance increases with time but in a non-uniform manner. The gradient changes with time. Thus the velocity is not constant during that time.

The velocity at an instant of time is the rate at which the distance is changing at that time and thus is the gradient of the graph at that time. Thus to determine the velocity at point D we draw a tangent to the graph curve at that instant. So at point D my estimate of the velocity is about (7.6 – 2.4)/(5 – 3) = 2.6 m/s.

At point E the distance–time graph shows a maximum. The gradient changes from being positive prior to E to negative after E. At point E the gradient is momentarily zero. Thus the velocity changes from being positive prior to E to zero at E and then negative after E. At E the velocity is zero.

From E to F the gradient is negative and so the velocity is negative. A negative velocity means that the object is going in the opposite direction and so is moving back to its starting point, i.e. the distance is becoming smaller rather than increasing. In this case, the object is back at its starting point after a time of 8 s.

9.2.3 Velocity–time graphs

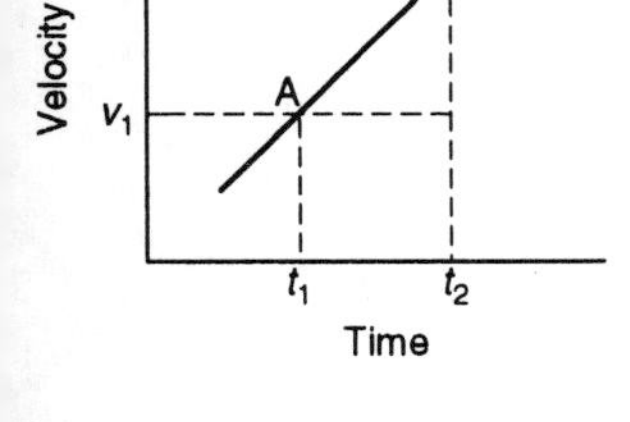

Figure 9.14 *Velocity–time graph*

If the velocity of an object is measured at different times then a velocity–time graph can be drawn. Acceleration is the rate at which the velocity changes. Thus, for the graph shown in Figure 9.14, the velocity changes from v_1 to v_2 when the time changes from t_1 to t_2. The acceleration over that time interval is thus

$$\text{acceleration} = \frac{v_2 - v_1}{t_2 - t_1}$$

But this is the gradient of the graph. Thus

acceleration = gradient of the velocity–time graph

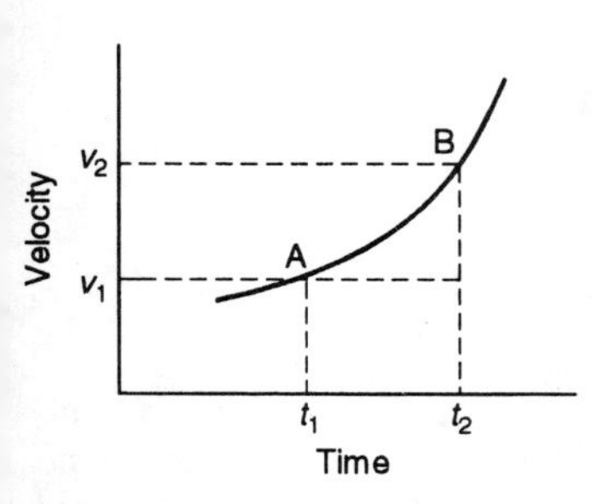

Figure 9.15 *Velocity–time graph*

In Figure 9.14 we have a straight line between the points A and B. Thus, since the gradient is the same for all points between A and B we have a uniform acceleration between those points. Consider the situation when the graph is not straight line, as in Figure 9.15. The velocity does not now change by equal amounts in equal intervals of time and the acceleration is no longer uniform. $(v_2 - v_1)$ is the change in velocity in a time of $(t_2 - t_1)$ and thus $(v_2 - v_1)/(t_2 - t_1)$ represents the average acceleration over that time. The smaller we make the times between A and B then the more the average is taken over a smaller time interval and more closely approximates to the instantaneous acceleration. An infinitesimal small time interval means we have the tangent to the curve. Thus if we want the acceleration at an instant of time then we have to determine the gradient of the tangent to the graph at that time, i.e.

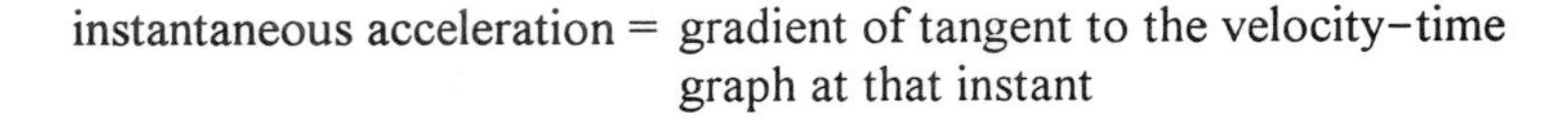

instantaneous acceleration = gradient of tangent to the velocity–time graph at that instant

The distance travelled by an object in a particular time interval is given by

distance = average velocity over the time × the time interval concerned

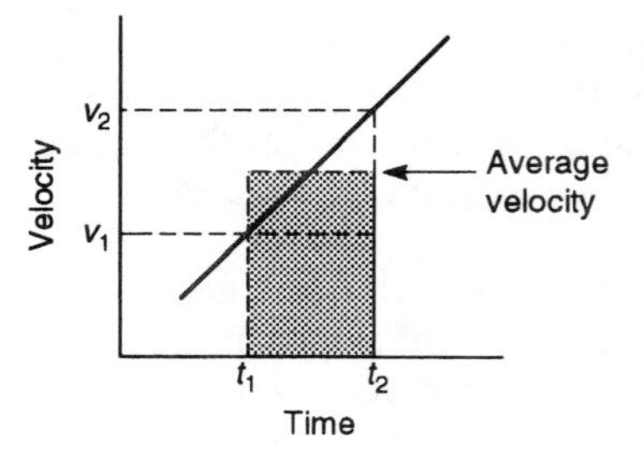

Figure 9.16 *Distance travelled*

Thus if the velocity changes from v_1 at time t_1 to v_2 at time t_2, as in Figure 9.16, then the distance travelled between t_1 and t_2 is represented by the area marked on the graph. But this area is equal to the area under the graph line between t_1 and t_2. Thus

distance travelled between t_1 and t_2 = area under the graph between these times

When the velocity is zero or there is uniform acceleration and so a straight line velocity–time graph, the areas can be easily calculated. However, if the acceleration is not uniform then we might estimate the area by counting the squares under the graph. Since the area of a square can be calculated, we can then obtain an estimate of the area under the graph.

Example

For the velocity–time graph shown in Figure 9.17, determine the acceleration at points A, B and C and estimate the distance covered in the 8 s.

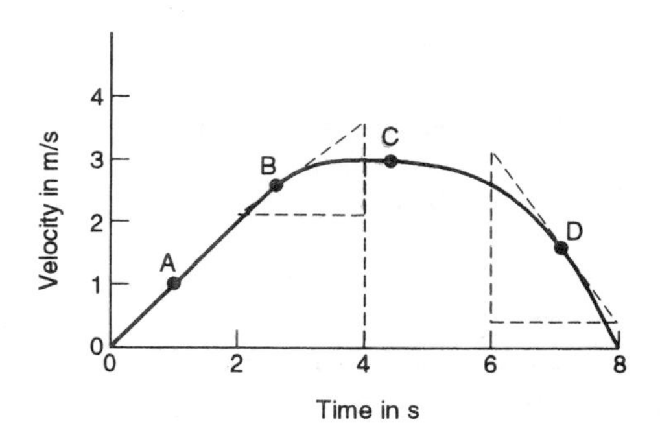

Figure 9.17 *Example*

Initially the graph is a straight line of constant gradient. The gradient, and hence the acceleration, at A is 2/2 = 1 m/s^2. At B the graph is no longer a straight line. The tangent to the curve at B has a gradient, and hence an acceleration, of about (3.7 – 2.2)/(4 – 2) = 0.75 m/s^2. At C the velocity is not changing with time. Thus there is a uniform velocity with zero acceleration. At D the velocity is decreasing in a non-uniform manner with time. The tangent to the curve at that point has a gradient of about (0.4 – 3.4)/(8 – 6) = –1.5 m/s^2.

The distance covered in the 8 s is the area under the graph. For 0 to 2 s we have the area of a triangle of height 2 and base 2, an area of $\frac{1}{2} \times 2 \times 2 = 2$. For the area between 2 s and 4 s we can consider the area to be about equivalent to a rectangle of 2.8 × 2, with 2.8 being the mean height of the area. This is thus an area of 5.6. For the area between 4 s and 6 s we can consider the area to be equivalent to a rectangle of 2.9 × 2 = 5.8. For the area between 6 s and 8 s we can consider this to be a reasonable approximation to a triangle and so an area of $\frac{1}{2} \times 2.8 \times 2 = 2.8$. Thus total area is 2 + 5.6 + 5.8 + 2.8 = 16.2. The unit of this area is (m/s) × m = m. Thus the distance is 16.2 m. With graph paper divided into squares, we could have estimated the area under the graph by counting squares.

Example

For the straight line motion described by the velocity–time graphs shown in Figure 9.18, which one shows (a) a constant velocity with no acceleration, (b) a constant acceleration, (c) a constant negative acceleration?

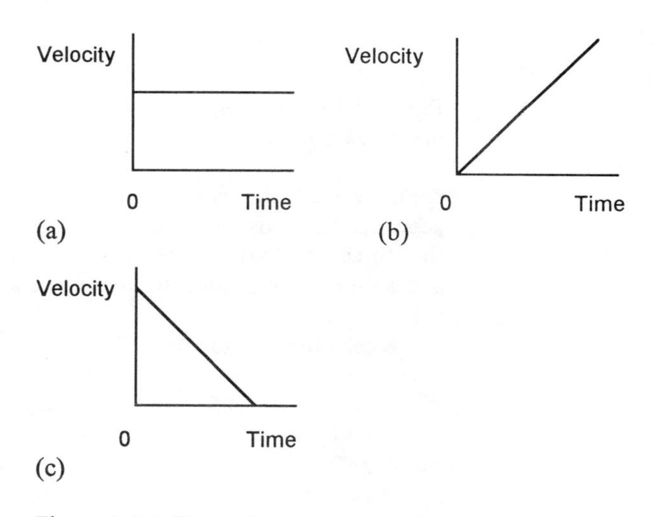

Figure 9.18 *Example*

(a) Figure 9.18(a) shows motion with constant velocity.
(b) Figure 9.18(b) shows motion with constant acceleration. The acceleration is positive because the velocity is increasing.
(c) Figure 9.18(c) shows motion with constant negative acceleration. The acceleration is negative because the velocity is decreasing.

9.3 Force and acceleration

What happens when we have a force acting on a body? Such a question can be investigated by applying a force to a trolley and studying its resultant motion by means of ticker tape (Figure 9.19) or perhaps electric timers. We need to be able to apply a constant force, then double that force in order to find what effect this has on the motion. We can do this by stretching a length of elastic. To stretch a length of elastic a force is required and if we stretch it always by the same amount it is reasonable to suppose that a constant force is being applied. Thus if we stretch the elastic shown in Figure 9.19(a) so that it is stretched always to the full length of the trolley a constant force is applied to the trolley. If we take two identical lengths of the same elastic in parallel then it is reasonable to suppose that if we stretch both of them we can apply twice the force (Figure 9.19(b)).

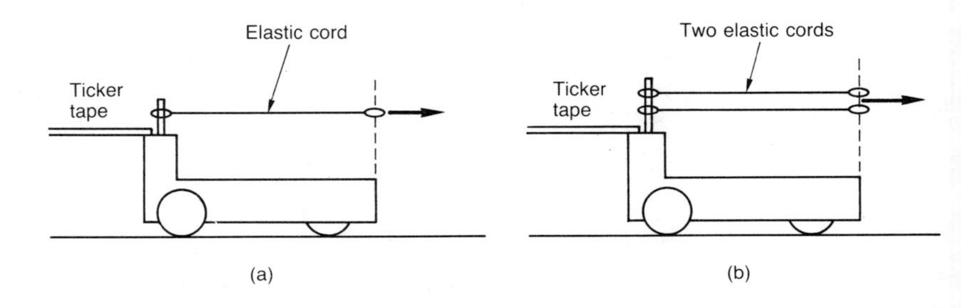

Figure 9.19 *Studying the motion of a trolley when (a) a constant force, (b) double that force, is applied*

Figure 9.20 shows the tape charts that result. The motion is constant acceleration. Thus a constant force produces a constant acceleration. When the force is doubled the acceleration is doubled. Thus we have the acceleration proportional to the force and can write

acceleration $\propto$ force

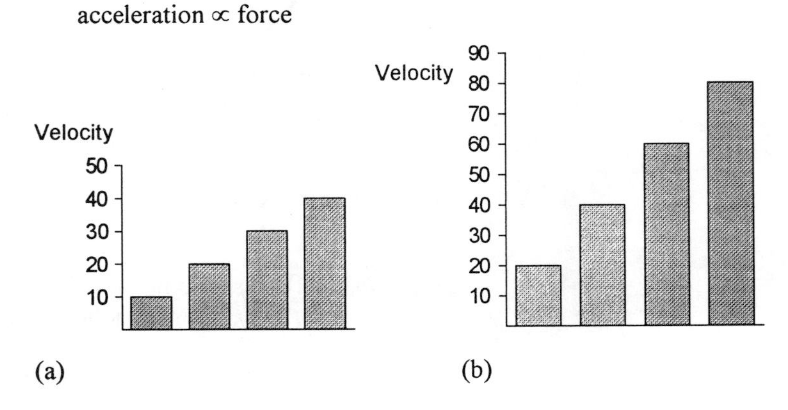

Figure 9.20 *(a) Single force, (b) double force*

There is another factor which affects the acceleration occurring when a force is applied and that is a factor we term the *mass* of the trolley. Suppose we stacked two trolleys, one on top of the other, and pulled them with the single length of elastic and then the double length. Doubling the trolleys halves the acceleration. The product of the number of the trolleys and the acceleration is a constant for the same force. Thus with a single force we have:

One trolley	Acceleration a	Product = $1 \times a = 1a$
Two trolleys	Acceleration $\frac{1}{2}a$	Product = $2 \times \frac{1}{2}a = 1a$

Thus we have a constant force producing a constant value of ma, where m is something called the mass which is doubled when we double the number of identical trolleys, and a the acceleration. *Mass* is a property we can term the inertia of an item, the bigger it is the more force is needed to accelerate it. We thus have

$$\text{mass} \times \text{acceleration} \propto \text{force}$$

With the unit of acceleration as m/s^2 and the unit of mass as the kg, we define a unit of force as the newton (N) so that

$$\text{mass} \times \text{acceleration} = \text{force}$$

This is usually written as

$$F = ma$$

where F is the force.

Example

A freely falling objects falls with a constant acceleration. What can be said about the force acting on it?

Because there is a constant acceleration there must be a constant force. This force is called the gravitational force. In a vacuum, when air resistance is removed, all objects fall with the same constant acceleration. This is because the gravitational force acting on an object is proportional to the mass of the object.

Example

What is the size of the force acting on an object of mass 2 kg if it causes an acceleration of 3 m/s^2?

Using $F = ma$ we have $F = 2 \times 3 = 6$ N.

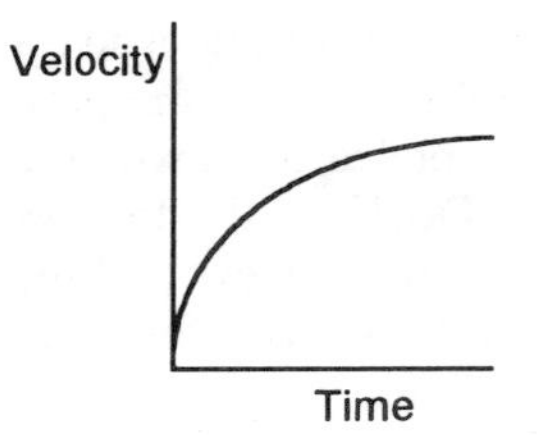

Figure 9.21 *Example*

Example

What can be said about the force acting on an object if the velocity–time graph is of the form shown in Figure 9.21?

Because the gradient of the graph is not constant, the acceleration is not constant. Hence the force is not constant.

9.4 Force and changes in length

Consider forces being applied to stretch a spring. If we measure the extension of the spring for different forces within the normal operating range of the spring and plot a graph then the result is similar to that shown in Figure 9.22. The graph is a straight line passing through the zero force – zero extension point. The extension is thus proportional to the force.

$$\text{extension} \propto \text{force}$$

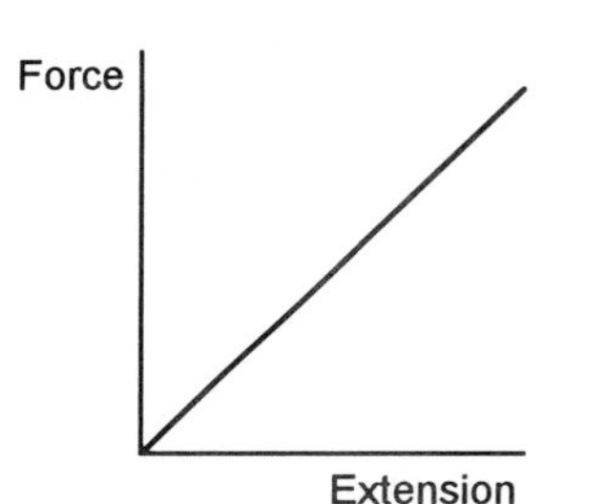

Figure 9.22 *Force-extension graph for a spring*

This relationship is known as *Hooke's law*. We can write this as an equation with an = sign, the usual form of writing this equation being as

$$F = kx$$

where F represents the force, k is a constant and x the extension.

We could carry out a similar experiment with compressing a spring, measuring the amount by which a spring is compressed for different forces within the normal operating range of the spring. The graph of force against the amount of compression is a straight line graph just like that for the force–extension and thus we have

$$\text{compression} \propto \text{force}$$

If we represent the amount of compression by x then we can write an equation of the same form as for the extension, i.e.

$$F = kx$$

Example

A spring, within its normal operating range, extends by 10 mm when a force of 50 N is applied. By how much will it be expected to extend when a force of 100 N is applied?

The phrase 'within its normal operating range' is used to indicate that the spring is expected to have the extension proportional to the force for the forces which fall within this range. For forces outside this range then the spring may behave non-linearly and become deformed. Thus, assuming the extension is proportional to the force, the force is doubled and so the extension is doubled. Hence the answer is 20 mm.

9.5 Force and static equilibrium

When a force acts on a body it accelerates (see section 9.3). However, if an object rests on a bench then there is the gravitational force acting on it in a downwards direction but it does not move (Figure 9.23). If you think of the bench as being rather like a trampoline, then when an object rests on its surface it becomes stretched, just like a spring, and supplies an upward directed force on the object. This upward directed force only arises because the object is exerting a force downwards. It is termed a *reaction force*. This reaction force is of the same size as the downwards directed force but acting in the opposite direction. The result of these two forces acting on the object is that they cancel each other out and so there is no resultant force acting on the object. Hence the object remains stationary on the bench. It is not only horizontal benches which supply reaction forces, whenever an object is in contact with a surface, whether horizontal or sloping, there is a reaction force. This reaction force is always at right angles to the surface.

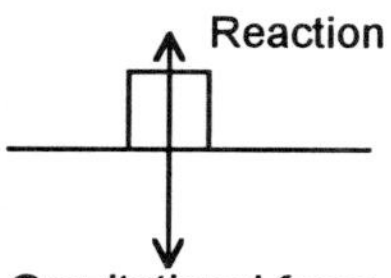

Figure 9.23 *No resultant force*

All freely falling objects in a vacuum fall with the same acceleration, this being termed the *acceleration due to gravity* g and having a value of about 9.8 m/s^2. When an object is restrained from falling then the gravitational force responsible for the acceleration in free fall has been cancelled out by some opposing force. Thus, for an object of mass m, the gravitational force must be mg. This is called the *weight*. Thus

$$\text{weight} = mg$$

9.5.1 Moments

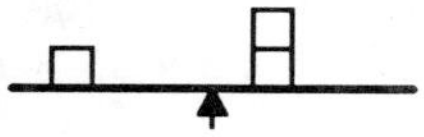

Consider an experiment involving a ruler balanced, like a seesaw, across a pivot (Figure 9.24). If we put an object, say a coin, on one side of the pivot point, the seesaw no longer balances. How far on the other side of the pivot point do we have to put a pile of two such coins to obtain balance? If we carry out such an experiment then the result is balance occurs when we have

Figure 9.24 *Balanced see-saw experiment*

number of coins on left-hand side × distance from pivot = number of coins on right-hand side × distance from pivot

Thus 1 coin on left-hand side a distance of 40 cm from the pivot is balanced by 2 coins on right-hand side a distance of 20 cm from the pivot.

The term *moment of a force* about an axis is used for the product of a force F and its perpendicular distance r from the axis (Figure 9.25).

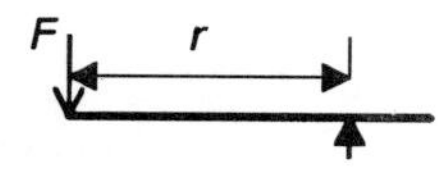

Figure 9.25 *Moment of a force*

$$\text{Moment} = Fr$$

The unit of the moment is the unit of force multiplied by the unit of distance. Thus with force in newtons (N) and distance in metres (m), the unit is N m. For the beam in Figure 9.24 to balance we must have the moment turning it about its pivot axis in an anticlockwise direction equal to the moment turning about the same axis in a clockwise direction (Figure

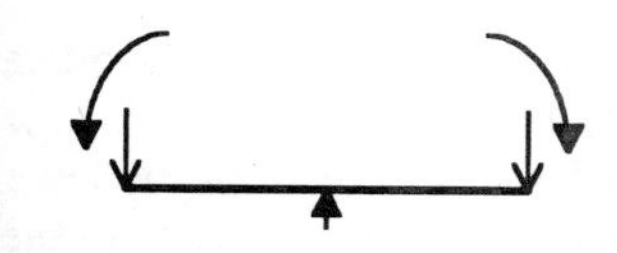

Figure 9.26 *Moments about the pivot*

9.26). Only then will there be no rotation. This is known as the *principle of moments*.

anticlockwise moments about an axis = clockwise moments about the same axis

Example

Determine the force F required to give balance for the pivoted beam shown in Figure 9.27.

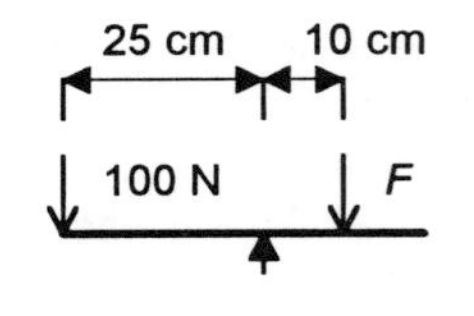

Figure 9.27 *Example*

The anticlockwise moment about the pivot is provided by the 100 N force and is 100×20 N cm. The clockwise moment is provided by the force F and is $F \times 10$ N cm. For balance to occur, the clockwise moment must equal the anticlockwise moment. Thus

$$F \times 10 = 100 \times 20$$

and so $F = 200$ N.

Example

Determine the force F required to give balance for the pivoted beam shown in Figure 9.28.

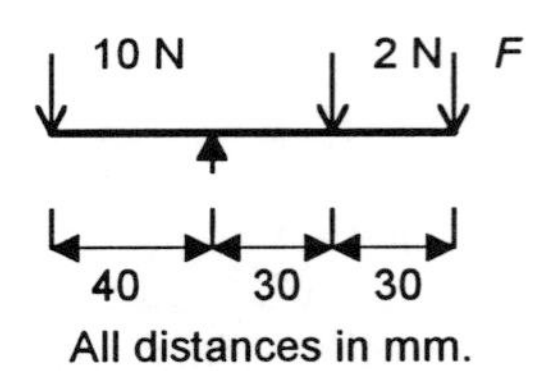

Figure 9.28 *Example*

Taking moments about the axis through the pivot, the clockwise moment is $10 \times 40 = 400$ N mm. The anticlockwise moments about the axis through the pivot is given by two forces and is thus 2×30 N mm plus $F \times 60$ N mm. Hence

$$60F + 60 = 400$$

Subtracting 60 from both sides of the equation, or transposing the 60 from the left- to right-hand sides, gives

$$60F = 400 - 60 = 340$$

Hence $F = \dfrac{340}{60} = 5.7$ N (to one decimal place).

9.5.2 Levers

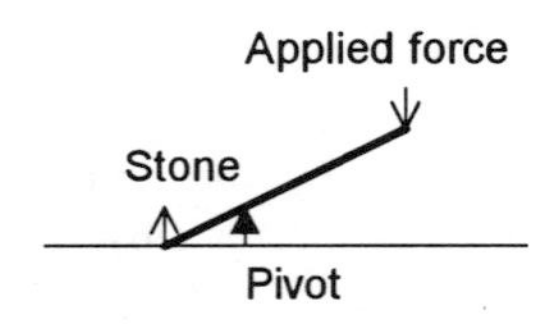

Figure 9.29 *Using a crowbar*

A crowbar is an example of a lever. Such a lever might be used to move a large boulder. The high force at one end of the lever needed to move the boulder is produced by a smaller force applied to the other end of the lever (Figure 9.29). The lever pivots about a point closer to the boulder than the applied force. The moment about the pivot produced by the applied force results in the same moment being applied to the stone.

Applied force × its distance from pivot	=	Force on stone × its distance from pivot

Because the distance of the stone from the pivot is much smaller than that of the applied force from the pivot, then the force applied to the stone is larger than the applied force.

In general, a *lever* can be used to change the size and/or the line of action of a force. The result depends on the position of the pivot, or *fulcrum* as it usually termed, and so the distances of the applied force, often termed the *effort*, and the force due to the load being moved.

Example

Figure 9.30 shows a pair of tinsnips. What effort will need to be applied if the force L required to shear through a metal sheet is 1600 N?

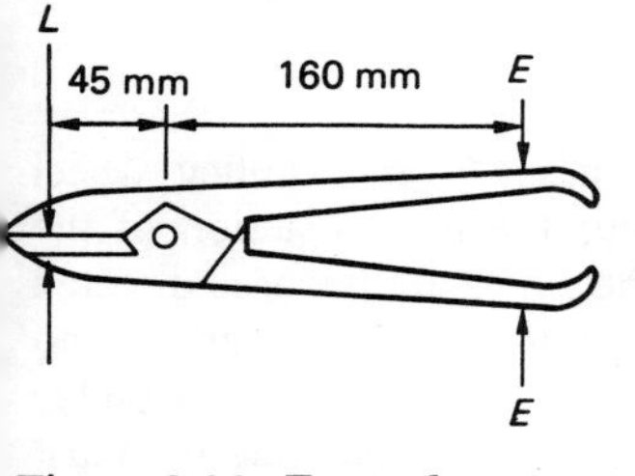

Figure 9.30 *Example*

Consider just one arm of tinsnips. What we have with the is a situation like that shown in Figure 9.29 for the crowbar. The moment about the pivot point applied to cut the sheet is $L \times 45 = 1600 \times 45$ N mm. This moment comes from the effort and so we must have

$$E \times 160 = 1600 \times 45$$

Thus, dividing both sides of the equation by 160, gives

$$E = \frac{1600 \times 45}{160} = 450 \text{ N}$$

9.5.3 Static equilibrium

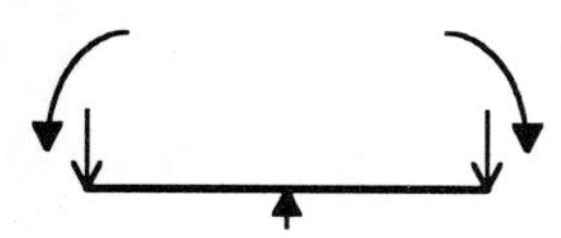
Figure 9.31 *See-saw*

An object is said to be in *static equilibrium* when there is no movement of the object or tendency to move in any direction. Thus it does not accelerate in any direction, i.e. there is no resultant force in any direction. It does not rotate about an axis in any direction. Thus if we consider a seesaw (Figure 9.31) in static equilibrium, for there to be no resultant force in any direction we must have:

sum of the upwards directed forces	=	sum of the downwards directed forces
sum of the forces to the left	=	sum of the forces to the right

In the example used in the figure there are no forces in the horizontal direction, but there are forces in the vertical direction. There are the two forces acting vertically down on the ends of the seesaw. This must, therefore, be balanced by a reaction force at the pivot. Only then will we have no resultant force. For no rotation we must have

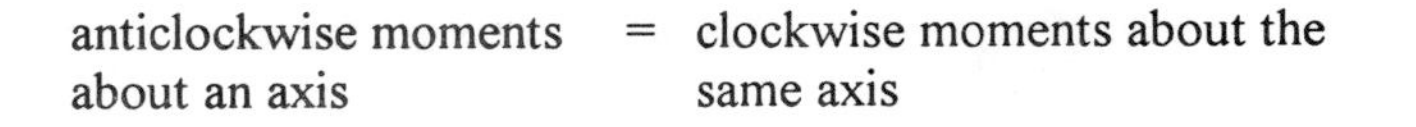
anticlockwise moments about an axis = clockwise moments about the same axis

This must apply whichever axis we choose to take moments about.

Example

For the balanced beam shown in Figure 9.32 (this was used in an earlier example as Figure 9.27), what is the reaction force at the pivot?

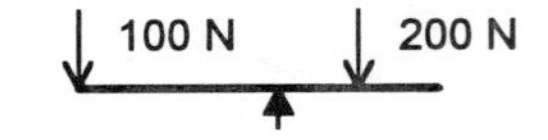

Figure 9.32 *Example*

The total of the downwards directed forces is $100 + 200 = 300$ N. Hence the reaction force at the pivot, since this is the only upwards directed force, must be 300 N.

9.5.4 Pulleys

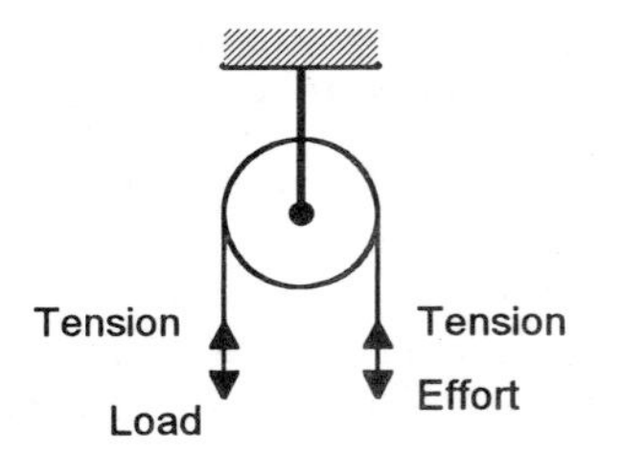

Figure 9.33 *Pulley*

Consider the forces involved when a cord passes over a pulley wheel (Figure 9.33). The single pulley wheel changes the line of action of the effort; what was a downwards directed force becomes an upwards directed force on the load. For equilibrium of the forces, when there is no acceleration, we must have for the cord on the right-hand side of the pulley the effort equal to the tension in the cord. On the left-hand side we must have the tension in the cord equal to the force exerted by the load. Since the cord is in equilibrium we must have no resultant force acting on it and so the tension at one end of it must be balanced by the tension at the other end. As a consequence, we must have the force on the load equal to the effort.

Now consider a pulley system having two wheel (Figure 9.34). When the system is in equilibrium and the effort is not causing the load to be accelerated, then we must have for the cord on the left-hand side of the system the effort equal to the tension T in the cord.

$$\text{Effort} = T$$

This must result in an equal tension acting on the central part of the cord. This likewise must result in an equal tension acting in the cord on the left-hand side of the system. If we consider the forces acting on the left-hand pulley, then we have upwards directed forces of $T + T = 2T$. If the pulley wheel has a weight of mg, then for that wheel to be in equilibrium we must have

$$\text{Load} + mg = 2T = 2 \times \text{Effort}$$

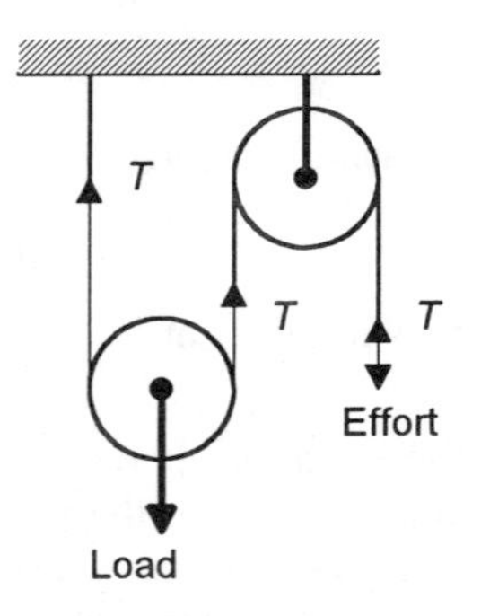

Figure 9.34 *Two pulley wheel system*

Thus, if the weight of the pulley wheel is small, we have virtually doubled the size of the effort force acting on the load. The system has magnified the effort force.

However, the distance through which the effort moves its end of the cord is twice that the load moves through, it is a movement reducer. Consider what happens if the effort moves its end of the cord through, say, 10 cm, then the 10 cm of cord must be provided by the rest of the cord shortening by that amount. This means 5 cm from the central cord and 5 cm from the left-hand cord. Thus the load is lifted by 5 cm when the effort moves through 10 cm.

Example

A two pulley wheel system, of the form shown in Figure 9.34, is used to lift a load of weight 1000 N through a distance of 1 m. What effort will be needed and through what distance must the effort move its end of the rope? Neglect the weight of the pulley wheels.

With such a system, the force applied to the load is twice the effort (see above discussion). Thus the effort needed will be 500 N. The distance moved by the effort will be twice that moved by the load. Thus the effort must move through 2 m.

9.6 Pressure

Consider a punch with a sharp point. When a force is applied to the punch the force is applied over just the small area of the point. However, if the punch is blunt, the force is applied over a larger area. With the sharp point the surface to which the punch is applied is readily marked. However, with the blunt punch this is not the case. A much larger force has to be applied. Thus the force per unit area is important. The force per unit area is called the *pressure*, i.e.

$$\text{pressure} = \frac{\text{force}}{\text{area}}$$

Example

What is the pressure applied by the surface of a point of area 1 mm^2 when a force of 500 N is applied to it?

The pressure is given by the above equation as

$$\text{pressure} = \frac{\text{force}}{\text{area}} = \frac{500}{1} = 500 \text{ N/mm}^2$$

9.7 Energy, work and heat

Consider the following situations. You lift an object from the floor onto a bench. This requires some effort and 'costs' you something. If you had to do it often you would feel tired. Now consider you use an electric motor to operate a hoist and lift the object from the floor onto the bench. It still 'costs' something in that 'electricity is used' and somebody has to pay the bill for it. The term used to describe the 'something' that is used is energy.

In science, *energy* is considered to be the something which enables people and machines to do useful jobs such a lifting or moving loads. Energy can appear in many different forms, e.g. potential energy, kinetic energy, electrical energy, heat energy, etc. These various forms of energy are discussed in the following pages.

9.7.1 Work

One way that energy can be transferred is as *work*. Work is said to be done when the energy transfer results in a force pushing something through a distance (Figure 9.35). The amount of energy transferred in such a situation is the work and is given by

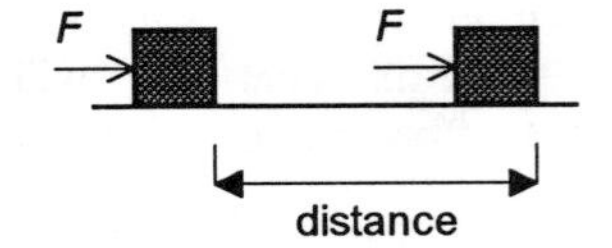

Figure 9.35 *Work*

$$\text{work} = \text{force} \times \text{distance moved}$$

When the force has the unit of the newton and the distance the metre, the unit of work is the joule (J).

Consider an object being lifted from the floor through a vertical height h onto a table (Figure 9.36). If the object has a weight mg then the force being moved is mg and the distance through which it is moved is h. Thus the work done is mgh. The object has gained, through this work, an amount of energy of mgh. It has this energy by virtue of its position on the table relative to the energy it had sitting on the floor. Energy an object has by virtue of its position is called *potential energy*. Thus the object gains potential energy of

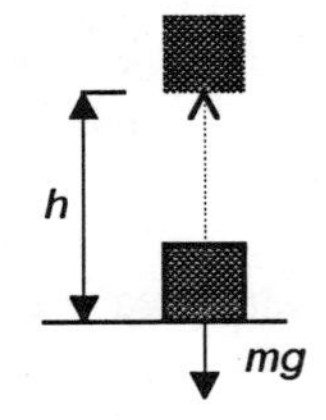

Figure 9.36 *Lifting an object*

$$\text{potential energy} = mgh$$

This form of potential energy is often called *gravitational potential energy* because it is the energy an object has by virtue of moving against a gravitational force. The unit of potential energy, indeed all forms of energy, is the joule (J).

Now consider the object falling off the table and down to the floor. It loses its potential energy of mgh and gains energy by virtue of its motion, this being termed *kinetic energy*. The potential energy has been transformed into kinetic energy. The kinetic energy gained is the work done. Thus

$$\text{KE gained} = F \times h$$

But $F = ma$ and since the acceleration a is the acceleration due to gravity g we have

$$\text{KE gained} = mgh$$

The kinetic energy gained is the potential energy lost. If the object starts from rest on the table and has a velocity v when it hits the floor, then the average velocity is $v/2$. The average velocity is the distance covered over the time taken. Hence

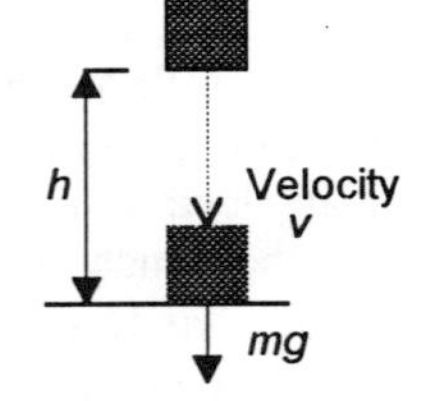

Figure 9.37 *A falling object*

$$\text{average velocity} = \frac{v}{2} = \frac{h}{t}$$

where t is the time taken to fall from the bench to the floor. If we multiply both sides of this equation by t we have

$$h = \frac{vt}{2}$$

The acceleration is the change in velocity divided by the time taken. Thus $g = v/t$. Hence

$$\text{KE gained} = mgh = m \times \frac{v}{t} \times \frac{vt}{2} = \tfrac{1}{2}mv^2$$

There are other forms of potential energy. Suppose we had the object attached to a spring and apply a force to the object which results in the spring being extended. Work is done because the point of application of the force is moved through a distance. Thus the object gains potential energy as a result of the work that has been done. This form of potential energy is termed *elastic potential energy or strain energy*. Suppose we had two permanent magnets and arranged them north pole facing north pole. If we push them closer together a force is required and so work is done. Thus the object gains potential energy, this form of potential energy being termed *magnetic potential energy*.

Example

What is the work done in using a hoist to lift a pile of bricks of mass 20 kg from the ground to the top of a building if the building has a height of 30 m? Take the acceleration due to gravity to be 9.8 m/s².

The work done is the force, i.e. the weight mg, multiplied by the vertical distance through which it is moved. Thus

$$\text{work done} = mg \times h = 20 \times 9.8 \times 30 = 5880 \text{ J}$$

Example

The locomotive of a train exerts a constant force of 120 kN on a train while pulling it at 40 km/h along a level track. What is the work done in 15 minutes?

In 15 minutes the train covers a distance of 10 km. Hence

$$\text{work done} = 120 \times 1000 \times 10 \times 1000 = 1200\,000\,000 \text{ J}$$

9.7.2 Heat

So far we have only considered work as a method of energy transfer. With work, energy is transferred when the point of application of a force moves through a distance. There is another method of energy transfer. *Heat* is the transfer of energy that occurs between two systems when there is a temperature difference between them. Thus if we observe an object and find that its temperature is increasing, then a transfer of energy, as heat, is occurring into the object. Heat transfer which results in a change of temperature is said to be *sensible heat*.

Consider what happens when we add, say, 100 g of water at 40°C to 100 g of water at 20°C, the water being contained in a light plastic container. The result is that the mixture comes to an equilibrium temperature of 30°C (the reason for specifying a light plastic cup is that we did not want to 'waste' any significant heat on raising the temperature of the cup). Now consider what happens when 50 g of water at 40°C is added to 100 g of water at 20°C. The mixture comes to an equilibrium temperature of 26.7°C. We can explain this if we consider the product of the mass of the water and its temperature change. Thus the 50 g of water cools from 40°C to 26.7°C and the 100 g of water increases in temperature from 20°C to 26.7°C. We have $50(40 - 26.7) = 100(26.7 - 20)$. If, instead of water at 40°C, we had added some other liquid or a solid at 40°C, the result would have been different. There is a factor associated with the materials being mixed. We thus think of the heat transfer from a hot material to a cold material as being given by

$$\text{heat transfer} = \text{mass} \times \text{specific heat capacity} \times \text{change in temperature}$$

The specific heat capacity is the factor which depends on the material concerned. The *heat capacity* of an object is defined as the amount of heat needed to change its temperature by 1°C, the *specific heat capacity c* being the amount of heat needed to change 1 kg by 1°C. The unit of specific heat capacity is J/ kg °C. Water, for example, has a specific heat capacity of about 4200 J/kg °C, aluminium 950 J/kg °C, copper 390 J/kg °C.

There are, however, situations where heat transfer occurs without a change in temperature but instead the material melts or changes into a vapour. Think of an ice cube at −10°C. An energy transfer to the ice results in its temperature rising to 0°C, but then the temperature stops rising, even though heat is continuing to flow into the ice. An energy transfer as heat into the ice cube at 0°C results in the ice melting to give water at 0°C. Only when the ice has melted does the heat start to raise the temperature. Energy is needed to change the state of the water from solid to liquid. Thus a graph of temperature against time during which there is a constant heat input is of the form shown in Figure 9.38. During the time that there is no change in temperature, energy is still being transferred to the solid but it is just being used to change the substance into a liquid. A similar graph occurs for when, for example, water is heated and changes into steam. Energy is needed to change the state of water from liquid to vapour. The heat which results in no temperature change is called *latent heat*.

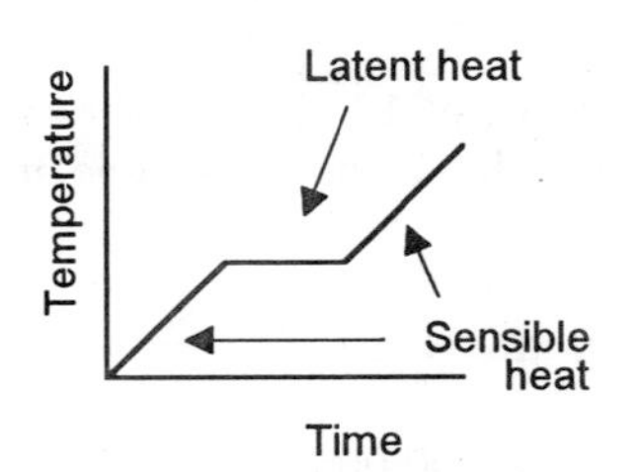

Figure 9.38 *Melting an ice cube*

The *specific latent heat L* of a substance is the amount of heat needed to change a mass of 1 kg of it from solid to liquid or liquid to vapour without any change in temperature occurring. It has the unit of J/kg. If we have a mass of m kg then the heat needed is mL.

Heat needed to change the state of m kg = mL

The latent heat of fusion of water is about 335 000 J/kg and of vaporization about 2257 000 J/kg.

A convenient method of carrying out experiments on the heating of materials is to use electrical heaters. See the section 9.7.1 for a discussion of such experiments.

Example

How much heat will an iron casting of mass 2 kg lose when temperature drops from 200°C to 20°C? The specific heat capacity of the iron can be taken as 480 J/kg °C.

The heat that has to be transferred out of the iron is

$$\text{heat} = mc \times \text{change in temperature} = 2 \times 480 \times 180 = 172\,800 \text{ J}$$

Example

How much heat is required to change 2 kg of water at 100°C to steam at 100°C? The specific latent heat of vaporization for water at 100°C is 2257 kJ/kg.

To change 2 kg of water at 100°C into steam at 100°C requires

$$\text{heat} = mL = 2 \times 2257 \times 1000 = 4514\,000 \text{ J}$$

9.7.3 Conservation of energy

Energy does not disappear when a useful job has been done, it just changes from one form to another. When you lift an object off the floor and on to a bench, work is done and energy transferred from you to the object which gains potential energy. The energy you have is provided by food. To lift an object of mass 1 kg through a height of 1 m uses the energy you gain from, for example, about 2.5 milligrams of sugar (this is typically about four grains of sugar). The energy lost by you is transferred via work into potential energy. When the object falls off the table and down to the floor it loses its potential energy but gains kinetic energy. When it is just about to hit the floor, all the potential energy that it has acquired in being lifted off the floor has been transformed into kinetic energy. When it hits the floor it stops moving and so all the kinetic energy vanishes. It might seem that energy has been lost. But this is not the case. The object and the floor show an increase in temperature. The kinetic energy is transferred via heat into a rise in temperature.

Energy is never lost, it is only transformed from one form to another or transferred from one object to another. This is the *principle of the conservation of energy*.

9.7.4 Power

Power is the rate at which energy is transferred, i.e.

$$\text{power} = \frac{\text{energy transferred}}{\text{time taken}}$$

But work and heat are methods by which energy can be transferred. Thus we can write

$$\text{power} = \frac{\text{work done}}{\text{time taken}}$$

and

$$\text{power} = \frac{\text{heat transfer}}{\text{time taken}}$$

When the unit of the energy transferred is joule (J) and the time seconds (s), then power has the unit of J/s. This unit is given a special name and symbol, the watt (W). Thus 1 W is a rate of energy transfer of 1 J per second.

Example

In an experiment to measure his/her own power, a student of mass 60 kg raced up a flight of fifty steps, each step being 0.2 m high and found that it took 20 s. What is the power?

The gain in potential energy of the student in moving his/her mass through a vertical height of 50×0.2 m is

$$\text{gain P.E.} = mgh = 60 \times 9.8 \times 50 \times 0.2 = 5880 \text{ J}$$

Hence the power is

$$\text{power} = \frac{\text{energy transfer}}{\text{time taken}} = \frac{5880}{20} = 294 \text{ W}$$

9.7.5 Efficiency

With a machine, energy is supplied to it and results in the machine transferring energy to a load. Not all the energy input to a machine is supplied by the machine as energy to the load, some of it is wasted (Figure 9.39). The *efficiency* of a machine may be defined as the percentage

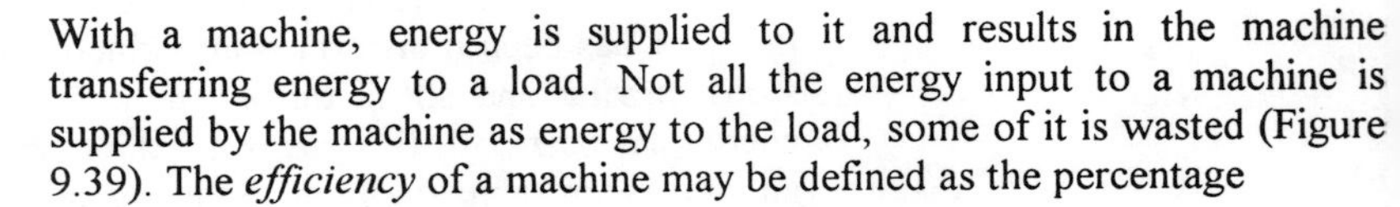

$$\text{efficiency} = \frac{\text{energy transferred from machine to the load}}{\text{energy transferred from effort to the machine}} \times 100\%$$

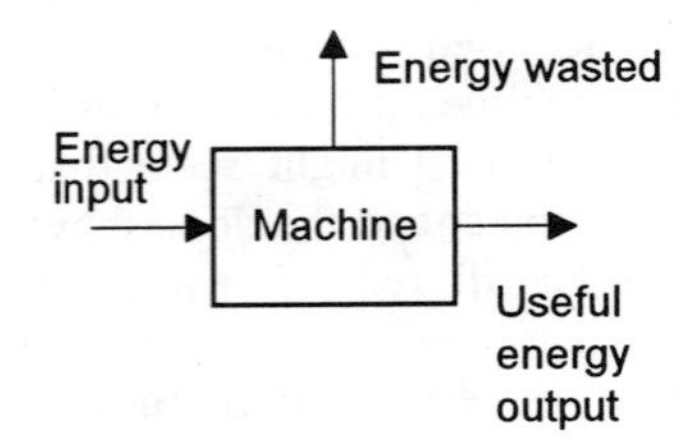

Figure 9.39 *Energy input and outputs with a machine*

This is often written as

$$\text{efficiency} = \frac{\text{useful energy output}}{\text{energy input}} \times 100\%$$

For example, human muscles are not 100% efficient but, at best, only about 25% efficient. This means that when you walk up stairs for every 100 J of energy needed to gain potential energy, another 300 J of energy has to be supplied by your body. This 300 J is wasted as heat.

Example

A machine is able to lift a mass of 50 kg through a vertical height of 5 m when it supplied with an energy input of 4000 J. What is the efficiency of the machine?

The useful energy output from the machine is work which results in a gain in potential energy of the mass. This gain in potential energy is

$$\text{gain in PE} = mgh = 50 \times 9.8 \times 5 = 2450 \text{ J}$$

The efficiency is thus

$$\text{efficiency} = \frac{2450}{4000} \times 100\% = 61.25\%$$

9.8 Electricity

Electricity is said to flow round a circuit when there is a *current*. We consider an electric current to be a flow of electric *charge*, the unit of charge being the coulomb (C). A current is something that is measured with an ammeter, the unit of current being the ampere (A). An electric circuit can be regarded as rather like water being pumped round a circuit of pipes. The water represents the charge and the rate of movement of the water the current. A current of 1 A is charge moving at the rate of 1 C per second. Water does not get lost on its way round the circuit and in electric circuits charge does not get lost. A flow meter at some point in the water circuit will show the rate at which water is passing through that part of the circuit, likewise an ammeter at some point in an electric circuit shows the rate at which charge is moving through that part of the circuit. To make the water flow round its circuit of pipes, a pump is used. This pump effectively lifts water up to a height and then allows it to flow downhill round the circuit and back to the pump again. A battery in an electric circuit fulfils a similar purpose. It is said to supply a *voltage*. Between two points in the water circuit we can measure the size of the hill down which the water is flowing. This hill means that is a potential energy difference between the water at the top of the hill and the water at the bottom. With an electric circuit we can use a voltmeter to measure the electric *potential difference* between the points and effectively the 'size of the hill' at that point down which the electric charge is flowing. The term voltage is often used instead of the term potential difference. The unit of voltage or potential difference is the volt (V).

To measure the current through some circuit element we put the ammeter in series with the element so the same current that flows through the element flows through the ammeter. To measure the voltage across a circuit element we put the voltmeter in parallel with it so that it measures the 'potential energy difference' between the two ends of the element. Figure 9.40 illustrates this.

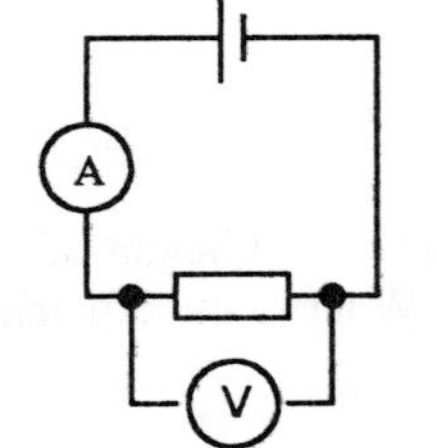

Figure 9.40 *Voltage and current measurement for a circuit element*

Suppose we use the circuit shown in Figure 9.40 to measure the voltage across an element and the current through it. The voltage V across the element divided by the current I through it is called the *resistance* R of the element.

$$R = \frac{V}{I}$$

When the voltage is in volts and the current in amperes, the resistance is in ohms (Ω). An element with a small resistance requires only a small voltage to give a sizeable current. An element with a higher resistance requires a larger voltage to give the same current.

Example

What is the resistance of a circuit element if there is a voltage of 4 V across it when the current through it is 0.1 A?

Using the equation given above

$$R = \frac{V}{I} = \frac{4}{0.1} = 40\,\Omega$$

9.8.1 Ohm's law

Figure 9.41 shows a circuit we could use to determine the relationship between the current through a circuit element and the voltage across it. The current through the element is varied by means of the variable resistor and readings taken of the current and voltage. Suppose the circuit element is a length of metallic wire, e.g. nichrome or constantin wire. The results might then be like:

Voltage in V	0	1	2	3	4
Current in A	0	0.1	0.2	0.3	0.4

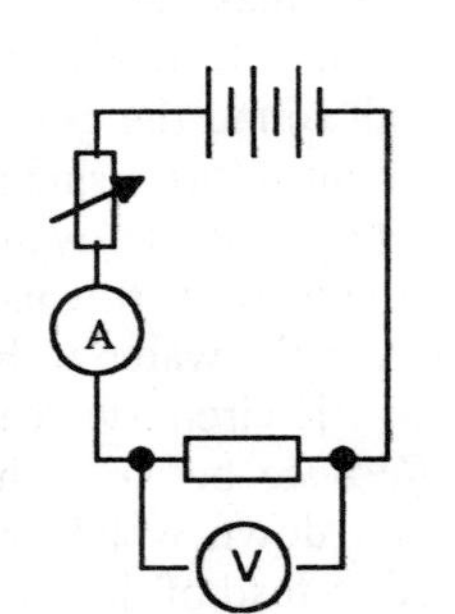

Figure 9.41 *Investigating current–voltage relationship*

When the results are plotted as a graph we have the result shown in Figure 9.42. The graph is a straight line graph passing through the zero voltage – zero current point. Thus the voltage is proportional to the current. Such a circuit element is said to be a linear device and obey *Ohm's law*.

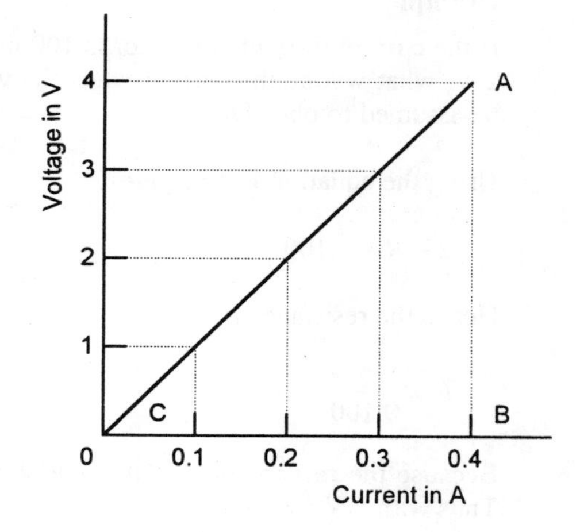

Figure 9.42 *Voltage – current graph*

Ohm's law can be stated as: provided physical conditions, such as temperature, do not change, then the voltage is proportional to the current, i.e.

$$\text{voltage} \propto \text{current}$$

Representing the voltage by V and the current by I we can write this as

$$V = \text{a constant} \times I$$

This constant is the resistance we defined in section 9.7. Thus

$$V = RI$$

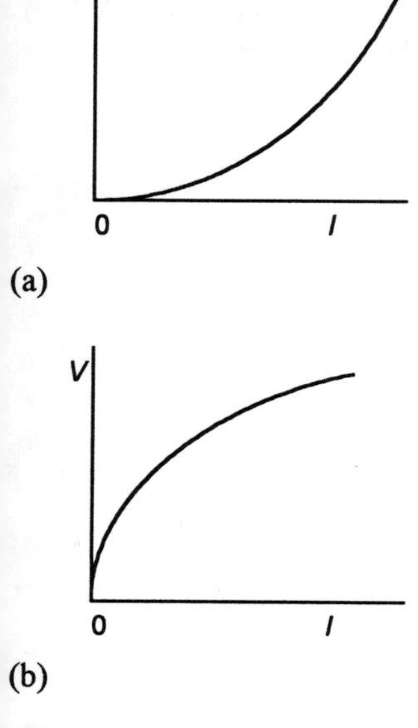

Figure 9.43 *(a) Lamp, (b) thermistor*

All that Ohm's law is stating is that the resistance is constant and does not change when the current is changed. The gradient of the voltage–current graph is the resistance. Thus for Figure 9.41, the gradient is AB/BC and so is

$$\text{gradient} = \frac{4-0}{0.4-0} = \frac{4}{0.4} = 10\,\Omega$$

Not all circuit elements obey Ohm's law. Figure 9.43 shows two examples, a filament lamp and a thermistor, which do not. They have graphs which are not linear. Thus the resistance depends on the current. With the lamp the resistance, i.e. the gradient of the graph, increases as the current increases, with the thermistor the resistance, i.e. the gradient of the graph, decreases as the current increases.

Example

If the current through a resistor is 100 mA when the voltage across it is 2 V, what will be the current when the voltage is 5 V? The resistor can be assumed to obey Ohm's law.

Using the equation $V = RI$ gives

$$2 = R \times 0.100$$

Hence the resistance R is

$$R = \frac{2}{0.100} = 20\ \Omega$$

Because the resistor obeys Ohm's law, the resistance will be constant. Thus with 5 V we have

$$5 = 20 \times I$$

Hence

$$I = \frac{5}{20} = 0.25\ \text{A} = 250\ \text{mA}$$

Example

Figure 9.44 shows a graph of how the voltage across a circuit element varies as the current changes. (a) Does the element obey Ohm's law? (b) What is the resistance of the element when the voltage is 2 V? (c) What is the resistance of the element when the voltage is 3 V?

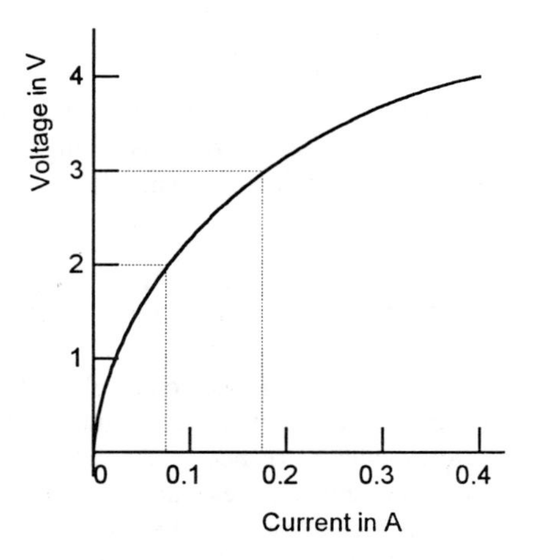

Figure 9.44 *Example*

(a) The graph is not a straight line passing through the zero voltage–zero current point. Thus the element does not obey Ohm's law.
(b) The resistance changes when the current or the voltage changes. Thus at a voltage of 2 V when the current is 0.075 A, the resistance is

$$\text{resistance} = \frac{2}{0.075} = 26.7\ \Omega$$

(c) At a voltage of 3 V when the current is 0.175 A, the resistance is

$$\text{resistance} = \frac{3}{0.175} = 17.1\ \Omega$$

9.8.2 Power

A voltmeter connected across a circuit element measures the energy transfer made by each coulomb of charge passing through that element. An energy transfer of one joule per coulomb is one volt. Thus a voltage of V volts means an energy transfer of V joules for each coulomb. Suppose we have a current of 3 A through some circuit element. This means that 3 coulombs of charge are passing through that element every second. If the current is I then I coulombs of charge are passing through the element each second. If the voltage across the element is V this must mean that for each coulomb, V joules of energy are transferred. Thus we must have VI joules of energy transferred every second. But the rate of energy transfer is the power. Hence

$$\text{power} = VI$$

The unit of power is the watt (W) if the current is in amperes (A) and the voltage in volts (V). Since $V = IR$, this equation can also be written as

$$\text{power} = (IR)I = I^2R$$

or

$$\text{power} = V\left(\frac{V}{R}\right) = \frac{V^2}{R}$$

Consider a situation where a lamp is connected across a battery and an ammeter shows that it takes a current of 3 A when the voltage across it is 6 V. The rate at which energy is being delivered to the map is

$$\text{power} = VI = 6 \times 3 = 18\ \text{W}$$

This is the rate at which energy is transferred between electrical and other forms, in this case mainly radiation and heat. With a lamp supplied by a battery the energy changes are from chemical energy in the battery to electrical energy and then to radiation and heat energies.

Now consider another situation in which electricity is used to run a motor. The input power to the motor is the voltage across the motor multiplied by the current through it. The motor may be used to lift a load through some vertical distance. We can then determine the output power in terms of the rate at which the load gains potential energy. If no energy is wasted anywhere in the motor, then the output power will equal the input power. However, generally there is some wastage, e.g. the motor gets hot, and so the motor is not 100% efficient. The motor efficiency is given by

$$\text{efficiency} = \frac{\text{mechanical output power}}{\text{electrical input power}} \times 100\%$$

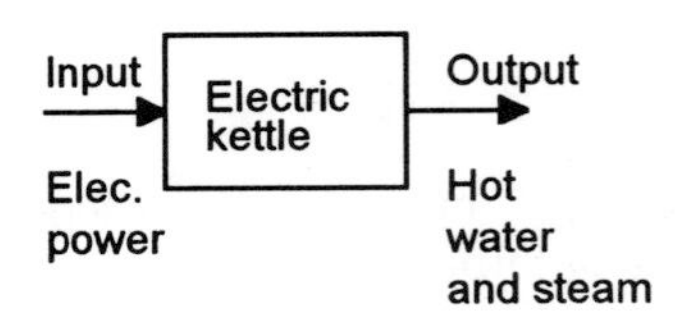

Figure 9.45 *Electrical kettle system*

As an illustration of an electrical system, consider an electric kettle (Figure 9.45). The kettle has an input of electrical power and an output of primarily boiling water and steam. We can explain what is happening from when the kettle is switched on with cold water to when it is switched off with hot water and steam being produced. Switching on the electricity allows a current to flow through the heating element and a voltage occurs across it. Electrical power is supplied to the heater and transformed into heat. This heats the water from room temperature, say 20°C, to 100°C. Thus suppose we have 500 g of water in the kettle and electrical power of 1 kW has to be supplied for 300 s to get the water to 100°C. The energy input is 1000 × 300 = 300 000 J. This is the energy which is needed to raise the temperature of 500 g of water through 80°C, plus the energy needed to raise the temperature of the kettle itself and lost to the surroundings. The specific heat capacity of water is about 4200 J/kg °C and so the energy needed to raise its temperature by 80°C is 0.500 × 4200 × 80 = 168 000 J. Thus of the 300 000 J supplied, just over half of it is used to heat the water. At 100°C, although energy is still being supplied, the temperature no longer increases but the water boils and the liquid changes into a vapour. Thus, suppose the kettle continued to boil the water for 50 s and afterwards it was found that the kettle only contained 480 g of water. The energy supplied during that 50 s is 1000 × 50 = 50 000 J. This is used to change 20 g of water at 100°C to steam at 100°C, plus lose heat to the surroundings. Thus the energy required per kilogram is 2500 000 J. If no heat was lost to the surroundings, this would be the latent heat of vaporization. A value obtained when no heat losses to the surroundings occur is 2257 000 J/kg.

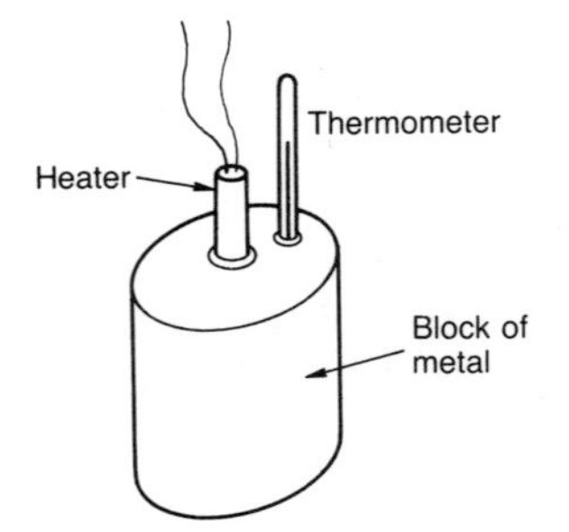

Figure 9.46 *Specific heat measurement*

The above illustrates the basis of experiments that can be used to determine specific heats and latent heats. To measure the specific heat capacity, electrical energy is supplied to an electric heater embedded in a block of the material or in the liquid concerned, the mass of the block or liquid having been measured (Figure 9.46). The current I and voltage V are measured to give the supplied electrical power. The energy is supplied for a measured length and time t and the resulting temperature rise determined. To minimize heat losses to the surroundings or to any container, the apparatus may be lagged with heat insulation and a very low mass, low specific heat capacity, material used for the container. The specific heat capacity c is then given by

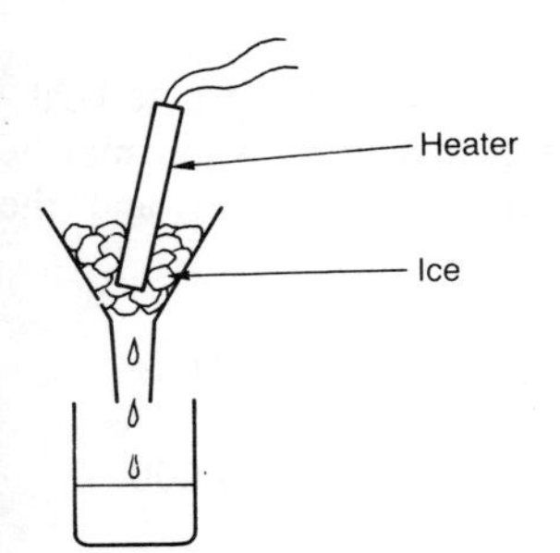

Figure 9.47 *Latent heat of fusion measurement*

$$IVt = mc \times \text{change in temperature}$$

To measure the latent heat of fusion, i.e. the latent heat involved when, say, ice changes to liquid water, the electrical heat is immersed in melting ice. This ensures that the ice is at 0°C. Then electrical energy is supplied for a measured time t and the amount of ice that is melted as a result is determined. This could involve the ice being in a funnel and the water that melts being collected in a beaker underneath (Figure 9.47). The energy supplied is IVt and if m kg is melted, the latent heat of fusion is

$$\text{latent heat} = \frac{IVt}{m}$$

To determine the latent heat of vaporization of, say, water, the amount m of water evaporated is determined when the electrical energy is being supplied for t s and the water is at its boiling point. The amount evaporated can be determined by finding the change in mass of the water from before the water is boiled off and afterwards. The latent heat is then given by the same equation as above.

Example

A 2-bar electric fire is rated as 2 kW when connected to the 240 V supply. What will be the current taken by the fire?

Using the equation $P = IV$, then

$$2000 = I \times 240$$

Hence

$$I = \frac{2000}{240} = 8.3 \text{ A}$$

For this reason a fuse that can take 8.3 A needs to be used in the plug connecting the fire to the voltage supply. Typically a 13 A fuse would be used in such a situation. This does not allow too large a current overload before fusing and breaking the circuit.

Example

A vacuum cleaner takes a current of 2 A when connected to the 240 V supply. What is the electrical power supplied?

Using the equation $P = IV$, then

$$P = 2 \times 240 = 480 \text{ W}$$

Example

For how long will a 50 W electric heater need to be run if it is to heat 1 kg of water through 40°C? The specific heat capacity of water may be taken as 4200 J/kg °C and heat losses to the container and the surroundings ignored.

The electrical energy transformed to heat in this time is $50t$. Thus

$$50t = mc \times \text{change in temperature} = 1 \times 4200 \times 40 = 168\,000$$

Hence

$$t = \frac{168\,000}{50} = 336 \text{ s or } 5.6 \text{ minutes}$$

Example

The current through a resistor of resistance 10 Ω is 0.2 A. What is the power?

Using the equation derived above

$$P = I^2R = 0.2^2 \times 10 = 0.4 \text{ W}$$

9.8.3 Alternating currents

The term *direct* voltage or current is used when the voltage or current is always in the same direction. For example, a battery supplies a direct voltage and a direct current through a resistor. Thus, for example, in the circuit shown earlier in Figure 9.40, the current and voltage readings remain steady indicating the current and voltage are not changing in value or in direction. The term *alternating* voltage or current is used when the direction alternates, continually changing with time.

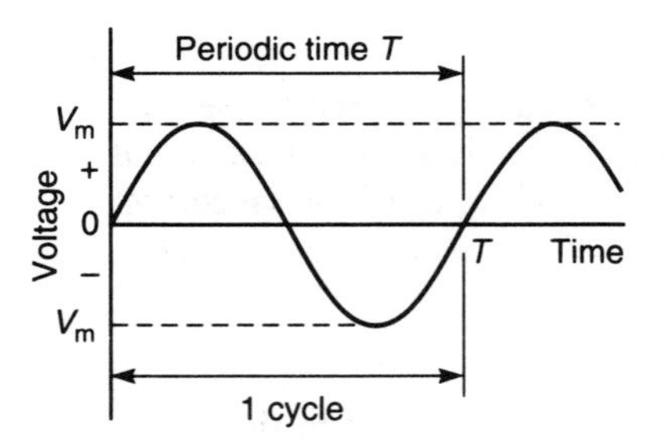

Figure 9.48 *Alternating voltage*

Figure 9.48 shows the form taken by the alternating voltages and current supplied by the mains supply. The alternating voltage or current oscillates from positive to negative values in a regular, periodic manner. In Figure 9.48 an alternating voltage is shown, the voltage value continually changing with time, oscillating from a maximum value in one direction $+V_m$ to a maximum value in the other direction $-V_m$. One complete sequence of such an oscillation is called a *cycle*. The time T taken for one complete cycle is called the *periodic time* and the number of cycles occurring per second is the *frequency* f. Thus $f = 1/T$. The unit of frequency is the hertz (Hz), 1 Hz being 1 cycle per second. In Britain the frequency of the alternating supply is 50 Hz. This means that 50 cycles occur every second.

The mains alternating current, or voltage, oscillates with time so that the average value over a cycle is zero. This is because during one cycle for every positive value of current or voltage there is a corresponding negative value, i.e. current or voltage in the opposite direction. Thus the average

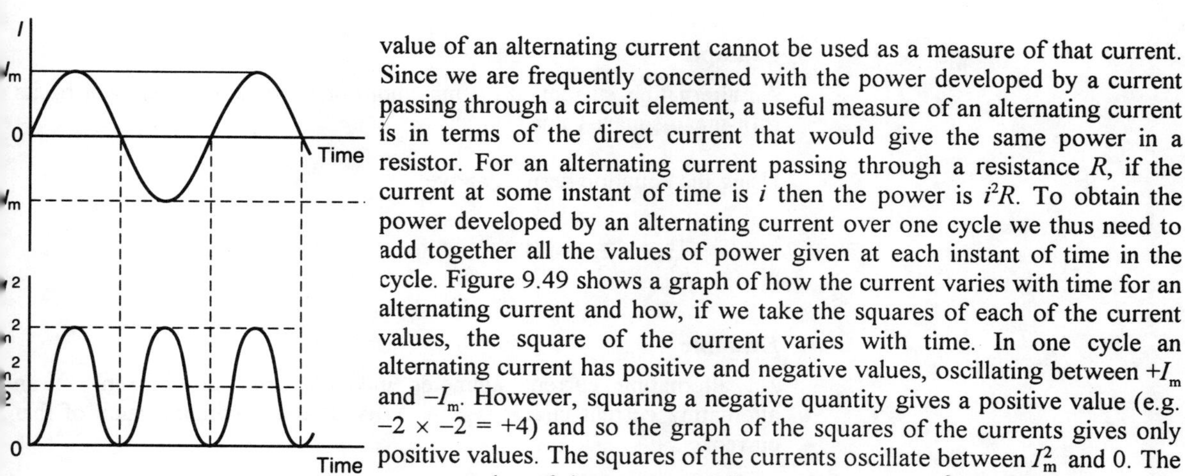

Figure 9.49 *Graphs of current with time and (current)^2 with time*

value of an alternating current cannot be used as a measure of that current. Since we are frequently concerned with the power developed by a current passing through a circuit element, a useful measure of an alternating current is in terms of the direct current that would give the same power in a resistor. For an alternating current passing through a resistance R, if the current at some instant of time is i then the power is i^2R. To obtain the power developed by an alternating current over one cycle we thus need to add together all the values of power given at each instant of time in the cycle. Figure 9.49 shows a graph of how the current varies with time for an alternating current and how, if we take the squares of each of the current values, the square of the current varies with time. In one cycle an alternating current has positive and negative values, oscillating between $+I_m$ and $-I_m$. However, squaring a negative quantity gives a positive value (e.g. $-2 \times -2 = +4$) and so the graph of the squares of the currents gives only positive values. The squares of the currents oscillate between I_m^2 and 0. The average value of the squares of the currents is thus $I_m^2/2$. Thus the average power is $RI_m^2/2$. The same power could have been produced by a d.c. current I if we have

$$RI^2 = \frac{RI_m^2}{2}$$

Dividing both sides of the equation by R and then taking the square root gives the equivalent of a d.c. current I as

$$I = \frac{I_m}{\sqrt{2}}$$

This equivalent current is called the *root-mean-square current*. We can likewise obtain an equation for the root-mean-square voltage

$$V = \frac{V_m}{\sqrt{2}}$$

Meters used to measure alternating currents and voltages generally indicate the root-mean-square values. If we measure the root-mean-square current through a resistor that obeys Ohm's law with direct current and the corresponding root-mean-square voltages across it then we obtain the Ohm's law relationship for the alternating current, i.e. $V = RI$.

Example

An alternating voltage has a frequency of 50 Hz. What is the time taken for one cycle of that voltage?

The alternating voltage completes 50 cycles every second. Thus

$$\text{time for 1 cycle} = \frac{1}{50} = 0.02 \text{ s}$$

Example

An alternating current has a maximum value of 2 A, what will be its root-mean-square value?

Using the equation derived above

$$I = \frac{I_m}{\sqrt{2}} = \frac{2}{\sqrt{2}} = 1.4 \text{ A}$$

Example

An alternating current ammeter indicates that the current in an alternating current circuit is 3 A. What is the maximum value of that current?

Using the equation developed above

$$I = \frac{I_m}{\sqrt{2}}$$

then $I_m = \sqrt{2}\,I = \sqrt{2} \times 3 = 4.2$ A

9.9 Expansion

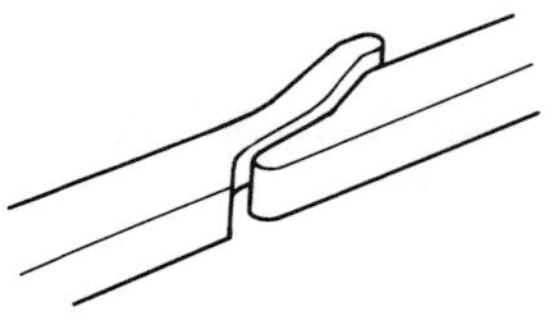

A consequence of increasing the temperature of objects and making them hotter is that they expand. Thus if a metal rod is heated it gets longer. If the rod is prevented from expanding, then the forces generated can be very large and the metal may deform or even break. Thus expansion can present problems to engineers. For example, railway lines expand on a sunny day and on a hot summer's day have been known to buckle. Thus methods have to be adopted to prevent such occurrences. Nowadays long lengths of welded rails are used and tapered at their ends (Figure 9.50) so that some movement is possible. The rails are also held very firmly by concrete sleepers which can withstand the large forces that can be generated by expansion, or contraction.

Figure 9.50 *Joint between railway lines*

Another example where expansion can cause problems is with bridges. Engineers have to allow for a bridge to expand when the temperature rises and contract when the temperature drops. This can be done by the bridge girders at one end of a bridge resting on rollers (Figure 9.51(a)). On the road surface over the bridge, the expansion and contraction can be allowed for by the use of a special joint (Figure 9.51(b)). Such a joint can consist of interleaved metal bars to give a metal grid within which expansion or contraction can occur and still leave a reasonable surface over which vehicles can travel.

Telephone wires and overhead electrical transmission cables have to be suspended between supports so that they are slack in summer. Otherwise when winter comes and the temperature drops, they contract and could become taut and break.

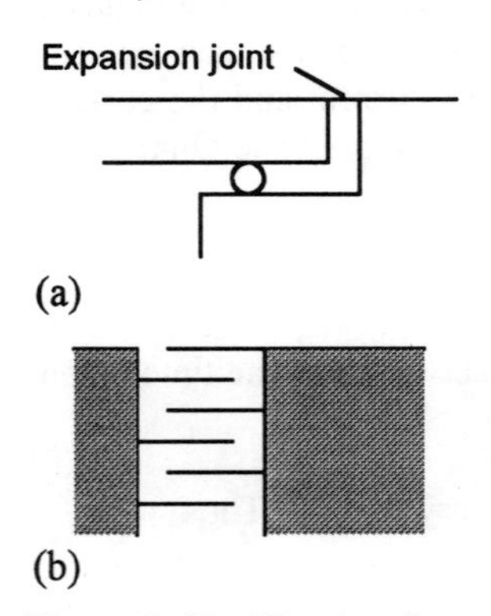

Figure 9.51 *Allowing for bridge expansion*

Riveting makes use of expansion and contraction. A hot, and so expanded, rivet is inserted through two holes in the metal plates being joined. The ends of the hot rivet are then hammered into a mushroom shape. When the rivet cools it contracts and pulls the two plates together.

It is not only solids that expand when the temperature rises, liquids and gases also expand. The mercury-in-glass thermometer is based on the property of mercury to expand when the temperature increases. Hot-air balloons rely on the expansion of air when heated. When the air expands the same mass of air occupies a larger volume, thus the density of the gas decreases. As a result the hot air in the balloon is of lower density than the surrounding colder air. The balloon thus rises like a cork rising in water.

9.9.1 Coefficient of expansion

For materials such as metals and many other materials, when expansion occurs the change in length is proportional to the change in temperature. Thus if we have a length L_0 at 0°C and a length L_t at a temperature of t°C, i.e. a change in temperature of t°C, then

$$(L_t - L_0) \propto t$$

The amount by which a rod expands for a given change in temperature also depends on the initial length of the material. For example, if we had a length of 2 m and changed the temperature by, say, 10°C then we would obtain twice the expansion that would occur for the same material of length 1 m when the temperature was changed by 10°C. Thus

$$(L_t - L_0) \propto L_0 t$$

We can write this as the equation

$$(L_t - L_0) = \alpha L_0 t$$

where α is a constant for a particular material. It is called the *coefficient of linear expansion*. The unit of the coefficient is per °C. This equation is generally rearranged to

$$L_t = L_0 + \alpha L_0 t$$

$$L_t = L_0(1 + \alpha t)$$

Typical values of coefficients are: aluminium 0.000 023 /°C, copper 0.000 017 /°C, mild steel 0.000 011 /°C.

The coefficient of linear expansion for a metal rod can be measured in the laboratory using apparatus (Figure 9.52) in which a micrometer (see chapter 10) is set so that its point touches the free end of the rod when it is at room temperature, this temperature being measured. The micrometer reading is then taken. The micrometer point is then withdrawn to allow for the expansion of the rod. Steam is then passed in a chamber round the rod

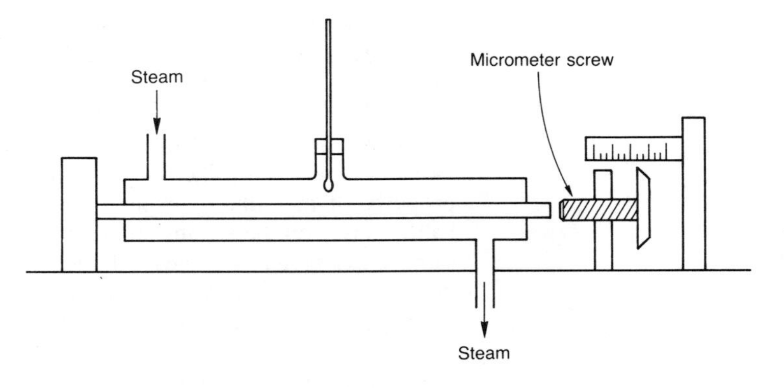

Figure 9.52 *Apparatus for measuring the coefficient of linear expansion*

to raise its temperature to 100°C. The micrometer point is then set to touch the free end of the rod again and the reading taken. The difference in the micrometer readings gives the amount by which the rod has expanded in going from room temperature to 100°C. This data, together with the initial length of the rod at room temperature, enables the coefficient of expansion to be calculated.

Example

A bar of copper has a length of 0.2 m at 0°C and is heated to 100°C. If the coefficient of linear expansion of copper is 0.000 017 /°C, by how much will the bar expand?

Using the equation $L_t = L_0(1 + \alpha t)$, then when we remove the brackets we have

$$L_t = L_0 + L_0\alpha t$$

Hence the expansion is

$$\text{expansion} = L_t - L_0 = L_0\alpha t = 0.2 \times 0.000\,017 \times 100 = 0.0034 \text{ m}$$

This is 3.4 mm.

Example

Telephone cables are fixed between two posts 40 m apart when the temperature is 25°C. How much slack should the installers allow if the cable is not to become taut until the temperature reaches 0°C? The coefficient of linear expansion of the cable is 0.000 017 /°C.

The slack is the amount by which the cable will contract when the temperature drops from 25°C to 0°C. Hence, using the equation $L_t = L_0(1 + \alpha t)$, then when we remove the brackets we have

$$L_t = L_0 + L_0\alpha t$$

Hence the contraction is

$$\text{contraction} = L_0 - L_t = L_0\alpha t = 40 \times 0.000\,017 \times 25 = 0.017 \text{ m}$$

Thus slack of 17 mm should be allowed. In the real world they would allow for a much lower temperature before breaking occurs.

Problems

Questions 1 to 30 have four correct answer options: A, B, C and D. You should choose the correct answer from the answer options.

Questions 1 to 3 relate to Figure 9.53 which shows four tape charts.

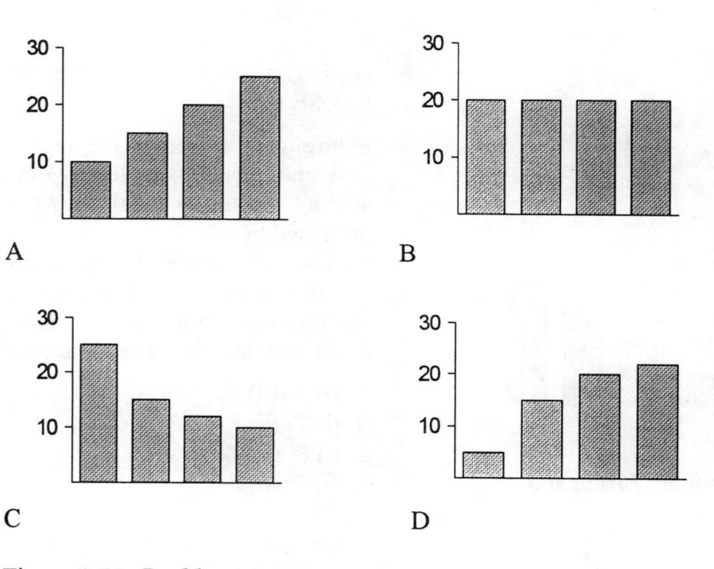

Figure 9.53 *Problems 1 to 3*

1 Which tape chart shows constant velocity motion?
2 Which tape chart shows constant acceleration motion?
3 Which tape chart shows a non-uniform deceleration?

4 Figure 9.54 shows a ticker tape produced by a moving object. The motion of the object is:

Start

Figure 9.54 *Problem 4*

A Constant velocity
B Constant acceleration
C Constant deceleration
D Variable acceleration

5 Figure 9.55 shows the ticker tape produced by a moving object. The motion of the object is:

Start

Figure 9.55 *Problem 5*

A Constant velocity
B Constant acceleration
C Constant deceleration
D Variable acceleration

6 Decide whether each of the following statements is TRUE (T) or FALSE (F).

Figure 9.56(a) shows the tape chart produced when a trolley on a horizontal bench is pulled by one length of elastic stretched a constant amount. Figure 9.56(b) shows the tape chart that could have been produced by:
(i) The same trolley being pulled by two parallel lengths of the same elastic, each being stretched the same constant amount.
(ii) Two of the same trolley stacked one on top of the other and pulled by a single length of the same elastic stretched the same amount.

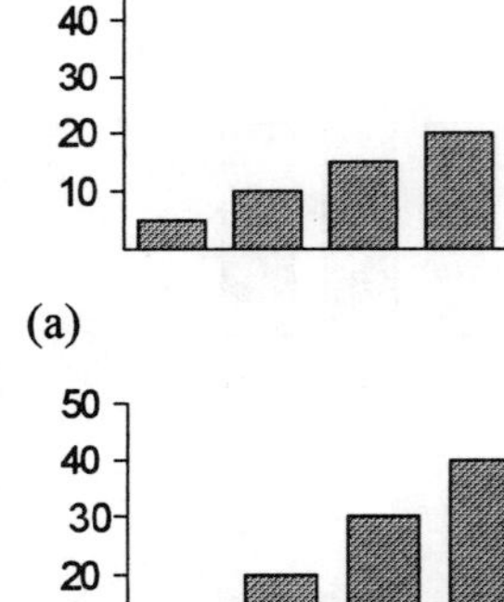

Figure 9.56 *Problem 6*

A (i) T (ii) T
B (i) T (ii) F
C (i) F (ii) T
D (i) F (ii) F

7 The graph in Figure 9.57 shows a motion for which:

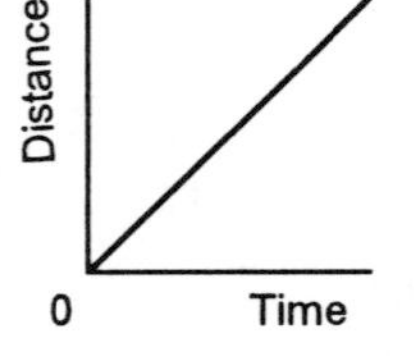

Figure 9.57 *Problem 7*

A The velocity is zero
B The velocity is constant at a non-zero value
C The velocity is changing at a constant rate
D The velocity is changing at a rate which varies

Questions 8 to 10 relate to Figure 9.58 which shows a graph will axes labelled as y and x. Decide whether each of the following statements is TRUE (T) or FALSE (F).

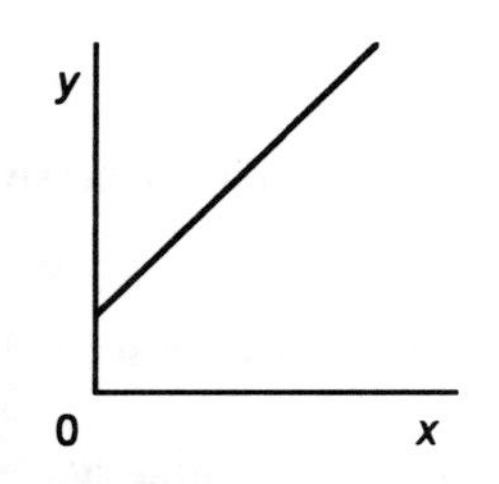

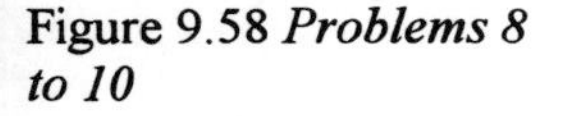
Figure 9.58 *Problems 8 to 10*

8 When y represents distance and x time then:
(i) The motion is constant velocity.
(ii) The motion starts at a distance which is not the zero distance mark.

A (i) T (ii) T
B (i) T (ii) F
C (i) F (ii) T
D (i) F (ii) F

9 When y represent velocity and x time then:
(i) The velocity is proportional to the time.
(ii) The acceleration is proportional to the time.

A (i) T (ii) T
B (i) T (ii) F
C (i) F (ii) T
D (i) F (ii) F

10 When y represents velocity and x time then:
(i) The velocity is constant and not zero.
(ii) The acceleration is constant and not zero.

A (i) T (ii) T
B (i) T (ii) F
C (i) F (ii) T
D (i) F (ii) F

11 Decide whether each of the following statements is TRUE (T) or FALSE (F).

When an unbalanced force acts on an object, it accelerates with a constant acceleration which is proportional to:
(i) Size of the force.
(ii) Size of the mass of the object.

A (i) T (ii) T
B (i) T (ii) F
C (i) F (ii) T
D (i) F (ii) F

12 When a force of 100 N acts on a mass of 2 kg the resulting acceleration is always:

A 50 m/s^2 in the direction of the force
B 50 m/s^2 downwards
C 200 m/s^2 in the direction of the force
D 200 m/s^2 downwards

13 When an object freely falls with an acceleration of 9.8 m/s^2 it:

A Covers the same distance in each second of its fall.
B Covers a distance which increases by 9.8 m in each successive second.
C Has a velocity which is the same in each second of its fall.
D Has a velocity which increases by 9.8 m/s in each successive second.

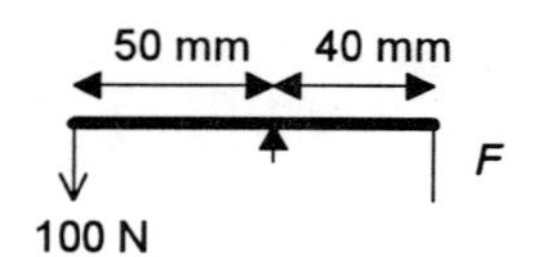

Figure 9.59 *Problem 14*

14 For balance to occur for the beam shown in Figure 9.59 resting across the pivot, the force F must be:

A 100 N in an upwards direction
B 100 N in a downwards direction.
C 125 N in an upwards direction.
D 125 N in a downwards direction

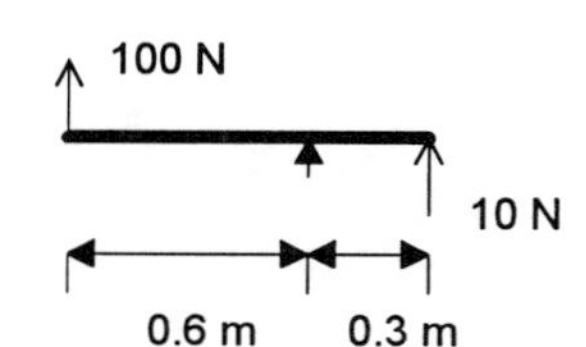

Figure 9.60 *Problem 15*

15 For the beam shown in Figure 9.60 there is a turning moment about the pivot axis of:

A 30 N m clockwise
B 30 N m anticlockwise
C 60 N m clockwise
D 90 N m anticlockwise

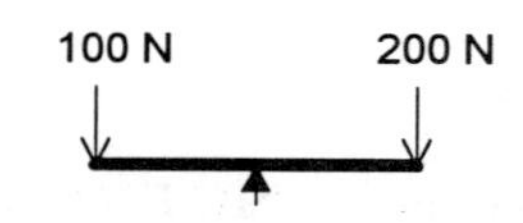

Figure 9.61 *Problem 16*

16 For the beam shown in Figure 9.61, the reaction at the pivot is:

A 100 N upwards
B 100 N downwards
C 300 N upwards
D 300 N downwards

17 Decide whether each of the following statements is TRUE (T) or FALSE (F).

When an object falls it:
(i) Loses potential energy.
(ii) Gains kinetic energy.

A (i) T (ii) T
B (i) T (ii) F
C (i) F (ii) T
D (i) F (ii) F

18 Decide whether each of the following statements is TRUE (T) or FALSE (F).

The potential energy transferred to an object when it is lifted through some height is doubled if
(i) The height through which it is lifted is doubled.
(ii) The mass of the object is halved.

A (i) T (ii) T
B (i) T (ii) F
C (i) F (ii) T
D (i) F (ii) F

19 Decide whether each of the following statements is TRUE (T) or FALSE (F).

The kinetic energy of an object moving with some velocity is doubled when:
(i) The velocity is doubled.
(ii) The mass of the object is doubled.

A (i) T (ii) T
B (i) T (ii) F
C (i) F (ii) T
D (i) F (ii) F

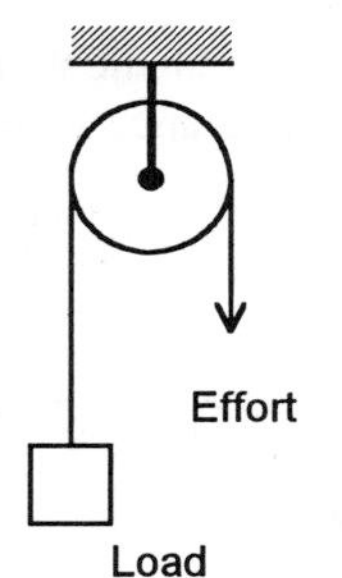

Figure 9.62 *Problems 20 and 21*

Questions 20 and 21 relate to Figure 9.62 which shows a pulley system. Decide whether each of the following statements is TRUE (T) or FALSE (F).

20 The tension in the rope over the pulley wheel is:
(i) The same size as the effort force.
(ii) The same size as the load force.

A (i) T (ii) T
B (i) T (ii) F
C (i) F (ii) T
D (i) F (ii) F

21 The work done by the effort is:
(i) Equal to the potential energy gained by the load.
(ii) The size of the effort multiplied by the distance it moves through.

A (i) T (ii) T
B (i) T (ii) F
C (i) F (ii) T
D (i) F (ii) F

22 The work done in lifting a mass of 3 kg through a vertical height of 2 m, where the acceleration due to gravity is 9.8 m/s^2, is:

A 0.6 J
B 6 J
C 29.4 J
D 58.8 J

23 A cube of mass 1 kg of side 10 mm rests flat on a horizontal table. where the acceleration due to gravity is 9.8 m/s^2. The pressure exerted on the table is:

A 0.01 N/mm^2
B 0.1 N/mm^2
C 0.98 N/mm^2
D 0.098 N/mm^2

24 Decide whether each of the following statements is TRUE (T) or FALSE (F).

When an ice cube melts:
(i) During the melting there is no change in temperature.
(ii) During the melting there is no heat flow into or out of the ice cube.

A (i) T (ii) T
B (i) T (ii) F
C (i) F (ii) T
D (i) F (ii) F

25 Figure 9.63 shows graphs of the voltage across a circuit element and the current through it. For which of the graphs is the resistance increasing as the current increases?

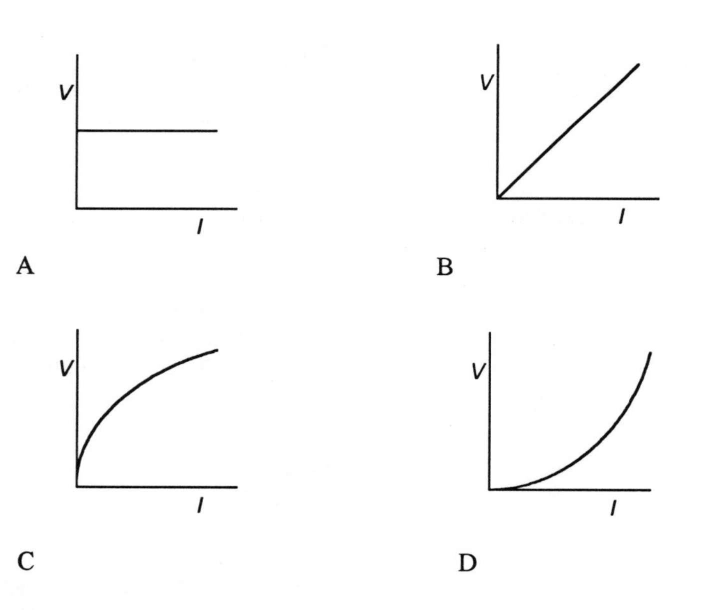

Figure 9.63 *Problem 25*

26 Decide whether each of the following statements is TRUE (T) or FALSE (F).

To double the rate at which electrical energy is transformed by a resistor:
(i) The current should be doubled.
(ii) The voltage should be doubled.

A (i) T (ii) T
B (i) T (ii) F
C (i) F (ii) T
D (i) F (ii) F

27 Decide whether each of the following statements is TRUE (T) or FALSE (F).

For the alternating current from the mains supply:
(i) The average current per cycle is zero.
(ii) The average power is zero.

A (i) T (ii) T
B (i) T (ii) F
C (i) F (ii) T
D (i) F (ii) F

28 Decide whether each of the following statements is TRUE (T) or FALSE (F).

For a resistor which obeys Ohm's law:
(i) The voltage across the resistor is proportional to the current through it.
(ii) The resistance is constant and does not change when the current changes.

A (i) T (ii) T
B (i) T (ii) F
C (i) F (ii) T
D (i) F (ii) F

29 Decide whether each of the following statements is TRUE (T) or FALSE (F).

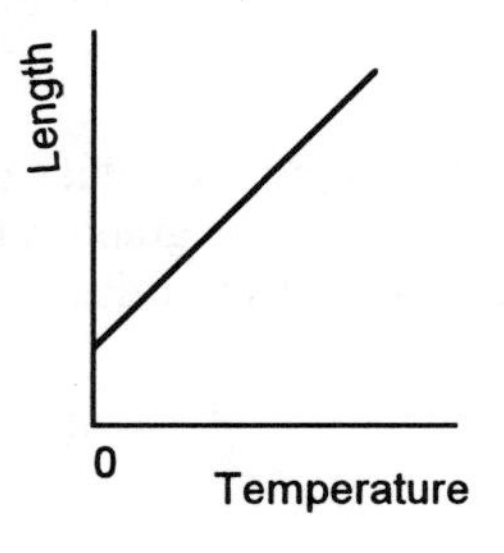

Figure 9.64 *Problem 29*

Figure 9.64 is a graph showing how the length of a metallic rod varies with the temperature. For the graph:
(i) The gradient of the graph is α, the coefficient of linear expansion.
(ii) The intercept of the graph with the length axis is L_0, the length at 0°C.

A (i) T (ii) T
B (i) T (ii) F
C (i) F (ii) T
D (i) F (ii) F

30 Decide whether each of the following statements is TRUE (T) or FALSE (F).

When the electrical power supplied to a heating element immersed in a fixed mass of liquid is doubled then, neglecting any heat losses from the liquid, we can have:
(i) The rate at which its temperature rises doubling.
(ii) The rate at which the liquid at its boiling point vaporizes is doubled.

A (i) T (ii) T
B (i) T (ii) F
C (i) F (ii) T
D (i) F (ii) F

31 An object moving in a straight line gives the following distance–time data. Plot the distance–time graph and hence determine the velocities at times of (a) 2 s, (b) 3 s, (c) 4 s.

distance in mm	0	10	36	78	136	210
time in s	0	1	2	3	4	5

32 An object moving in a straight line gives the following distance–time data. Plot the distance–time graph and hence determine the velocities at times of (a) 5 s, (b) 8 s, (c) 11 s.

distance in m	0	1	15	31	49	75	110
time in s	0	1	4	6	8	10	12

33 An object starts from rest and maintains a uniform acceleration of 0.5 m/s^2 for 4 s. It then moves with a uniform velocity for 3 s before a uniform retardation which brings it to rest in 2 s. Plot the velocity–time graph and hence determine the retardation during the last 2 s and the total distance covered.

34 A car starts from rest and travels along a straight road. Its velocity changes at a uniform rate from 0 to 50 km/h in 20 s. It then continues for 50 s with uniform velocity before reducing speed at a constant rate to zero in 10 s. Sketch the velocity–time graph and hence determine (a) the initial acceleration, (b) the final retardation and (c) the total distance travelled.

35 The velocity of an object varies with time. The following are values of the velocities at a number of times. Plot a velocity–time graph and hence estimate the acceleration at a time of 2 s and the total distance travelled in the 5 s.

velocity in m/s	0	1.2	2.4	3.6	3.6	3.6
time in s	0	1	2	3	4	5

36 The velocity of an object changes with time. The following are values of the velocities at a number of times. Plot the velocity–time graph and hence determine the acceleration at a time of 4 s and the total distance travelled during the 20 s.

velocity in m/s	0	8	10	16	20	20
time in s	0	4	8	12	16	20

37 The velocity of an object changes with time. The following are values of the velocities at a number of times. Plot the velocity–time graph and hence determine the acceleration at a time of 2 s and the total distance travelled between 1 s and 4 s.

velocity in m/s	25	24	21	16	9	0
time in s	0	1	2	3	4	5

38 State whether the distance–time graphs and the velocity–time graphs will be straight line with a non-zero gradient or zero gradient or curved in the following cases:
(a) The velocity is constant and not zero.
(b) The acceleration is constant and not zero.
(c) The distance covered in each second of the motion is the same.
(d) The distance covered in each successive second doubles.

39 Determine the values of the forces labelled as F in Figure 9.65. All the distances are in metres.

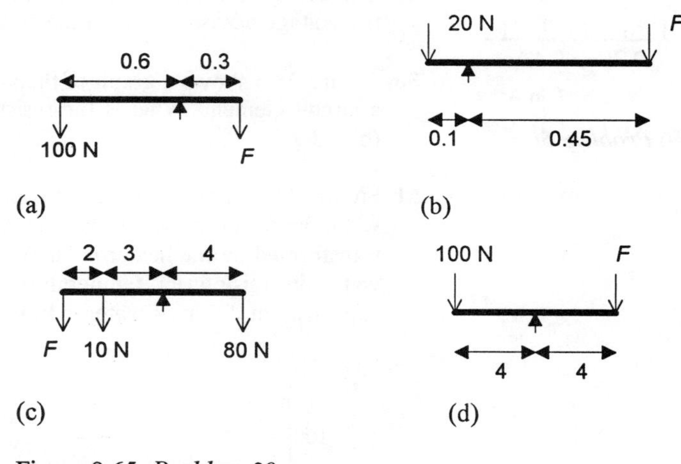

Figure 9.65 *Problem 39*

40 State the sequence of energy transformations that occurs with:
(a) An object falling off the bench to the floor.
(b) An electric immersion heater used to produce hot water.
(c) A student running up some stairs.
(d) An electric lamp.
(e) An electric motor.

41 Calculate the work done when a hydraulic hoist is used to lift a car of mass 1000 kg through a vertical height of 2 m. Take the acceleration due to gravity to be 9.8 m/s^2.

42 Calculate the work done in pushing a broken-down car a distance of 20 m if a constant force of 300 N is applied.

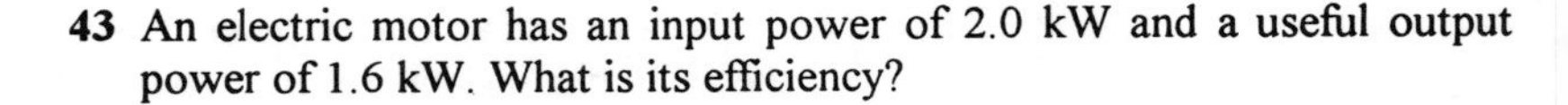

43 An electric motor has an input power of 2.0 kW and a useful output power of 1.6 kW. What is its efficiency?

44 Calculate the heat required to raise the temperature of 2 kg of copper by 60°C if it has a specific heat capacity of 390 J/kg °C.

45 Calculate the heat required to change 2 kg of ice at 0°C into liquid at 0°C if the specific latent heat of fusion for water is 335 000 J/kg °C.

46 A resistor, which obeys Ohm's law, has a resistance of 10 Ω. What will be the current through it when the voltage across it is (a) 2 V, (b) 3 V?

47 Describe the form of the voltage–current graph for a resistor that obeys Ohm's law.

48 Describe the form of the force–extension graph for a spring if the extension is proportional to the force.

49 The current through a resistor of resistance 10 Ω is 0.2 A. What is (a) the voltage across it and (b) the electrical power transformed?

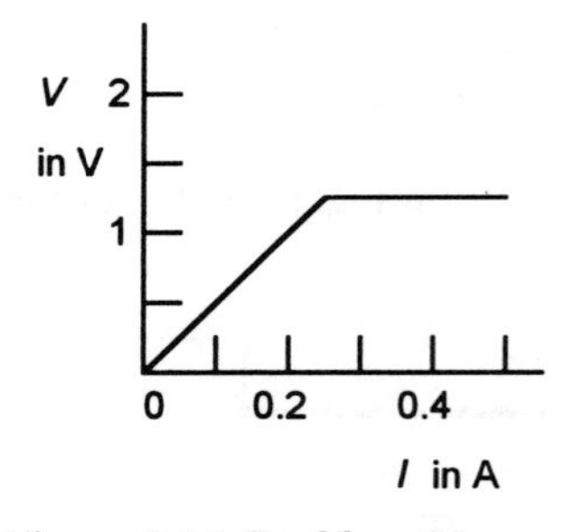

Figure 9.66 *Problem 50*

50 Figure 9.66 shows a graph of the voltage plotted against the current for a circuit element. What is the resistance when the current is (a) 0.1 A, (b) 0.3 A?

51 Figure 9.67 shows how the temperature of water changes with time when an electric heater is immersed in it. If the electrical power transformed by the heater is 50 W, estimate (a) the heat supplied to the water in raising its temperature to 100°C, (b) the latent heat of vaporization if 1 g of water is transformed into steam.

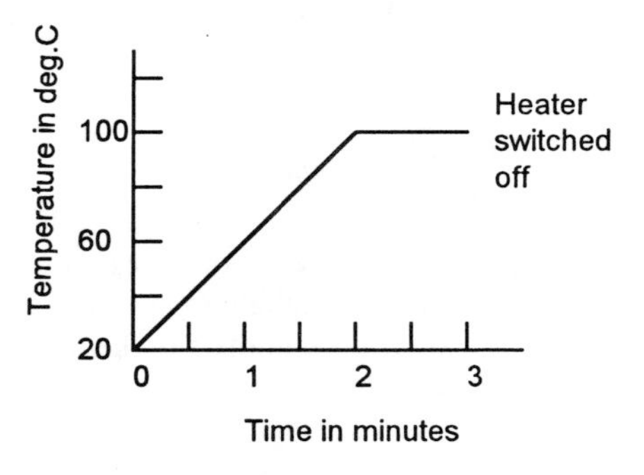

Figure 9.67 *Problem 51*

52 Figure 9.68 shows how the voltage across a resistor varies for resistors A and B as the current through them changes. (a) Which of the resistors obeys Ohm's law and (b) estimate their resistances when the current is 1 A.

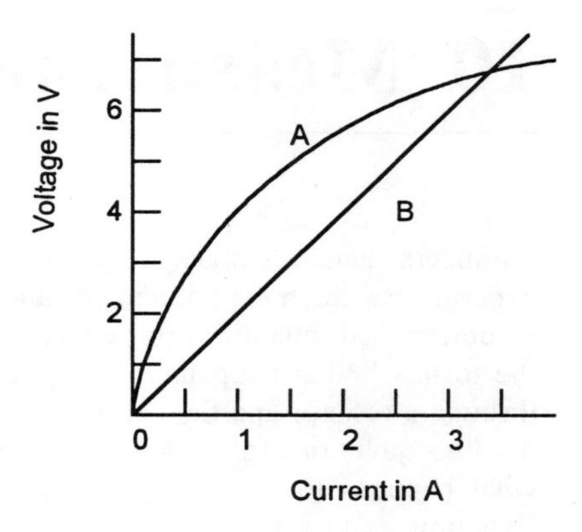

Figure 9.68 *Problem 52*

53 A cable has a length of 100 m at 0°C. How long will it be at 20°C if the material has a coefficient of expansion of 0.000 017 /°C.

54 What is the linear expansion of a steel bridge of span 100 m when the temperarure rises by 20°C? The coefficient of linear expansion of the steel is 0.000 011 /°C.

10 Measurement

10.1 Experiments and measurements

Engineers and scientists tend to spend a lot of time carrying out experiments. Such experiments enable theories to be tested, relationships to be determined, quantities to be measured, answers obtained to questions of the form - 'What happens if?'. For example, measurements of the current through a resistor and the voltage across it enables the relationship between the two quantities to be determined. We might investigate the question of what happens to the resistance of a resistor if the temperature changes. Experiments in a course of instruction provide the opportunity for knowledge, skills and understanding to be acquired through investigations involving the 'real world'. For example, an investigation of an electric kettle used to boil water through measurements made of temperature and how it varies with time can lead to the acquisition of understanding of such concepts of sensible heat and latent heat and the consequences of such for the boiling of liquids.

The ability to make measurements is also needed by engineers in order to perform such tasks as marking out a piece of metal for machining, measuring the temperature to which a piece of metal has been heated for a heat treatment process, measurement of the current and/or voltage occurring in an electrical circuit.

This chapter is a consideration of the basic skills involved in carrying out experiments and making measurements. This section is concerned with the stages likely to be involved in an experiment, the handling of the results of measurements and the writing of reports of experimental work.

10.1.1 Experiments

The stages involved in devising an experiment are likely to be:

1 *Aim*

Define the aim of the experiment. What is the purpose of the experiment, what is to be found out? For example, the aim of an electrical experiment might be to find out whether a particular resistor obeys Ohm's law. Another experiment might be to find out how the extension of a spring is related to the forces used to stretch it.

2 *Plan*

When the aim is clear, the experiment needs to be planned. This means making decisions about such matters as what measurements are to be made, how the measurements are to be made and what instruments are needed. Thus with an experiment to find out whether a resistor obeys Ohm's law the measurements to be made are current and voltage. What size currents and voltage are likely to be involved and so what meters should be selected?

3 *Preparations*
Once the experiment is planned, preparations must be made to carry out the experiment. This involves collecting and assembling the required instruments and making certain you know how to operate them. For example, if the experiment requires the use of an instrument with a vernier, do you know how to read such a scale. Are there any health-and-safety factors which need to be taken into account?

4 *Preliminary experiment*
In some cases a preliminary experiment might be needed to find out whether the method you propose or the instruments you have selected are suitable. With the use of an ammeter and a voltmeter for the measurement of resistance, a preliminary experiment might be used to determine whether the current and voltage ranges of the instruments are adequate to enable the full range of measurements to be made.

5 *Doing the experiment and making measurements*
The next stage is to carry out the experiment, making such measurements as are required. All measurements should be recorded and details of the instruments used also recorded. For example, in the measurement of a temperature the instrument used needs to be recorded because it could have an impact on the data obtained. A mercury-in-glass thermometer, because it is relatively slow reacting, is likely to lag behind a fast-changing temperature change and so affect the results obtained whereas a thermocouple is much faster reacting. Thus the temperatures readings indicated by these could differ.

6 *Repeating measurements*
Often there can be a need to repeat some measurements in order to verify that the first set of results obtained can be reproduced and can thus be relied on. It is often worthwhile doing at least some preliminary analysis of the data before putting all the equipment away in order to check whether there is some oddity about perhaps some measurement or set of measurements and repetition is necessary. For example, with a multi-range ammeter you might have misread the range for some readings.

7 *Analysis of the data*
Following the results collection is analysis of the data to find out what the data tells you. For example in an experiment involving the measurement of the extension of a spring when different forces are used to stretch it, the analysis might consist of plotting a graph of extension against force in order to establish whether the extension is proportional to the force. See sections 10.1.2 and 10.1.3 for further discussion.

8 *Report*
Finally there needs to be a report of the experiment in which the findings are communicated to others. See section 10.1.4 for a discussion of report writing.

10.1.2 Measurement records

Consider a measurement of a quantity which is not expected to vary. This might be perhaps the measurement of the time taken for an object to fall through some fixed distance. All measurements are subject to errors. In this case if we suppose that a stopwatch was used, then the stopwatch button has to be pressed at the instant the object starts to fall and again pressed at the instant it has covered the specified distance. There is invariably some variability about the speed of response, or anticipation, of the person pressing the stopwatch button and so some error associated with the measurement. Thus one measurement might give a time of 3.1 s while a second measurement when the experiment is repeated gives 3.3 s. We might expect that with such a measurement the result will sometimes be too high and sometimes too low. If we take a number of such measurements and calculate the average value then we might assume that the average value is a better indication of the time than just one measurement. Thus for just the two readings of 3.1s and 3.3 s, the average is

$$\text{average} = \frac{\text{sum of the readings}}{\text{number of readings taken}} = \frac{3.1+3.3}{2} = 3.2 \text{ s}$$

Thus 3.2 s is more likely to be the true value than 3.1 s or 3.3 s. This assumes that the 3.1 s result was to low a time and the 3.3 s was too high a time. We cannot be sure of this and so it is wise to take more readings and the average of them. The more readings that are taken the more likely it is that the random variation that occurs between the readings will cancel out and the average value give an indication of what might be called the true value. Suppose we obtain ten measurements of the time, the table of results might look like:

Time in s	3.1	3.3	3.1	3.0	3.3	3.0	3.4	3.2	3.4	3.0

The average value is the sum of these measurements, i.e. 3.1 + 3.3 + 3.1 + 3.0 + 3.3 + 3.0 + 3.4 + 3.2 + 3.4 + 3.0 = 32.0 divided by the number of readings, in this case 10. Thus

$$\text{average} = \frac{32.0}{10} = 3.2 \text{ s}$$

This 3.2 s is a more reliable value for the time than any one of the readings taken.

A convenient way to present data, whether for repeated measurements of a single quantity or for where one quantity is related to another, is in the form of a table. This applies to both the rough notes made during the experiment and in the written-up report of the experiment. Tables must have clear headings which indicate the quantity for which the data is given and the units of that data. Thus for the time measurements given above, the table consists of a single row with the single row heading of time in the unit of seconds. Consider another measurement of the length of a rod. The results might be of the form:

Length in mm	20.1	20.2	20.0	20.3	20.1	19.9	20.3	20.1	20.2	20.0

This table indicates that all the readings are of length and in the unit of millimetres.

Now consider an experiment where the current through a resistor is measured for a number of voltages. The results might then be of the form:

Current in A	0	0.1	0.2	0.3	0.4
Voltage in V	0	1	2	3	4

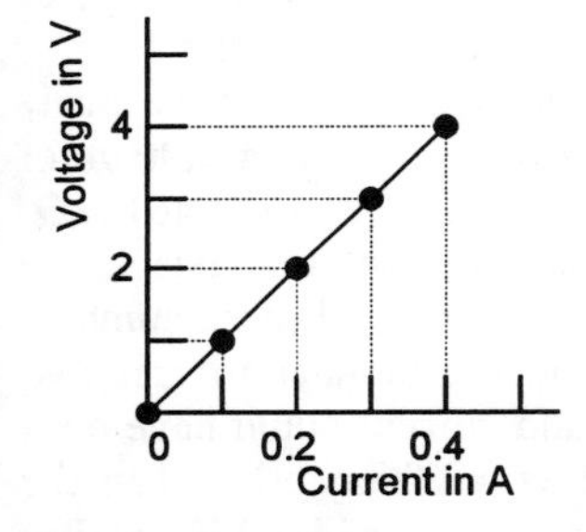

Figure 10.1 *Voltage–current graph*

The first row is of current readings and all the results are in units of amperes. The second row is of voltage readings and all the results are in units of volts. When the current is 0 the voltage is 0, when the current is 0.1 A the voltage is 1 V, etc.

To see the relationship between the current and voltage data in the above table, a graph can be used. A graph gives a pictorial representation of the data and relationship or trends in the data can be often more easily seen with a graph than by looking at a table of data. Figure 10.1 shows such a graph. It is also possible with a graph to more easily see if some data points do not follow the general trend shown by the majority of the points. With Figure 10.1 all the data points fall on the graph line. But suppose we had the following data:

Current in A	0	0.15	0.25	0.50	0.60
Voltage in V	0	1	2	3	4

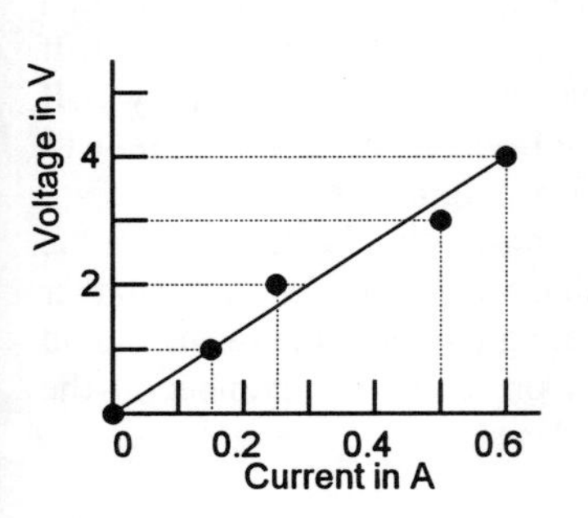

Figure 10.2 *Voltage–current graph*

The points would be as shown in Figure 10.2. In that situation, since we anticipate the result being a straight line, we can draw the 'best' straight line through the points. The best line is the one for which the data points are reasonably equally scattered either side of the line. With the line drawn, the data point at 0.25 A is about as far above the line as the data point at 0.50 A is below it. The data point at 1 V is about as far to the left of the line as that at 3 V is to the right. The other points fall on the line.

Graphs enable data values to be obtained by interpolation and extrapolation. *Interpolation* is when we use a graph to find data points on the graph line within the range of the data. Thus for Figure 10.2 if we wanted a likely value for the current when the voltage is 2.5 V, then we can look up that value on the graph and obtain a value of 0.375 A. *Extrapolation* is when we extend a graph line beyond the range of the data in order to obtain an estimate of what a data point might be if it can be assumed that the same trend continues. Thus if we take the graph in Figure 10.1 we might use it to estimate what the current might be when the voltage is 5 V. This is a voltage value outside the limit of the data giving

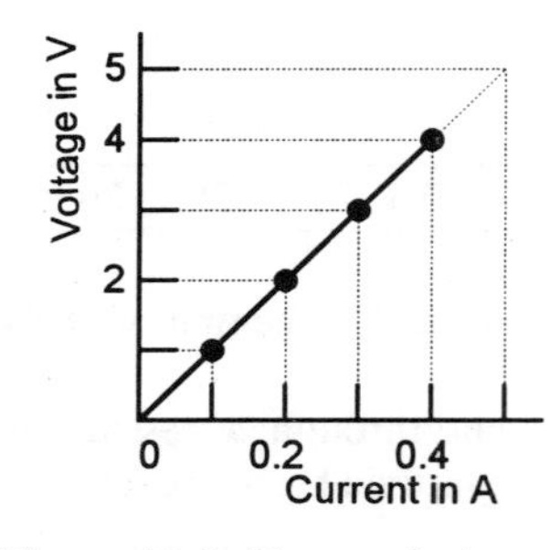

Figure 10.3 *Extrapolation*

the graph and so we have to extrapolate the line to that value. Figure 10.3 shows the extrapolation. The result is a current of 0.5 A.

When quoting measurement values it is important to realise that when a measurement is quoted as, say, 2.0 V, that this implies that the actual value lies between 1.95 V and 2.04 V. The 0 is assumed to have significance in that it implies we could have had a reading of perhaps 1.9 V or 2.1 V. The value of 2.0 indicates that it is nearer to 2.0 than either of the possible readings on either side of it. Writing a value as 2.0 is to give it to two *significant figures*. A value of 2.12 V lies between 2.115 V and 2.125 V and is being quoted to three significant figures. Data should only be quoted to the number of significant figures with which the measurement can be made. Thus if a meter can only give a reading to one decimal place, e.g. 1.9 V, then it should not be quoted as 1.90 because this would imply a greater accuracy than the measurement justifies. When a value is quoted as 0.00102 A then the number of significant figures is the number of figures between the first non-zero figure and the last figure, i.e. the 1 to the 2.

When the result of a calculation produces a number, it may have many figures. Thus for example, the resistance obtained by dividing a voltage of 2.0 V by a current of 0.13 A is, using my calculator, 15.384615 Ω. However, because we only had the voltage and current readings to an accuracy of two significant figures, it is pointless quoting all these numbers in the resistance value and it is better to round the number to just two significant figures. This is because the voltage and current could have been slightly different without them being recorded as anything other than the values given. Thus they might really have been 2.03 V and 0.129 A. The resistance would then have been given by my calculator as 15.736434 Ω. Rounding to two significant figures involves looking at the third figure and deciding what clue it gives as to the possible number that should be quoted for the second figure. With 15.384615 the third significant figure is the 3. If the third figure is 5 or greater then the second figure is rounded up by 1. If the third figure is less than 5 the second figure is left alone. With 15.384615 this is the case and so the result to two significant figures is 15 Ω.

When multiplying or dividing numbers, the result should be given to the same number of significant figures as the value which has the least number of significant figures. When adding or subtracting numbers the result should be rounded to the same number of decimal places as the number in the calculation with the least number of decimal places.

Example

What are the number of significant figures in the distance measurements of (a) 2.01 m, (b) 1.0 m, (c) 0.123 m?

(a) The number of significant figures is 3.
(b) The number of significant figures is 2.
(c) The number of significant figures is 3.

Example

Write down the answer of the following calculations to an appropriate number of significant figures:
(a) 1.1×2.1, (b) 2.0×4.103, (c) $12 + 2.5$.

(a) Using a calculator gives 2.31. Since each of the values is quoted to two decimal places then the result should also be quoted to that number. Hence the result is 2.3, there being no rounding up.
(b) Using a calculator gives 8.206. Since the value with the least number of significant figures has two then the result should be quoted to two. Hence the result is 8.2, there being no rounding up.
(c) Addition gives 14.5 The number with the least number of decimal places is the 12. So the result should be quoted to the same number and be, with rounding up, 15.

10.1.3 Spreadsheets

The term *spreadsheet* is used for computer programs that organize data in the form of a table. Essentially a spreadsheet consists of a large number of boxes, or *cells* as they are commonly termed, arranged in vertical columns and horizontal rows. Each cell in the table has a unique address. Columns are identified by the letters of the alphabet A, B, C, etc. and rows by numbers 1, 2, 3, 4, etc. After 26 columns the identifications continues AA, AB, AC, etc. Figure 10.4 shows the start of such a table with the addresses of some cells. Cell addresses are always specified by first the letter of the column and then the number of the row. Thus, for example, B2 refers to the cell in the B column and row 2. A spreadsheet is thus just a table of cells with unique addresses for each cell. The importance of this table is that data can be entered into cells and formulas written for the manipulation of the data.

	A	B	C	D	etc.
1	A1	B1	C1	D1	
2	A2	B2	C2	D2	
3	A3	B3	C3	D3	
4	A4	B4	C4	D4	
5	A5	B5	C5	D5	
6	A6	B6	C6	D6	
7	A7	B7	C7	D7	
etc.					

Figure 10.4 *Cell addresses*

The data that can be entered into a cell may be an item of text, a numerical value or a formula involving numerical values held in other cells. The formula may involve such processes as addition, subtraction, multiplication and division. Thus, for example, a formula might be entered in cell B5 that it should display the sum of the numerical values in cells B1, B2, B3 and B4. The methods used to write such a formula depends on the software used. With my software I can write it as:

sum (B1:B5) *or* = sum (B1..B5) *or* @ sum (B1..B5)

or B1 + B2 + B3 + B4 *or* = B1 + B2 + B3 + B4

The : or the .. is used to indicate that it is all addresses in the column between B1 and B5. For subtraction of, say, B2 from B1 I can write

B2 – B1

For multiplication of, say, B1 and B2 I can write

B1*B2

The * sign is used to indicate multiplication. If I want a percentage of some value, say B1, I can write

B1*40%

If I want to divide, say, B2 by B1 then I can write

B2/B1

Brackets can also be used in formulas to indicate an order of operations. Thus, for example,

(10*B2)–B1

means that the data at address B2 is multiplied by 10 and then has the data at address B1 subtracted from it.

Consider an example of where a number of measurements are made of some quantity and the average is required. What we might then have is a spreadsheet like that shown in Figure 10.5. Column A is used to enter text giving labels indicating which measurement we are referring to. Column B contains the data. Cell B1 is used for a label. Then cell B2 contains the results of measurement 1, cell B3 the result of measurement 3, cell B4 the result of measurement 4, etc. Cell B7 has the formula sum(B2:B6) entered into it using, with my software, a special pull-down box called Edit Formula. In other programs a special key might be used to indicate that a formula is being entered in a cell. When the data is entered into those cells the sum appears in B7. If the data in any one of the cells is changed, the sum changes. Cell B8 has the formula B7/5 entered, using the Edit Formula

(a)

	A	B
1		Length in mm
2	Measurement 1	B2
3	Measurement 2	B3
4	Measurement 3	B4
5	Measurement 4	B5
6	Measurement 5	B6
7	Sum of measurements	sum(B2:B6)
8	Average	B7/5

(b)

	A	B
1		Length in mm
2	Measurement 1	10.1
3	Measurement 2	10.2
4	Measurement 3	10.4
5	Measurement 4	10.3
6	Measurement 5	10.1
7	Sum of measurements	51.1
8	Average	10.22

Figure 10.5 *(a) The form of the spreadsheet with the addresses for data and the formulas that are to be entered into cells, (b) the result when data is put in the cells.*

box, into it. It thus displays the average of the five readings. The formulas for B7 and B8 could have been combined in a single formula so that the average appears without any preliminary total appearing.

Figure 10.6 shows another example involving measurements being made of the voltage across a resistor and the current through it at a number of currents. The resistance is determined for each of the sets of values. Row 1 is used for labels to indicate which column is voltage, which current and which resistance and the units of these quantities. Column A is the voltage, Column B the current and column C the resistance obtained by dividing the voltage value by the corresponding current value. Row 2 then contains the voltage and current values for the first current measurement. Row 3 contains the voltage and current values for the second current measurement. The formula used is the address is column A divided by the corresponding address, i.e. the same row, in column B. As the result indicates, the resistance changes as the current changes.

(a)

	A	B	C
1	Voltage in V	Current in A	Resistance in Ω
2	A2	B2	A2/B2
3	A3	B3	A3/B3
4	A4	B4	A4/B4
5	A5	B5	A5/B5
6	A6	B6	A6/B6

(b)

	A	B	C
1	Voltage in V	Current in A	Resistance in Ω
2	0.5	0.04	12.5
3	1.2	0.14	8.57
4	1.7	0.24	7.08
5	2.2	0.32	6.88
6	2.8	0.55	5.09

Figure 10.6 *(a) The addresses and formulas, (b) the results with data.*

Such programs also tend to offer the facility of being able to extract data from a spreadsheet and turn it automatically into a graph or chart. Thus for Figure 10.6 the spreadsheet program could produce a graph of the data in column C plotted against that in column B. For the spreadsheet shown in Figure 10.5 we might want a chart showing the frequency with which the various length values occur in measurements.

10.1.4 Report writing

The aim of a report of an experiment is to communicate to readers what the experiment was about, how it was carried out and the findings. A report should be easy to read and be complete but concise. The layout of a report can considerably affect its clarity and ease of reading. A structured layout, with headings, is thus generally used. The following are typical report sections and a form of structure. In some cases, it may be appropriate to combine some sections under a single heading.

1 *Title*
The report needs a clear and concise title which indicates what the experiment was about. For example, a title might be: The determination of the relationship between force and extension for a spring.

2 *Introduction*
This gives the background to the experimental investigation and the particular aims of the investigation. For example we might have the following for the experiment on the determination of the relationship between force and extension for a spring. When forces are applied to a spring it increases in length. The aim of this experiment is to investigate the relationship between the extension and the force and, in particular, to determine if the extension is proportional to the force.

3 *Method*
This section is devoted to a clear statement of how the experiment was carried out and the materials and instruments used. Any problems encountered should be given. Diagrams can be used to make the explanation clearer.

4 *Theory*
A statement of any theory or equation relevant to the investigation.

5 *Results*
Details of the data obtained.

6 *Interpretation of the results*
This section deals with the analysis of the data. Thus in the case of the investigation to find the relationship between the force and extension of a spring, the data might be plotted as a graph to see whether a straight line graph through the origin is obtained.

6 *Conclusions*
This section refers back to the purpose of the experiment and indicates how far the experiment has gone to meet that purpose. Thus in the case of the spring, it would indicate whether a relationship has been found between the force and the extension and what it was.

For a long report there may be an abstract immediately following the title, giving an overview of the experiment and its findings so that a reader can quickly determine what was done, why it was done, and the main findings.

In writing a report the normal style is to write in, what is termed, the third person. This means writing in the form – 'The current was measured for voltages of 1 V, 2 V, 3 V and 4 V' – rather than in the first person as – 'I measured the current at voltages of 1 V, 2 V, 3 V and 4 V'. Reports are generally written for the most part in the past tense, i.e. a statement of what was done and not as though it had been written while you were doing the experiment. Thus there might be – 'Measurements were made of the current at a number of voltages'. However, in the interpretation of the results we might have – 'The graph indicates that the current is proportional to the voltage'. When abbreviations or symbols are first used, they should be explained.

10.2 Measurement of length

The basic tools used by engineers for the measurement of length are the engineer's rule, vernier callipers and the micrometer screw gauge.

The *engineer's rule* is made from stainless steel or chrome-plated carbon steel with graduations which are engraved for accuracy and clarity. The edges of the rule are ground so that it can also be used as a straight edge. The end from which measurements are made, i.e. the 0 end, is also ground. Such rules are made to high standards of precision and need careful looking after. They are, however, only used for quick measurements where high accuracy is not required. For higher accuracy vernier instruments or micrometer screw gauges are used.

All *vernier instruments* rely on very accurately engraved scales. The main scale is marked in standard increments like the engineer's rule. The vernier scale which slides along the main scale is marked with divisions slightly smaller than the main scale divisions. For example, the main scale of a vernier may be marked in 1 mm increments and the vernier scale in increments of 0.9 mm. This enables the instrument to be used to read to 0.1 mm, the difference between these two scale divisions (Figure 10.7).

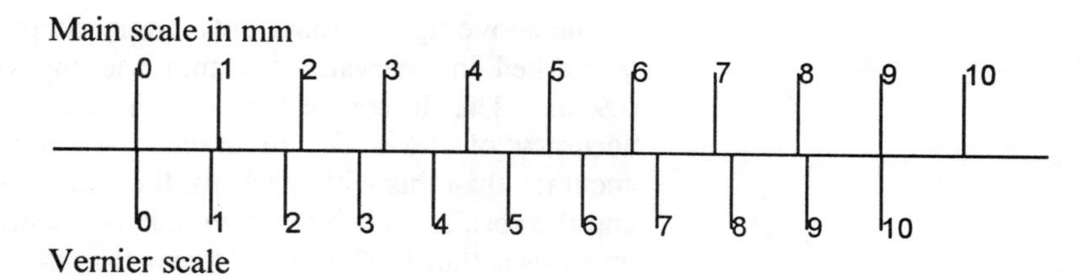

Figure 10.7 *A vernier scale to read to 0.1 mm.*

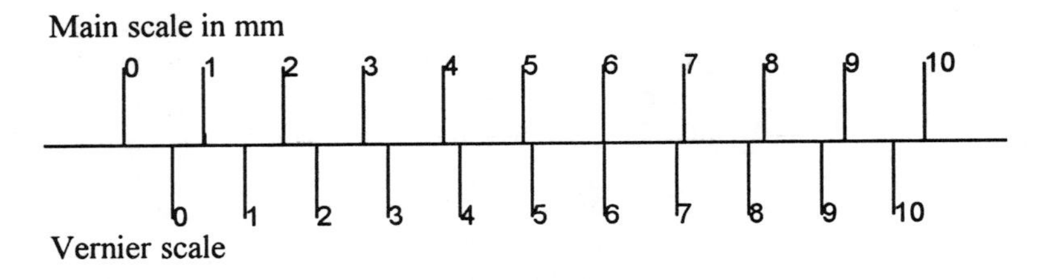

Figure 10.8 *Scale reading 0.6 mm*

With Figure 10.7 the 0 on the vernier scale coincides with the 0 on the main scale. The reading of such a scale is 0.0 mm. Figure 10.8 shows the scales when the 0 on the vernier scale is between the 0 and the 1 on the main scale. This means that the length is between 0 and 1 mm. The mark on the vernier scale that is directly opposite one on the main scale is the 6. Thus the reading of such a scale is 0.6 mm.

To read a vernier scale the procedure is:

1. Read the main scale reading immediately prior to the length being measured.
2. Add to that total the vernier reading indicated by the mark on the vernier scale which is exactly opposite a division on the main scale.

As a further illustration, Figure 10.9 shows a reading given by another length. The 0 on the vernier scale is between the 11 and 12 on the main scale. The vernier mark of 2 is exactly opposite a mark on the main scale. Thus the reading is 11.2 mm.

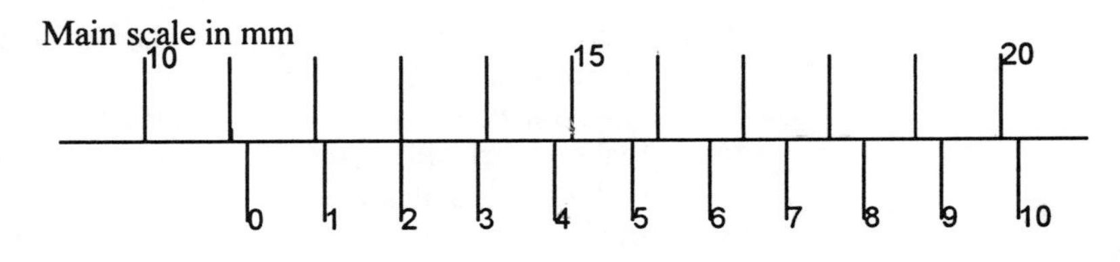

Figure 10.9 *Vernier scale reading 11.2 mm*

The above figures show only a very simple vernier scale. The main scale is marked in intervals of 1 mm and the vernier scale in increments of 0.9 mm. The difference between them is 0.1 mm and this is the reading accuracy of the scale. In practice we are more likely to have greater accuracy than this with, perhaps, the main scale marked in 1 mm increments and the vernier scale marked in 0.98 mm increments. The difference in scale intervals is thus 0.02 mm and this is thus the reading accuracy of the scale.

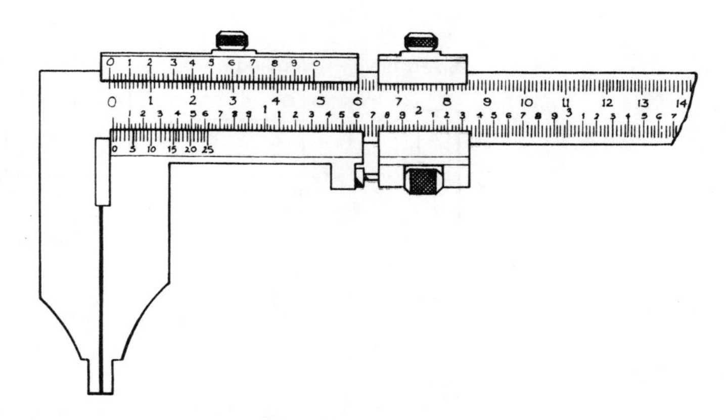

Figure 10.10 *Vernier callipers*

Vernier callipers (Figure 10.10) are an example of an instrument for the measurement of length which uses a vernier scale. The main scale is marked on the main body of the instrument and the vernier scale on the sliding jaw assembly. There are two main scales and vernier scales on the instrument shown, one metric and the other in inches. When used for external measurements (Figure 10.11(a) the jaws are positioned either side of the object to lightly grip it and then the reading taken. When used for internal measurements, e.g. the internal diameter of a tube, the jaws are positioned so that their external surfaces lightly press against the inside of the object (Figure 10.11(b)). The reading taken must have added to it the total jaw thickness.

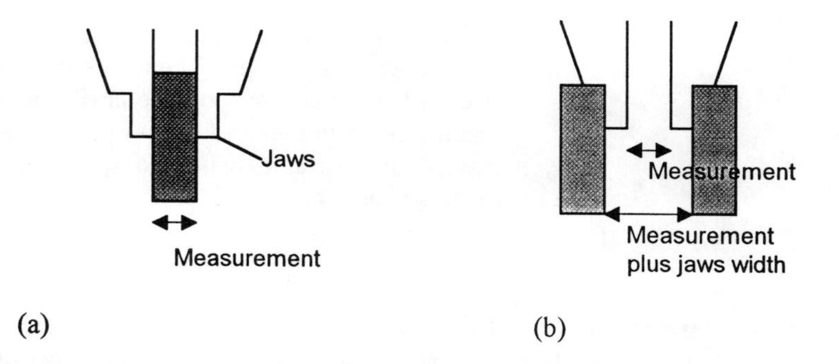

Figure 10.11 *(a) External, (b) internal measurements*

Vernier scales also occur on many other measurement instruments. They are used, for example, with vernier protractors for the measurement of angles, vernier height gauges for the measurement of heights, machine tools for the measurement of linear and angular movements, etc.

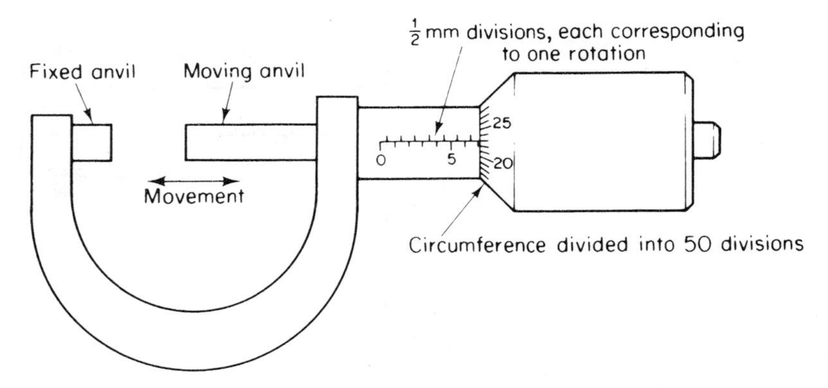

Figure 10.12 *Micrometer screw gauge*

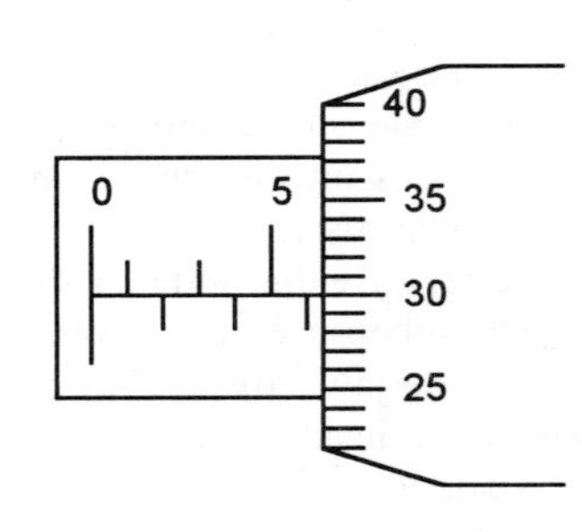

Figure 10.13 *Micrometer reading*

The *micrometer screw gauge* (Figure 10.12) depends on the measurement of the angular rotation of a precision screw. In a metric micrometer, the screw thread has a pitch of 0.50 mm. Thus one revolution of the screw thread results in a linear movement of 0.50 mm. The barrel of the micrometer is marked in 0.50 mm increments, the marking being placed alternatively above and below a line for ease of reading. The thimble is marked so that its circumference is divided in 50 divisions. One rotation of the thimble corresponds to 0.50 mm so one such division is 0.01 mm. Thus with the scales in the positions shown in Figure 10.13, the reading is 6.30 mm. When using a micrometer, care must be taken to avoid straining the frame of the instrument by tightening up too tightly the anvils against the object being measured. The micrometer movable anvil should be rotated up against the object until the ratchet, at the end of the micrometer gives two or three clicks.

The micrometer is a precision instrument and should be kept clean and in its case when not in use. The anvils should be kept slightly open. Before it is used for a measurement, the anvils should be closed and the instrument reading checked taken. If it does not indicate zero then the barrel should be rotated in its frame, with the special C-spanner provided, until a zero reading is obtained.

10.3 Measurement of weight and mass

This section is concerned with the basic instruments used to measure the weight, or mass, of an object. An object of mass m is acted on by a force, generally termed its weight, of mg as a result of gravity. g is the acceleration of free fall due to gravity. Basic instruments used for the measurement of weight are the spring balance and the lever arm balance. While such instruments are generally given scales in mass units, they are really measuring the weight and a value for the acceleration due to gravity has been assumed. For the measurement of mass an analytical balance is commonly used. Such an instrument does not assume a value for the acceleration due to gravity.

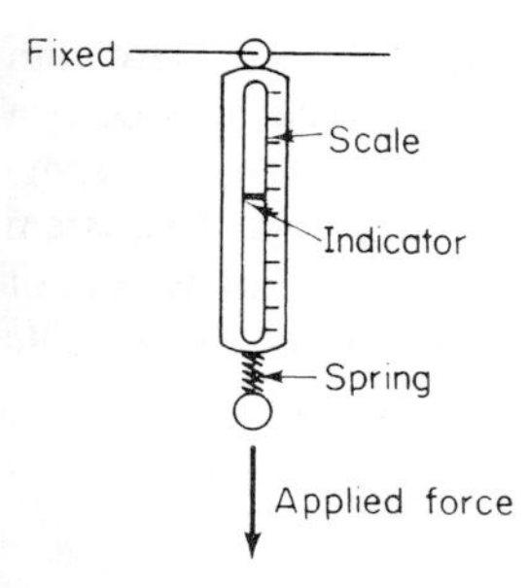

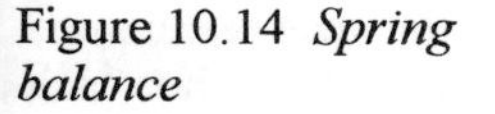

Figure 10.14 *Spring balance*

The *spring balance* (Figure 10.14) is a simple instrument that can be used for the measurement of weight, i.e. the gravitational force pulling down on a mass. It depends for its action on the extension of a spring being proportional to the force used to extend it. Direct reading spring balances are not capable of high accuracy as the extensions are relatively small.

The *lever arm balance* (Figure 10.15) is, as the name implies, a lever system in which the unknown mass is placed in the balance pan and the gravitational force acting on it causes the lever arm to swing out over the scale until the moment supplied by the counterweight on the lever arm balances the moment due to the mass on the balance pan. As the lever arm rotates the counterweight is moved further and further away from the pivot. Different ranges can be selected by moving the counterweight.

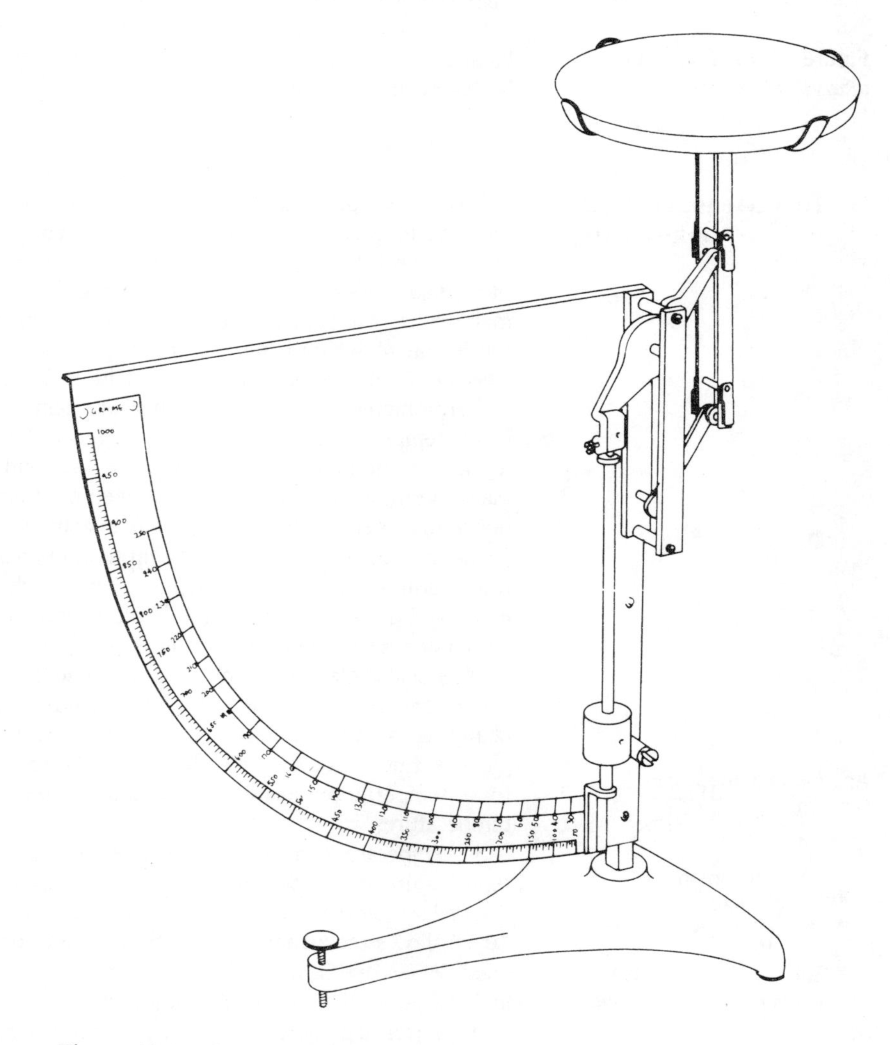

Figure 10.15 *Lever arm balance*

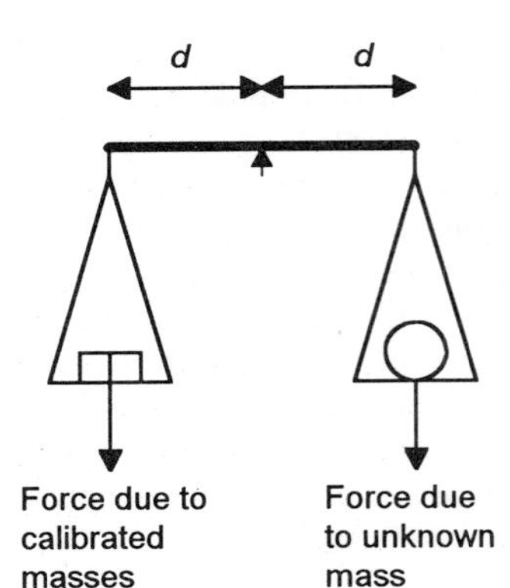

Figure 10.16 *The basic analytical balance*

The basic *analytical balance* (Figure 10.16) is essentially a seesaw with the object, for which the mass m_u is required, placed on one balance pan and then calibrated masses m_c placed on another balance pan, the same distance from the pivot as the unknown mass on its pan, until the system balances. When this occurs then the clockwise moment due to the weight of the unknown mass is balanced by the anticlockwise moment of the weights of the standard masses in the other pan. We thus have

$$m_u g \times d = m_c g \times d$$

Hence $m_u = m_s$ and its does not depend on the value of the acceleration due to gravity. Such balances are capable of determining masses with a high degree of accuracy.

Spring balances tend to be used for rough measurements, lever arm balances for measurement of greater accuracy and analytical balances for higher accuracy measurements.

10.4 Measurement of temperature

Basic methods used for the measurement of temperature are the mercury-in-glass thermometer and thermocouples.

The *mercury-in-glass thermometer* depends for its action on the volume of a fixed amount of mercury increasing as the temperature increases. The mercury is constrained to expand along a length of capillary tube and thus the height to which it moves up the tube is a measure of the temperature. The thermometer will have a range within the limits −35°C to +600°C. Such a thermometer, for use in the measurement of temperatures round about room temperature, might have a range of 0 to 40°C with scale divisions marked in intervals of 0.1°C. Because the readings given by a mercury-in-glass thermometer depend on how deep the thermometer is immersed in the hot liquid, thermometers are usually specified as having been calibrated for partial immersion, i.e. immersion up to some particular point on the stem of the thermometer, total immersion in which the thermometer is used immersed up to the level of the liquid in the thermometer stem or complete immersion when the thermometer is completely immersed.

Mercury-in-glass thermometers are widely used in laboratories for the measurement of the temperatures of liquids when the temperatures are not changing rapidly. Because of the bulk of the thermometer and the fact that glass is a poor conductor of heat, such thermometers take a while to react to a temperature change. They thus cannot be used for fast-changing temperatures.

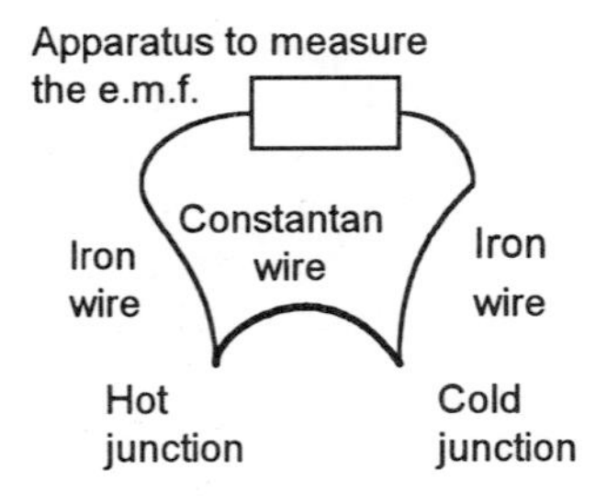

Figure 10.17 *Thermocouple*

A temperature measuring device that can be used for fast-moving temperatures is the *thermocouple*. A thermocouple consists basically of two dissimilar metal wires connected together to form two junctions. Figure 10.17 shows the arrangement with iron and constantan being the two wires used. When one of the junctions, the so-termed cold junction, is kept at 0°C and the other, the so-termed hot junction, is at some different temperature a voltage (termed the thermoelectric e.m.f.) is produced which is related to the temperature of the hot junction. This e.m.f. can be measured by suitable apparatus and displayed as perhaps a digital display or a pointer moving across a scale. The voltage to be measured is rather small and so a microvoltmeter is required (note, 1 microvolt is one millionth of a volt). The thermocouple is monitoring the temperature at the point where the two wires form a junction and so can be used for the measurement of highly localised temperatures. Also, because the mass of material at a junction is very small and metals are good conductors of heat, it does not take much to bring them up to temperature and so they can respond very quickly to temperature changes.

Depending on the materials used for the thermocouple wires, the range covered can vary within about −200°C to +1700°C. For example, copper with constantan can be used for −100°C to +400°C, iron and constantan for −100°C to +800°C.

10.5 Measurement of time

The basic instrument used for the measurement of time is the *stopwatch*. The mechanical stopwatch relies on the regular oscillations back and forth of the watch balance wheel, each oscillation being used to rotate, by a slight amount, the watch hand round the watch face. The stopwatch can be made to start and stop by a button being manually pressed, hence a time interval can be measured. Electronic clocks are available which can be started or stopped by either manual means or suitable electrical signals. Such clocks rely for their operation on the use of the mains alternating voltage or the generation of a suitable alternating voltage by an electronic circuit in the clock.

One of the problems of using manual methods to start and stop a clock is the fact that people have reaction times. Thus when an event occurs, a significant amount of time can occur before a person reacts and starts, or stops, the clock. The reaction time depends on a number of factors, not least the individual concerned and their state of alertness at the time.

10.6 Measurement of electrical quantities

The measurement of electrical quantities discussed in this chapter are the basic methods for the measurement of current, voltage and resistance, with a discussion of the cathode ray oscilloscope and some of the measurements it can be used to make.

10.6.1 Measurement of current and voltage

The basic instrument used for the measurement of current is the *moving coil galvanometer*. This consists of a coil of wire in a magnetic field provided by a permanent magnet (Figure 10.18). When there is a current through the coil it rotates and in doing so moves a pointer over a scale. This basic meter movement is likely to give a full scale deflection (fsd) for a current of a few millamperes or less. In order to enable larger currents to be used, a resistor is connected in parallel with the instrument so that only a percentage of the current passes through the meter, the rest passing through

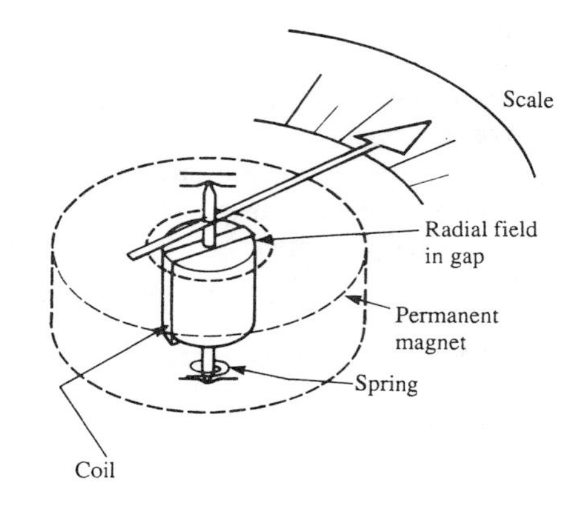

Figure 10.18 *Basic meter movement*

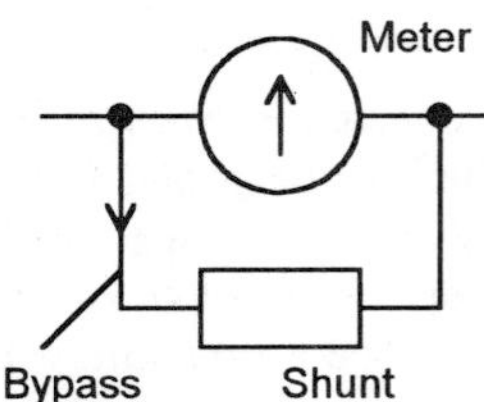

Figure 10.19 *A shunt*

a bypass resistor (Figure 10.19). Such a resistor is called a *shunt*. A multi-range meter will have a number of such shunts and they can be selected by rotating a dial on the instrument or using a particular pair of input terminals.

For example, with a basic meter movement which has a full range deflection of 1 mA, a shunt that takes 99% of the incoming current will give a full scale deflection with 100 mA since only 1 mA will actually pass through the meter. With a shunt that takes 99.9% of the incoming current, the meter will have a full scale deflection with 1000 mA, i.e. 1 A, since only 1 mA will pass through the meter.

The moving coil meter is a direct current meter and cannot be used for the measurement of alternating current without some additional circuitry. Such circuitry rectifies the alternating current, i.e. transforms it into an equivalent direct current.

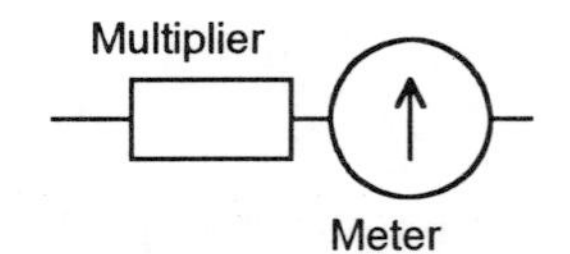

Figure 10.20 *A multiplier*

The moving coil meter can be used as a direct voltage voltmeter by using it to measure the current through a fixed resistor. Since the voltage is proportional to the current, then the meter gives a measure of the voltage. Different ranges for the meter are given by using different resistances, such resistors being placed in series with the instrument (Figure 10.20). Such resistors are called *multipliers*. Such instruments are typically used for voltage ranges from about 0–50 mV to 0–100 V. For alternating voltages, the meter can be used with a rectifier circuit to convert the alternating current into an equivalent direct voltage.

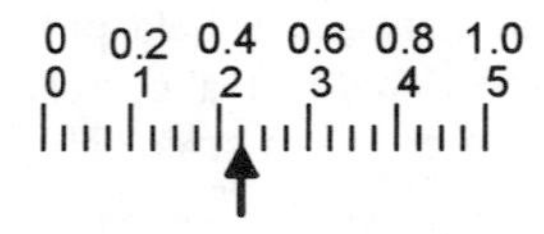

Figure 10.21 *Example*

Example

A multimeter is used to measure current. The meter has two current ranges 0 to 1 A and 0 to 5 A. The selection switch is set to 5 A. With the reading indicated in Figure 10.21, what is the value of the current?

The reading is 2.25 A.

10.6.2 Measurement of resistance

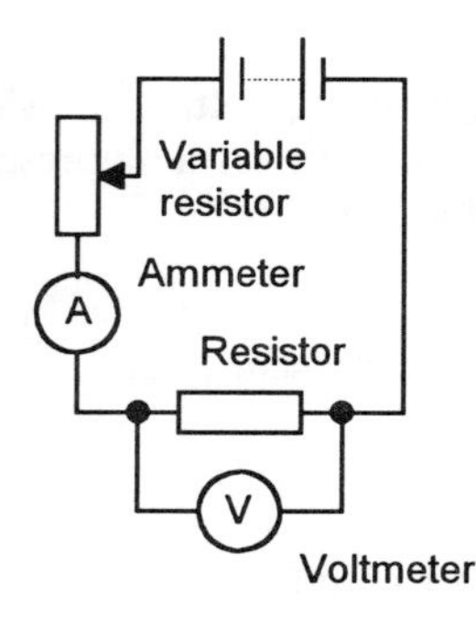

Figure 10.22 *The determination of voltage-current relationships*

Figure 10.22 shows the basic circuit that can be used to determine the relationship between the voltage across a resistor and the current through it. For a resistor that obeys Ohm's law, the voltage is proportional to the current and so the result is a straight line graph passing through the zero voltage–zero current point. The gradient of the graph, i.e. V/I, is the resistance.

When a resistor does not obey Ohm's law then the result is not a straight line through the origin. For such a situation, the results of the voltage and current values might be tabulated in columns A and B and column C give the resistance calculated for the voltage–current values. The outcome may then be shown as a graph of the resistance plotted against the current.

When a rough value of the resistance of a resistor is required, a multi-range meter might be used with the range switch set to resistance. Before the reading is taken, the zero resistance condition must be set for the instrument. This involves connecting the two terminals together by a piece of copper wire so that there is effectively no resistance between them (Figure 10.23(a)). Then the zero adjustment knob on the meter is rotated until the meter reading is zero ohms. Now when the resistor is connected between the terminals of the instrument (Figure 10.23(b), the reading displayed on the appropriate scale, is the resistance. The value given is not to any great accuracy and is often used to determine the order of magnitude of the resistance, or even whether there is a short circuit in some appliance or some resistance.

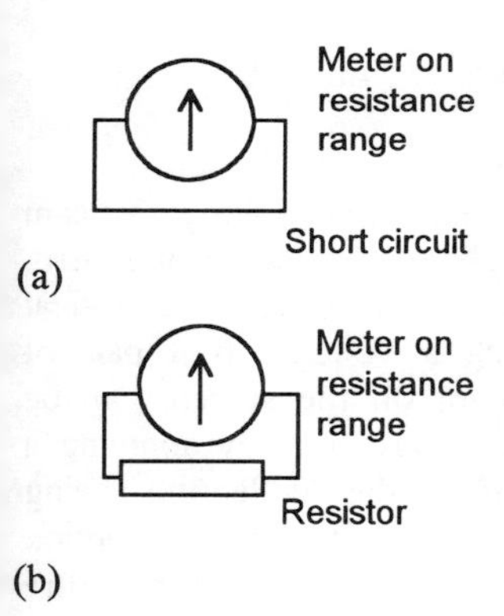

Figure 10.23 *Measuring resistance*

The resistance measurement with the multi-range meter works on the principle that there is a battery inside the instrument which supplies a current through the resistor. The meter monitors this current. The bigger the resistance of the resistor the smaller will be the current. Thus the size of the current is a measure of the resistance.

Example

A rough measurement of a resistor by means of a multi-meter indicates that the resistance is about 20 Ω. If you have to make measurements using the circuit shown in Figure 10.22 of the voltage across the resistor and the current through it for a number of currents, what range voltmeter and ammeter will be needed if the d.c. voltage supply is 3 V?

If the variable resistor is set to zero then the maximum current will occur in the circuit and all the voltage from the supply will be across the resistor. Thus the maximum voltage is 3 V and so a voltmeter should be selected which covers that range. Ideally the meter should have a range of 3 V so that all the voltage readings can be spread over the full width of the scale. The maximum current occurs when the voltage is a maximum. Thus the maximum current is $V/R = 3/20 = 0.15$ A = 150 mA. Thus a meter should be selected to cover this range. Ideally the meter should have a range of 150 mA so that all the current readings can be spread over the full width of the scale.

10.6.3 The cathode ray oscilloscope

The *cathode ray oscilloscope (CRO)* is essentially a voltmeter which displays a voltage as the movement of a spot of light on a fluorescent screen. Figure 10.24 shows the basic structure of a CRO.

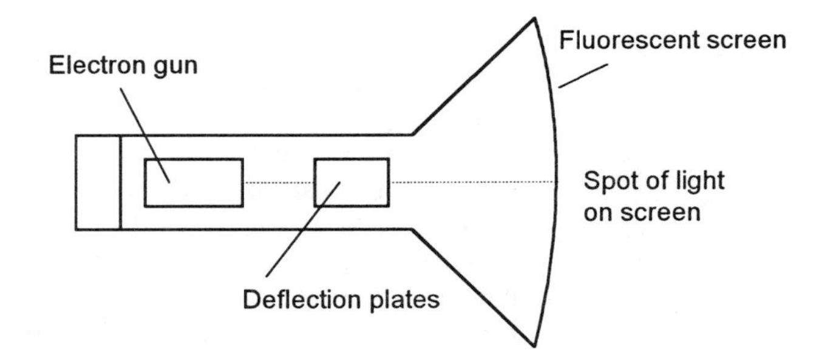

Figure 10.24 *Basic structure of the cathode ray oscilloscope*

Electrons are produced in an electron gun assembly which directs a beam of them along the tube to produce a spot of light on a fluorescent screen. The beam, and hence the spot on the screen, can be deflected in a vertical direction, termed the Y-direction, by applying a voltage to a pair of deflection plates. The beam, and hence the spot on the screen, can be deflected in a horizontal direction, termed the X-direction, by applying a voltage to another pair of deflection plates, the deflection being proportional to the voltage. Thus with a voltage applied to the Y-deflection plates we might move the spot on the screen in the way shown in Figure 10.25(a), with the voltage applied to the X-deflection plates the movement might be in the way shown in Figure 10.25(b). The screen is marked with a grid of centimetre squares to enable such deflections to be easily measured. Thus in (a), if each square corresponds to a voltage of 1 V then the Y-input was 3 V, in (b) if each square corresponded to, say, 2 V then the X-input was 6 V.

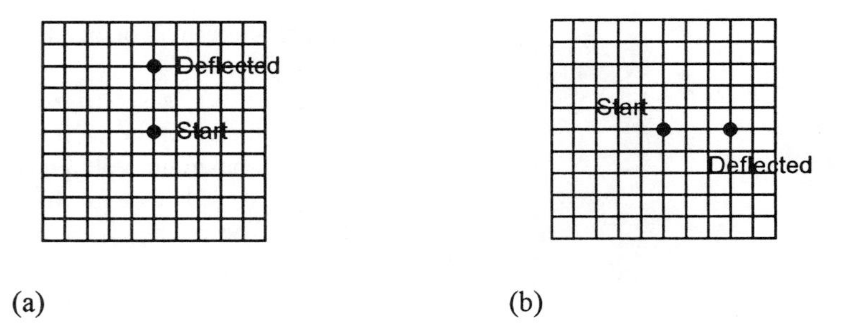

(a) (b)

Figure 10.25 *Spot deflected in (a) the vertical direction, (b) the horizontal direction.*

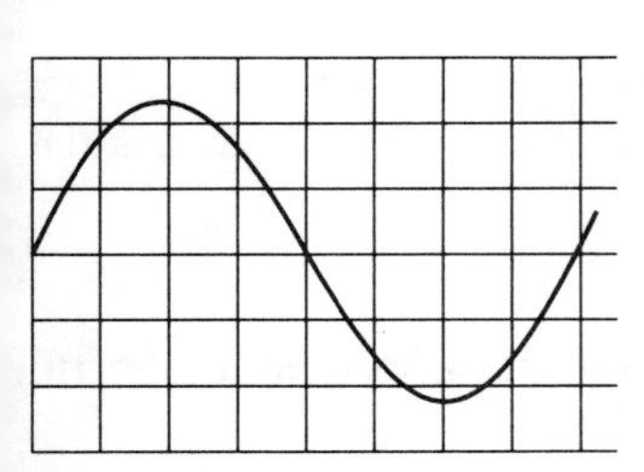

Figure 10.26 *Alternating voltage displayed*

If an internal signal, called the *time base signal*, is applied to the X-deflection plates then the spot on the screen can be made to move from left to right across the screen so that each centimetre takes the same time. At the end of its travel across the screen it flies back to the left at high speed in order to start its motion across the screen again. Thus with, say, an alternating voltage applied to the Y-deflection plates and a time base signal to the X-deflection plates, the screen displays a graph of input voltage against time. With an alternating voltage, and a suitable time base signal so that it takes just the right time to move across the screen and get back to the start on the left to always start at the same point on the voltage wave, the screen might look like that shown in Figure 10.26. From measurements of the displacement in the Y-direction to give the maximum displacement from the zero axis we can determine the value of the maximum voltage. By using a calibrated time base so that we know the time taken to cover each centimetre in the horizontal direction, we can determine the time taken to complete one cycle of the alternating voltage.

Figure 10.27 shows the controls likely to be found on a very basic cathode ray oscilloscope.

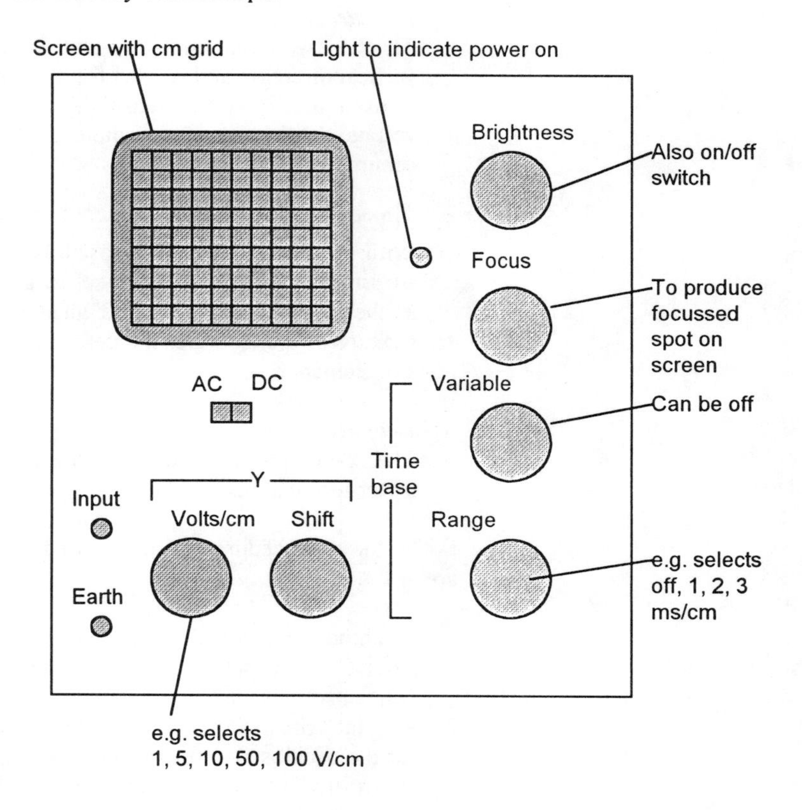

Figure 10.27 *Controls on a basic cathode ray oscilloscope*

The controls on the basic cathode ray oscilloscope are:

1 *Brightness*
This control is also generally used as an off/on switch. It adjusts the brightness of the spot on the screen.

2 *Focus*
This adjusts the electron gun so that is produces a focused, i.e. sharp, spot or trace on the screen.

3 *Time base*
This has two controls. With the variable knob set to off, the calibrated range switch can be used to select the time taken for the spot to move per cm of screen in the horizontal direction from left to right. The variable knob can be used to finely tune the time. Also, when the time base is switched off, the variable control can be used to shift the spot across the screen in the X-direction.

4 *Y-controls*
This has two controls. The Y-shift is to move the spot on the screen in a vertical direction and is used for centring the spot. The volts/cm control is used to adjust the sensitivity of the Y-displacement to the input voltage. With it set, for example, at 1 V/cm then 1 V is needed for each centimetre displacement.

5 *AC/DC switch*
With switch set to d.c. the oscilloscope will respond to both d.c. and a.c. signals. With the switch set to a.c. a blocking capacitor is inserted in the input line to block off all d.c. signals. Thus if the input was a mixture of d.c. and a.c. the deflection on the screen will be only for the a.c. element.

6 *Y-input*
The Y-input is connected to the oscilloscope via two terminals, the lower terminal being earthed.

The basic procedures to be adopted in order to operate an oscilloscope are:

1 Brightness control off. Focus control midway. Y-shift control midway. AC/DC switch set to DC. Y volts/cm control set to 1 V/cm. Time base control set to 1 ms/cm.
2 Plug into the mains socket. Switch on using the brightness control. Wait for the oscilloscope to warm up, perhaps a minute. Move the brightness control to full on. A bright trace should appear across the screen. If it does not, the trace may be off the edge of the screen. Try adjusting the Y-shift to see if it can be brought on to the screen.
3 Centre the trace using the Y-shift control. Reduce the brightness to an acceptable level and use the focus to give a sharp well defined trace.

4 Switch the time base control to the required range for the signal concerned. Apply the input voltage. Select a suitable number of volts/cm so that the trace occupies a reasonable portion of the screen and measurements can be made.

Example

A cathode ray oscilloscope is being used to determine the value of a direct voltage. With the Y volts/cm control set at 5 V/cm, connection of the voltage to the Y-input results in the spot on the screen being displaced through 3.5 cm. What is the applied voltage?

Since the spot displaces by 1 cm for every 5 V, a displacement of 3.5 cm means a voltage of $3.5 \times 5 = 17.5$ V.

Example

A cathode ray oscilloscope has an alternating voltage input. With the Y control set to give 1 V/cm and the time base at 1 ms/cm, the trace shown in Figure 10.28 was found. What is (a) the maximum voltage and (b) the frequency of the signal?

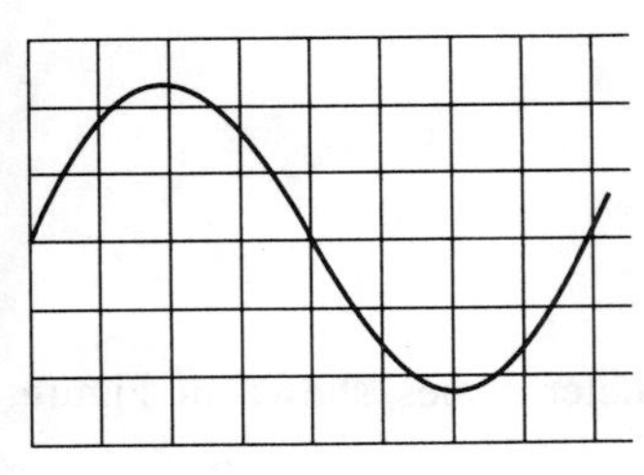

Figure 10.28 *Example*

(a) The alternating voltage oscillates between $+V_m$ and $-V_m$ about the central axis. The trace on the screen is of a wave which oscillates with a displacement which varies from +2.2 cm to −2.2 cm about the central line through the wave. Thus, since the control is set at 1 V/cm, the maximum voltage is 2.2 V.
(b) The number of centimetres needed for the wave to complete one cycle is 8 cm. This means one cycle takes 8 ms, i.e. 0.008 s. The frequency is the number of cycles completed per second and is thus $1/0.008 = 125$ Hz.

Problems

Questions 1 to 12 have four correct answer options: A, B, C and D. You should choose the correct answer from the answer options.

1 The following are a set of readings, in millimetres, of the width of a bar:

10.1, 10.2, 10.0, 10.4, 10.2, 10.1, 10.3, 10.0, 10.4, 10.3

The average value is:

A 10.1 mm
B 10.2 mm
C 10.3 mm
D 10.4 mm

Questions 2 to 4 relate to the following formulas used with spread-sheets.

A sum(B1:B4)
B B1*B4
C B1/B4
D B1 + B4

Select the formula that could be used to:

2 Multiply data at address B1 by that at address B4.
3 Divide data at address B1 by that at address B4.
4 Add the data at address B1, B2, B3 and B4.

5 The reading of the vernier scale shown in Figure 10.29 is:

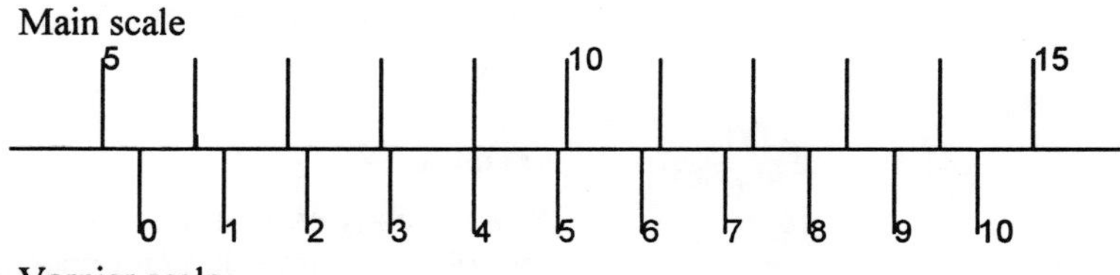

Figure 10.29 *Problem 5*

A 4.9
B 5.4
B 6.4
C 9.4

6 The reading of the screw gauge micrometer scales shown in Figure 10.30 is:

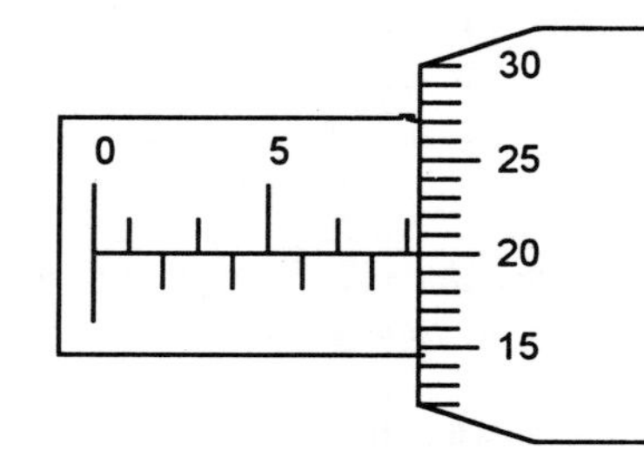

Figure 10.30 *Problem 6*

A 7.20
B 9.20
C 20.7
D 20.9

7 When vernier callipers are used to determine the internal diameter of a tube, the jaw width:

A Is ignored, having no effect on the value of the diameter
B Is subtracted from the reading given by the scale.
C Is added to the reading given by the scale.
D Is a scaling factor and multiplies the reading given by the scale.

8 Decide whether each of the following statements is TRUE (T) or FALSE (F).

The analytical balance:
(i) Depends for its action on the clockwise moment of the weight of the unknown mass balancing the anticlockwise moment of the weight of calibrated masses.
(ii) Can be used for the accurate measurement of mass.

A (i) T (ii) T
B (i) T (ii) F
C (i) F (ii) T
D (i) F (ii) F

9 Decide whether each of the following statements is TRUE (T) or FALSE (F).

A thermocouple has the advantages over a mercury-in-glass thermometer of being:
(i) Faster to respond to changing temperatures.
(ii) Able to measure the temperature at virtually a point.

A (i) T (ii) T
B (i) T (ii) F
C (i) F (ii) T
D (i) F (ii) F

10 Decide whether each of the following statements is TRUE (T) or FALSE (F).

A shunt is used with a moving coil galvanometer to:
(i) Enable the meter to be used as a voltmeter.
(ii) Enable the meter to be used for the measurement of larger current.

A (i) T (ii) T
B (i) T (ii) F
C (i) F (ii) T
D (i) F (ii) F

11 A basic moving coil galvanometer has a meter which gives a full scale deflection with 1 mA. When used with a shunt which enables 99.99% of the current to bypass the meter, the meter will give a full scale deflection for a current of:

A 0.01 A
B 0.1 A
C 1 A
D 10 A

12 With the Y volts/cm control of a cathode ray oscilloscope set to 5 V/cm a direct voltage deflects the trace on the screen by 2 cm. The voltage is thus:

A 0.4 V
B 2 V
C 5 V
D 10 V

13 With the time base control of a cathode ray oscilloscope set to 0.01 ms/cm a trace on the screen is found to take 4 cm to complete one cycle. The frequency of the voltage input is thus:

A 0.000 04 Hz
B 0.04 Hz
C 25 Hz
D 2500 Hz

14 The vertical position of a trace on the screen of a cathode ray oscilloscope is adjusted by using the:

A The variable time base control
B The range time base control
C The Y-shift control
D The Y volts/cm control

15 An alternating voltage has a frequency of 50 Hz. To obtain a trace showing two complete cycles on the screen of a cathode ray oscilloscope of width 10 cm, the time base control should be set to give:

A 20 ms/cm
B 25 ms/cm
C 40 ms/cm
D 50 ms/cm

16 The following are experiments which can be carried out involving the measurement techniques, data analysis and report writing techniques discussed in this chapter and also involve the science and mathematics concepts introduced in chapters 8 and 9. In each case, plan the experiment, identify the appropriate devices for making the measurements, carry out the measurements, record the results and write a report.

(a) Take a number of samples of different diameter bare copper wire. Measure the length of each sample, its diameter and its mass. Because the diameter may show some variations along its length you will need to take a number of diameter measurements and obtain an average.

Tabulate the results and hence determine the relationship between the mass of a sample and its volume. Note that the volume of a cylinder or length L and diameter d is $(\pi d^2/4)L$.

(b) Construct a simple pendulum by taking a length of string, about 1 m, attaching a small, but heavy, mass to one end and tying the other end to some support so that the pendulum can freely swing back and forth when the mass is displaced about 100 mm to one side. Determine the frequency of oscillation of the pendulum by taking a number of repeated measurements of the time taken for ten complete oscillations. Hence obtain an average time for ten oscillations and so obtain a value for the frequency.

(c) Determine, using a voltmeter and ammeter and a d.c. supply of about 3 or 4 V, the resistance of a fixed radio-type resistor (use one of about 25 Ω). Take readings of the voltage across the resistor at a number of different currents through it and use a graph of voltage against current in order to determine the resistance. Does the resistor obey Ohm's law?

(d) Determine, using a voltmeter and ammeter and a d.c. supply of about 3 or 4 V, how the resistance of a thermistor varies with the current through it. Take readings of the voltage across the thermistor at a number of different currents through it. Tabulate your results and determine the resistance at each current. Hence plot a graph of the resistance against the current. Does the thermistor obey Ohm's law?

(e) Determine, using a voltmeter and ammeter and a d.c. supply of about 3 or 4 V, how the resistance of a flashlamp bulb (e.g. a 1.25 V, 0.25 A bulb) varies with the current through it. Take readings of the voltage across the thermistor at a number of different currents through it. Tabulate your results and determine the resistance at each current. Hence plot a graph of the resistance against the current. Does the bulb obey Ohm's law?

(f) Take a beaker containing hot water (e.g. water at about 60°C) and, using a suitable thermometer, make measurements of the temperature of the water at different times and so plot a graph showing how the temperature of the water changes with time. Is the temperature drop from the initial value proportional to time?

(g) Take a beaker containing crushed ice and, using a suitable thermometer, make measurements of the temperature of the ice/water at different times and so plot a graph showing how the temperature of the ice/melted ice changes with time. Offer an explanation for the results.

(h) Take a beaker containing cold water and into it insert a 12 V electrical immersion heater and a thermometer. Connect the immersion heater via a switch to a suitable circuit so that the voltage across it can be measured and the current through it measured, selecting suitable range meters. Start a clock at the time the electrical heater is switched on and measure how the temperature changes with time for a time of about 5 minutes. Then switch the heater off but continue to monitor the temperature for another 2 minutes. Plot the results as a graph showing how the temperature varies with time. Explain the form of the resulting graph.

(i) Determine the average thickness of a sheet of metal from measurements taken at a number of points using a micrometer.

(j) For a batch of resistors, determine the resistance of each and hence obtain the average resistance value and the maximum spread of resistance values about the average. For a large number of resistors it could be convenient for the work to be divided among a number of teams and all enter their results into the same spreadsheet, then using the spreadsheet to determine the average and give a suitable graphical display showing how the values vary.

11 Engineering in society

11.1 Engineering technology

This chapter is a consideration of three key aspects of engineering technology, namely materials, information technology and automation, and their applications in home and leisure activities, industry and commerce, and health and medicine. The chapter can do no more than indicate some of the very numerous applications and their impact on our lives. Chapter 12 is concerned with the impact of engineering activities on the environment.

What do we mean by technology? Before attempting an answer to that question, consider what is meant by science. Science is interested in trying to answer the questions – 'What is nature like and why is it like that?' For example, science is trying to find out facts about materials and why they behave as they do. Technology on the other hand is concerned with how to use the resources of nature more effectively in order to satisfy the needs and desires of people. Thus technology is concerned with using materials to make products that people want. Science helps technology by supplying new information about nature, with technology aiding science by providing new instruments and techniques. For engineers science is a means to an end.

11.2 Materials

This section is broken down into two sections: a discussion of how materials have evolved over the years and the effects of their evolution on people, and a discussion of some examples of materials in use now.

11.2.1 The evolution of materials

The early history of the human race can be divided into periods according to the materials that were predominantly used. Thus we have the stone age, the bronze age and iron age.

In the *stone age* (about 8000 BC to 4000 BC) the people could only use the materials occurring around them such as stone, wood, clay, animal hides, bone, etc. The tools they made were limited to what they could fashion out of these materials. Thus tools were limited to those that could be made from stone, flint, bone and horn.

By about 4000 BC people in the Middle East were able to extract *copper* from its ore and it rapidly became an important material. Copper is a ductile material which can be hammered into shapes, thus enabling a greater variety of items to be fashioned that was possible with stone. Because the copper ores contained impurities which were not completely removed by the smelting, alloys were produced. It was soon realised that the deliberate adding of additives to copper could produce materials with improved properties. About 2000 BC it was found that when tin was added to copper an alloy was produced which had an attractive colour, was easy to form and harder than copper alone. This alloy was called *bronze*. Thus we have the *bronze age*.

About 1200 BC the extraction of *iron* from its ores signalled another major development and the *iron age*. Iron in its pure form was, however, inferior to bronze but by heating items fashioned from iron in charcoal and hammering them, a tougher material, called *steel*, was produced. Plunging the hot metal into cold water, i.e. quenching, was found to improve the hardness. It was also found that reheating and cooling the metal slowly produced a less hard but tougher and less brittle material, this process now being termed tempering. Thus *heat treatment processes* were developed.

The large-scale production of iron can be considered an important development in the evolution of materials in that it made the material more widely and cheaply available for products. Large-scale iron production with the first coke fuelled blast furnace started in 1709. Cast iron was used in 1777 to build a bridge at the place in England now known as Ironbridge. The term *industrial revolution* is used for the period that followed as the pace of developments of materials and machines increased rapidly and resulted in major changes in the industrial environment and the products generally available to people. 1860 saw the development of the Bessemer and open hearth processes for the production of steel and this date may be considered to mark the general use of steel as a constructional material. Aluminium was extracted from its ores in 1845 and produced commercially in 1886. In the years that followed, many new alloys were developed. The high strength aluminium alloy Duralumin was developed in 1909, stainless steel in 1913, high strength nickel-chromium alloys for high temperature use in 1931. Titanium was first produced commercially in 1948. High strength maraging steels were developed in 1961.

While naturally occurring *plastics* have been used for many years, the first manufactured plastic, celluloid, was not developed until 1862. In 1906 Bakelite was developed. The period after about 1930 is often termed the *plastic age*. This period saw a major development of plastics and their use in a wide range of products. Polyethylene is an example of a scientific investigation yielding a surprise result. In 1931 a Dutch scientist, Michels, was given approval by ICI to design apparatus which could be used to carry out research into the effects of high pressure on chemical reactions. In 1932 the investigation began and in March 1933 a surprise result was obtained. The chemical reaction between ethylene and benzaldehyde was being studied at a pressure of 2000 times the normal atmospheric pressure and a temperature of 170°C when a waxy solid was found to form. The material that had been formed was *polyethylene*. The experiment had not been designed to develop a new material but that was the outcome. The commercial production of polyethylene started in England in 1941. The development of *PVC* was, however, an investigation where a new material was sought. In 1936 there was no readily available material which could replace natural rubber. In the event of a war, the Second World War followed in 1939, Britain's natural rubber supply would be at risk since it had to travel by sea from the far east. Thus a substitute was required and research was initiated. In July 1940 a small amount of PVC was produced, but there was to be many problems before commercial production could occur in 1945 of PVC with suitable properties.

The evolution of materials over the years has resulted in changes in the life style of people. Thus when tools were limited to those that could be fashioned out of stone there was severe limitations on what could be achieved with them. The development of metals enabled finer products to be fashioned. For example, bronze swords were far superior weapons to stone weapons. Consider what the world would be like today if plastics had not been developed.

In 1930 the world production of steel was about 300 000 million kg, 2 000 million kg of copper and zinc and there was virtually no aluminium or plastics. By 1950 there had been only a slight increase in the production of steel, zinc and copper, but aluminium has risen to some 1000 million kg and plastics to 1000 million kg. Another twenty years later, 1970, steel, zinc and copper, had showed only a small increase but aluminium had now risen to equal the production of copper and zinc, at about 10 000 million kg and plastics had overtaken them to become 20 000 million kg. By 1990 the amount of plastic used had increased even more.

Materials production industries can be classified as operating in three stages:

1 *Primary*
This industry is involved in the extraction of the necessary raw materials from the earth, e.g. mining, quarrying, forestry, oil and gas extraction.

2 *Secondary*
This industry is involved in the conversion of the raw materials into the required basic material, e.g. ingots of iron or sheet steel.

3 *Tertiary*
This industry is concerned with the fabrication of the material into the products required by individual users, e.g. engineering components, car bodies.

11.2.2 Materials in use

Consider some examples of materials in use in the home and leisure activities, in industry and commerce, and in health and medicine.

Until comparatively recently, if you wanted a *tennis racket* then there was no choice of material for the racket. It had to be wood. Now there can be other materials used. Professional players are very likely to use a racket made from a composite material. The racket is made by injecting a melt of a polymer containing carbon fibres into a racket shaped mould. This would give a racket with a solid composite for the frame and the handle. However, the properties of the racket are improved if, while the racket is still in the mould and only the outer skin of the composite has solidified, the liquid core is poured out so that the handle solidifies as a hollow core. The handle can then be filled with a polyurethane form. The effect of this form of construction is to give a racket in which the vibrations that occur when a ball is hit are rapidly damped out and not transmitted through to the arm of the player.

If you have to go to the dentist for a *dental filling*, until quite recent times that filling would inevitably have been a silvery material which did not match your teeth. This material was made by mixing mercury with a silver–tin alloy, the material setting hard as a result of a chemical reaction. Now dentists can use a polymer-based material to which a filler can be added to enable the dentist to produce a filling which matches the colour of your teeth. The polymer is a thermosetting one involving the mixing of two chemicals which then sets to give a hard, rigid, product.

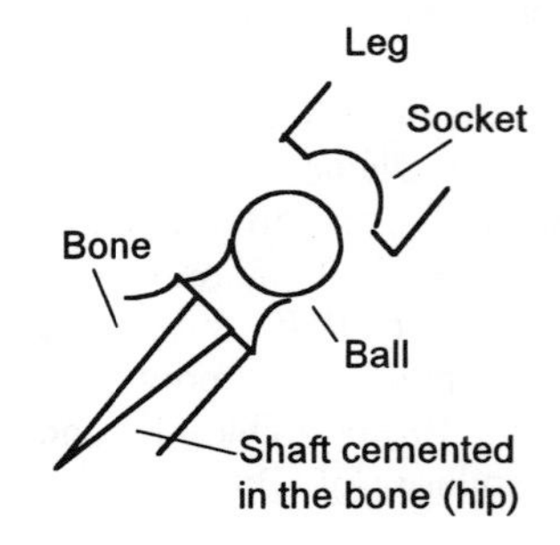

Figure 11.1 *Hip joint implant*

You might be even more unfortunate and need a *hip joint replacement*. The material used to replace such a joint must be accepted by the body as though it was its own bone and have properties very similar to bone. It would, for example, present problems if the implant was of a material which had a significantly different coefficient of expansion than human bone. The material has also to be resistant to attack by body fluids, it must not become rusty. For the ball part of the joint (Figure 11.1) stainless steel or titanium has been the material used. Developments are now resulting in a carbon fibre composite being used which has properties more like that of bone than metals.

Consider the materials that have been used in *car construction* over the years and the changes that have occurred. The early designers of cars were more concerned with the mechanical components than the body of the car. Thus the early cars were little more than a simple frame, a chassis, to which the engine, gearbox and wheels were attached; the engineers were not interested in the body. Coach builders used to building horse-drawn vehicles were then used to erect a body on the chassis. The result was a wooden-framed body covered with either thin wooden sheets or fabric. The bodywork was thus carried out quite independently of the mechanical components mounted on the chassis. With mass production came the use of sheet steel which could be formed into the required shapes by machines. The result was the replacement of wooden frames with infill sheets by a self-supporting body formed by steel sheets. It was soon recognised that a body could be built which took on the role of the chassis as the mounting bed for the mechanical components. Essentially the structure became a box on which the mechanical components could be mounted. Thus the modern form of construction of cars was developed.

Until relatively recently *car exhausts* were something that had to be replaced relatively frequently since they corroded so badly as a result of exposure to the hot exhaust gases and the action of water and solids thrown up from road surfaces, in particular salt used for de-icing roads in winter. The material used for these exhausts was mild steel. During the 1970s British Steel initiated a major effort to find a better material for exhausts. A significant improvement was found using steels with about 12% chromium. As a result a steel called Hyform 409 with the composition: 0.02% carbon, 0.6% silicon, 0.3% manganese, 11.4% chromium and 0.4% titanium was developed. With such a steel, or similar steels, exhausts now last much longer.

When running, cars emit from their exhausts carbon monoxide, a product of incomplete combustion which it is difficult to disperse in air, unburnt hydrocarbons, and nitrous oxides. Nitrous oxides are major participants in the chemical reactions that occur with sunlight to produce smog that, like carbon monoxide, lead to respiratory problems. In 1976 legislation in the United States concerning the maximum amounts of such gases which were permitted from cars led to the use of *catalytic converters* on American cars. Now legislation in Europe is leading to catalytic converters becoming standard on cars here. Such converters reduce the level of carbon monoxide, unburnt hydrocarbons and nitrous oxides in the exhaust gases. A catalytic converter is essentially a tube in which a platinum catalyst is mounted. The term catalyst is used for a substance which changes the rate of a chemical reaction without itself undergoing any permanent chemical change. The catalyst is on a ceramic mesh element supported within a stainless steel tube. The temperatures these materials have to withstand are up to about 900°C. Additionally they have to have good resistance to corrosion as a result of the exposure to the hot exhaust gases. Stainless steels are used because of good corrosion resistance, such steels containing 12% or more of chromium.

11.3 Information technology

The term *information* is used for items of knowledge. Information surrounds us in our everyday lives. For example, you might listen to the weather forecast on the radio or watch it on the television. The information about the weather is represented in the form of a mixture of words, numbers and, in the case of TV, graphical images. If you look at a clock or watch in order to determine the time you are receiving information. The term *information technology* is used for the processing of information. Such processing involves the collection of information, its storage, its retrieval from store, its processing and transmission to the point where it is to be used (Figure 11.2).

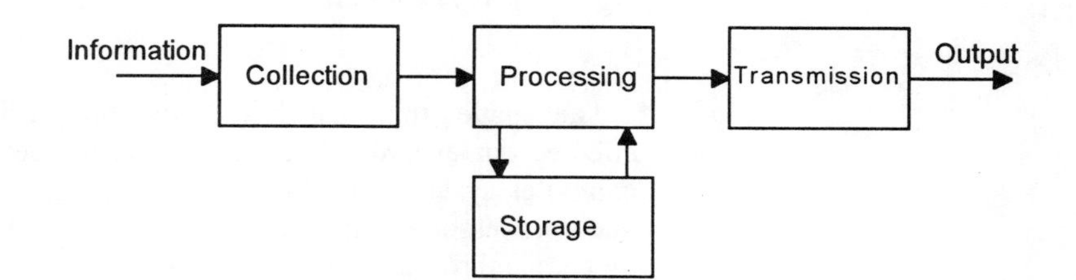

Figure 11.2 *The information technology system*

In order to process information, a number of devices, termed hardware, have to be connected together and controlled to achieve the processing. The instructions used to control the hardware are termed the software. With modern information technology systems, the arrangement of hardware and software together constitute a computerised information system. As a simple example, your watch, if modern, is likely to be a computerised information system involving a single semiconductor chip which is programmed by a set of instructions to output a signal at regular time intervals to a liquid crystal display. As another example, consider the information processing in weather forecasting (Figure 11.3). Information is gathered by weather satellites in orbit about the earth, transmitted to receivers on the earth, communicated by radio or cable links to a computer centre where the computer gathers together all the information to produce charts that show what is happening. This information is then interpreted by weather forecasters to produce weather forecasts for the area concerned. This is then presented by weather forecasters, using words and numbers and, in the case of TV, graphical presentations. This is then transmitted by radio waves through the air to be picked up by the radio or TV aerial and, after some processing, transmitted as words through your radio or as words through your TV speakers and a picture on the TV screen.

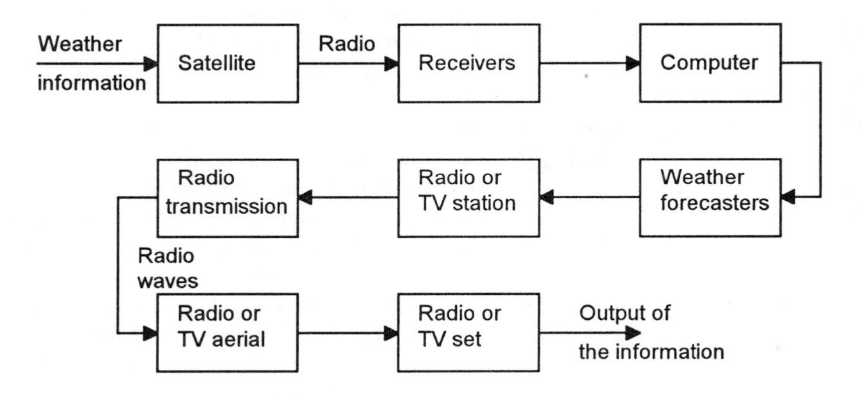

Figure 11.3 *The weather forecasting system*

The single item that has revolutionized information technology and affected our everyday lives is the microprocessor. This is the heart of any computer system. The following is a discussion of the evolution of the microprocessor and the effects of its use in information technology systems on people through a discussion of some examples of it applications.

11.3.1 The development of the microprocessor

The development of the microprocessor can be considered to have started in 1947 with the invention of the *transistor* by Brattain, Bardeen and Shockley. This consisted of a small piece of germanium, a semiconductor, with two closely spaced wire contacts on one of its surfaces. This small piece of germanium was termed a *chip*. Developments occurred very rapidly and transistors very rapidly replaced the large power-consuming vacuum tubes that had, up to then, been used in electronic circuits. However, with large circuits there was still the problem of making the connections between all the transistors and other components used. The solution to this was the invention of the *integrated circuit* by Kilby in 1958 and Joyce in 1959. This involves the production of a circuit as a single entity; transistors, resistors and capacitors, on a single chip of semiconductor, silicon in this case. This production of a circuits as a single entity reduced the costs per circuit and improved reliability. Further progress involved the construction of more and more complex circuits in an increasingly small chip of material. In 1959 the integrated circuit had just one transistor on its chip, by 1970 there could be about 1000 transistors, by 1980 about a million, by 1990 many millions.

As the number of components per chip increased so the capability of a chip increased. Thus in 1971 the *microprocessor* was invented by Hoff at Intel. The microprocessor is an integrated circuit which is capable of being programmed so that it can react in different ways according to its inputs, i.e. it is the heart of a computer. Instead of having to design integrated circuits for each specific application, the same microprocessor could be used in a wide variety of devices, only the program need be changed. Thus watches, cash registers, washing machines, etc. need not have a special integrated circuit designed for each but could use the same microprocessor, just the program being changed.

The first microprocessor, the Intel 4004, had about 2000 transistors on a single chip and was originally deigned for use in a calculator. Developments since then have been very rapid, not only with the number of transistors and hence computing power possible with a chip but also with the applications of such chips. By the 1970s about 1000 to 10 000 transistors could be put on a single chip. This gave the Intel 8080 and 8085 microprocessors. By the late 1970s and early 1980s, 10 000 to 100 000 transistors were being put on a single chip, giving the Intel 80886 and 8088 microprocessors. Later developments then gave chips with more than 1000 000 transistors per chip. This trend to increasing computing power on a chip, coupled with the reduction in cost, is generating more and more applications for microprocessors.

11.3.2 Information technology in use

Consider some examples of information technology in use in the home and leisure activities, in industry and commerce, and in health and medicine.

The ever decreasing cost of microprocessors means that they are becoming very cheap ways of controlling a wide variety of devices. Consider for example a *domestic washing machine*. Such a machine is required to carry out a number of tasks in sequence, e.g.

1 Check that the door is closed.
2 Open a valve to start filling the machine with water.
3 Close the value when the water reaches a predetermined level.
4 Switch on the drum motor to rotate the drum, and washing, for a predetermined time. This is the prewash cycle.

5 Switch on the discharge pump for a predetermined time to empty the water from the drum.
6 Open the water valve again to start filling the drum with water.
7 Close the valve when the water reaches a predetermined value.
8 Switch on the heater element.
9 Switch off the heater element when the water reaches a predetermined temperature.
10 Switch on the drum motor to rotate the drum for a predetermined time. This is the main wash cycle.
11 Switch on the discharge pump for a predetermined time to empty the water from the drum.
12 Open the water valve again to start filling the drum with water.
13 Close the valve when the water reaches a predetermined value.
14 Switch on the drum motor to rotate the drum for a predetermined time. This is part of the rinse cycle.
15 Switch on the discharge pump for a predetermined time to empty the water from the drum.
16 Repeat steps 12, 13, 14 and 15 for a number of times.
17 Switch on the drum motor to fast speed, together with the discharge pump, for a predetermined time. This is the spin dry cycle.
18 Stop the machine and, after a time of usually 60 s, release the door lock so that the washing can be removed.

Until comparatively recently, such a sequence of operations would be controlled by a set of cams which rotated and, as a result of their shape, operated electrical contacts at specific times. Figure 11.4 shows the form such a cam might take. The program is stored in the cam shape. A flat portion opens the switch, a circular portion closes it. By determining the lengths of such parts of the cam, when it rotates it switches on and off an electrical switch for predetermined times. However, a microprocessor system can replaces such a system of cams and offer greater reliability and greater variations in the program of instructions that can be followed.

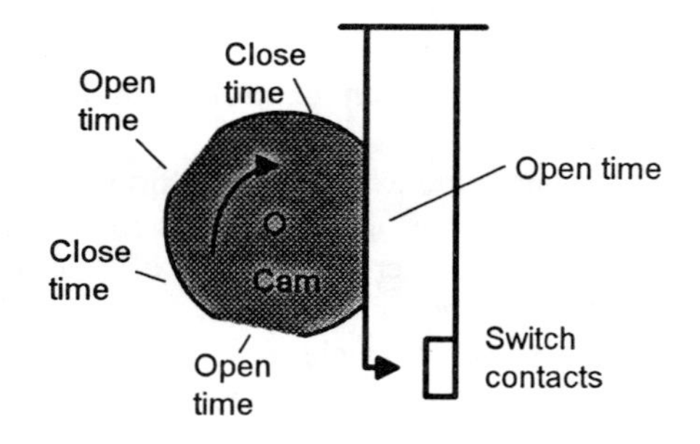

Figure 11.4 *A cam operated switch*

As another example of where microprocessors are replacing mechanical devices, consider the simple *bathroom scales*. Until comparatively recently they generally involved the weight of the person on the scales causing a cantilever to bend and in so doing, by a system of levers and gear wheels, cause a pointer to move across a scale. Figure 11.5 shows the type of arrangement that has been used. Now the system is being replaced by a system where, when a person stands on the scale, the cantilever bends but now the bending is used to produce an electrical signal which is processed by the microprocessor to give the weight in a digital read out. The electrical signal is produced by strain gauges attached to the cantilever. These are flat lengths of wire stuck on it like postage stamps. When the wire stretches its resistance increases and so can be used to give an electrical signal. The microprocessor based system has thus no mechanical moving parts, hence its greater reliability.

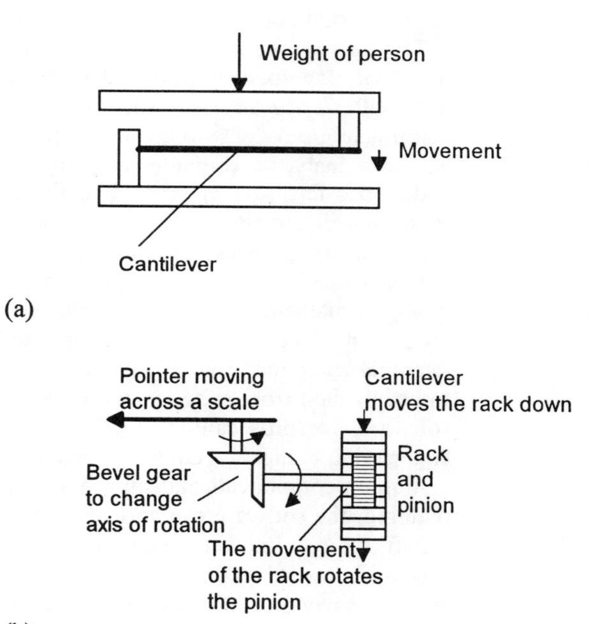

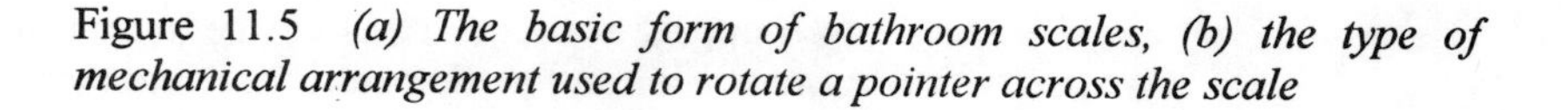

Figure 11.5 *(a) The basic form of bathroom scales, (b) the type of mechanical arrangement used to rotate a pointer across the scale*

Another example of where microprocessors have had an impact is with *cameras*. It is not that long ago that you had to estimate how far off the subject being photographed was and manually set the focus. You also had to estimate what the exposure should be, perhaps using a meter to measure the light intensity. You then has to set the aperture and shutter speed. Now automatic focus and automatic exposure cameras are common, all you have to do is point the camera at the subject and press the trigger, the microprocessor in the camera does the rest.

With *cars* we now have active suspension, antiskid brakes, engine management control, cruise control, fuel/trip computers and many other systems all based on the use of microprocessors to process signals and take control of systems and make driving safer and easier. For example, the active suspension, or adaptive damping (ADS) as it is often termed, is a microprocessor controlled suspension system which adjusts the damping of the suspension, switching between hard and soft damping with great speed, to ensure that the suspension reacts almost instantly to prevailing road or driving conditions. The ABS electronic anti-lock braking system helps the driver maintain steering control, even under emergency braking and on wet or slippery surfaces. The microprocessor system is continually carrying out a braking system check and making adjustments to ensure maximum performance. Engine management systems adjust the running of the engine to ensure optimum operation for the conditions prevailing.

The microprocessor has meant that the *personal computer* becomes an item that many people have in their homes and that virtually all companies use. With *word processing* software it has replaced typewriters as the main instrument used for producing typed information. The word processing software enables a computer to carry out all the functions of a typewriter and also offers several other facilities such as the ability to copy, move, insert and delete single characters, complete lines or entire blocks of text. They can spell check, grammar check, sort lists alphabetically, compile contents lists and indexes. A problem with the ordinary typewriter is the typing of mathematical equations with all the mathematical symbols that are used. With suitable word-processing software this presents no problem. Personal computers can be used with *spreadsheet* software (see chapter 10) to enable data from experiments to be logged and analysed. With *database* software, records can be stored, sorted and selected. For example, an engineering company could maintain a database of components with information on components and suppliers. Then when a particular component is required, the software can sort through the data to extract details of those that fit the requirement. The personal computer can be used as a means to access via the telephone line vast banks of information on the *information superhighway* or world-wide net. The computer is used to dial an ordinary telephone number that puts it in contact with a web of networks of computers round the world. Once connected, the computer user has access to shopping services, travel advice, technical advice, data on a wide range of topics, or on-line chat sessions with other computer users.

In *medicine* computers are being used for more efficiently storing *medical records* than in paper files in cabinets. Such records are easily called up by a doctor in any part of the hospital. They are also used to enable doctors to carry out *clinical diagnosis* in a way which would not previously have been possible. X-rays can be used to penetrate the body and give a picture of the internal structure. For example, they might be used to determine where a bone fracture is or some change that has occurred as a result of disease, e.g. a cancerous growth. The normal X-ray picture, however, gives a view of everything that is in the body between the point at which the X-rays enter and the point at which they emerge to be recorded. When detail is required of small elements within the body the conventional

X-ray can present problems. This problem can be overcome by using a *CT scanner* (computerised axial tomography scanner). A fine beam of X-rays is passed through the body and detected by a detector on the other side which is only able to detect the X-rays passing straight through in a fine beam. The incident beam of X-rays and the fine beam detector are then scanned across the body so that readings are taken from a wide range of angles at which the beam passes through. A vast number of readings might be taken and all the results are fed to a computer which analyses all the data and can produce a picture on a computer screen which enables fine detail of any feature within the body to be examined and the angles at which it is viewed changed.

11.4 Automation

The *domestic washing machine* referred to in the previous section is an example of an automated machine (Figure 11.6). The machine takes in dirty clothes, washing powder, water and electrical power and out comes clean clothes and dirty water without any human having to intervene and carry out any task.

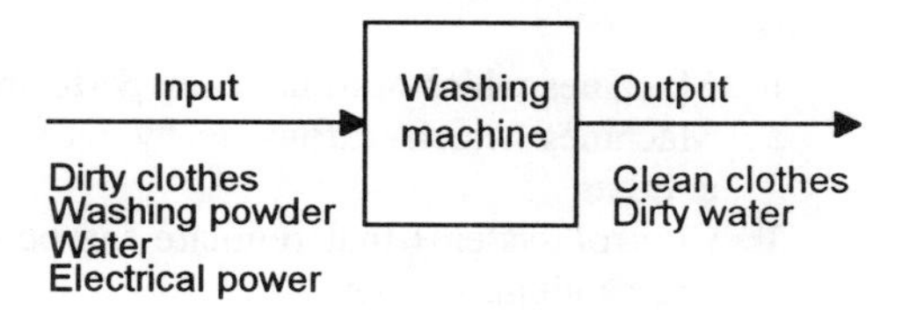

Figure 11.6 *The automatic washing machine*

Another example of an automatic machine is the *bank cash dispenser* (Figure 11.7). You insert into the dispenser a cashcard carrying a magnetic strip with a unique code to identify the account holder. The card user has then to key in a personal identification number. If this number tallies with the number given by the magnetic strip then cash withdrawal can proceed. Following the card user keying in the amount of money to be withdrawn, the information about the person making the withdrawal and the amount requested is transmitted to the bank's computer. This checks that there are sufficient funds in the account and, if so, deducts the money from the account and sends back a signal authorizing the cash machine to pay out the cash. The system is automatic in that no person is involved in checking the identification of the bank customer, the account details, entering the withdrawal in the account and paying out the cash.

The above are just two examples of automation. *Automation* can be defined as the handling or material or information in some process without direct and continuing human operation. The word automation probably first appeared in print in 1948 in an article in the American Machinist about the Ford motor factory and its production processes for cars. The definition of

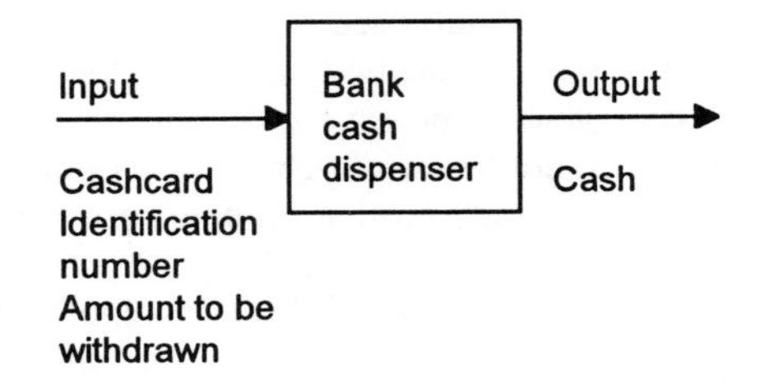

Figure 11.7 *Automatic bank cash dispenser*

automation given there was the application of mechanical devices to manipulate work pieces into and out of equipment, turn parts between operators, remove scrap, and to perform tasks in timed sequence with the production equipment so that the production line can be wholly or partially under push-button control. Automation began with the automatic handling of parts in the production process in the metal working industry. It has, however, spread much further than that now.

Automatic production can be considered to involve three main aspects:

1 Machines which automatically perform operations.
2 Machines which automatically move materials from one machine to another.
3 Control systems that regulate the performance of both the production and handling systems.

The following discussion illustrates how these various aspects have developed over the years.

Automatic machines can be traced back to the early 1800s with an example of such a machine being the machinery devised for the British Admiralty by Marc Brunel to manufacture pulley–blocks. The machinery consisted of a number of machines, each of which was set to carry out a specific machining job. There was no automatic transfer of parts between machines and workmen were used for that part of the process. As a result, from an input to the machinery of timber emerged completed pulley blocks, the entire operation enabling 10 men to replace the 110 men that had previously being required to make 130 000 pulley blocks per year. A major advance in automation was the automatic transfer machine which automatically moved workpieces between machines and fed them in proper orientation to machines. Such a type of machine was probably first used by the Waltham Watch Company in 1888, and first applied in the car industry at Morris Motors in England in about 1924 for the production of engines. Such types of machines became commonplace in the car industry and many other industries involved in mass production in the 1930s.

A major advance in automation was the advent of *numerical control* (NC) in the early 1950s. Information about successive positions of tools, machine tables, speeds and feed rates were indicated by the punch holes on

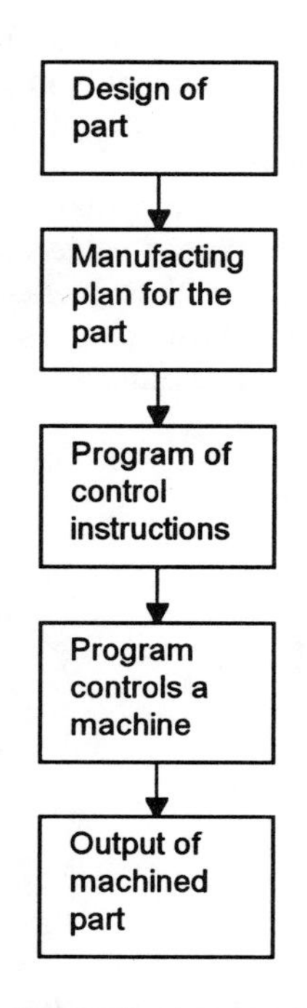

Figure 11.8 *Basic principle of NC*

a punched paper tape. This tape thus enabled machinery to be programmed to carry out a sequence of operations with the program capable of being altered for other products with the same machinery. Figure 11.8 illustrates the principle of numerical control. The design of the product leads to the planning of the manufacturing which is then presented as a program consisting of holes on a punched tape. The tape is then fed into the control system of the NC machine and organises the tools, speeds, cutting actions, etc. necessary to produce the product. The development from numerical control using punched tape was *computer numerical control* (CNC) in which a computer used its program to control machines and the sequencing of operations. Such a development first occurred in the early 1970s. The program instructions were no longer a sequence of punched holes on paper tape but put into the memory of a computer. Now, a modern computer controlled machine will have sensors able to supply the computer with information about what is happening in the machining, the computer to consider the information and make decisions about how to proceed on the basis of that information in relation to the required product, and actuators to carry out its instructions (Figure 11.9).

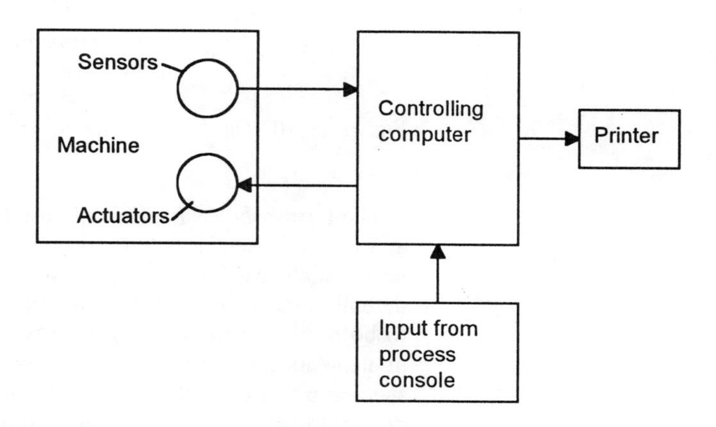

Figure 11.9 *The basic form of a CNC system*

Many control situations occur where sequencing of operations is required. For example, in the automatic washing machine referred to earlier in this section a sequence of operations involving opening valves, closing valves, switching on motors, etc. is required. There are many other situations where such a control mechanism is required. In the late 1960s the motor car manufacturer General Motors faced such a problem. Up till then many sequence operations were controlled by relays. These are electrical switches which can be actuated by a current passing through a coil in the switch. Thus by arranging the sequencing of currents a sequence of switches can be operated. From this problem emerged the *programmable logic controller*. This is a computer system which has as its input essentially signals from switches, i.e. on/off signals (Figure 11.10). On the basis of this information the computer carries out a program to supply as output further on/off signals. Thus, in the case of a washing machine, the inputs might be signals from a valve as to whether it is open and so water entering the drum and a signal from the machine drum as to whether the water is up to the required mark. When both signals indicate on, then the computer triggers the next stage in its operation. This might be to switch the valve off and start the pump to empty the machine. The controller thus works in a number of stages, assessing the signals at each stage and using logic such as if A *and* B on or if A *or* B on, etc. to determine the course of its action.

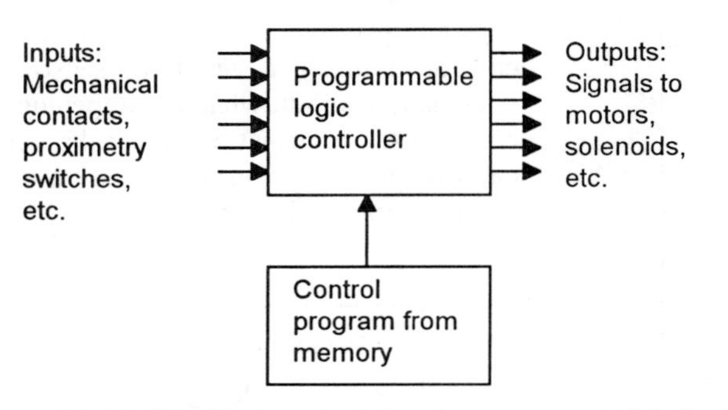

Figure 11.10 *The basic principle of a programmable logic controller*

The term *robot* was originally used for an automated humanoid machine and was an essential feature of much science fiction. The term is, however, now widely used for devices used in the production process. The usual definition used for an industrial robot is a reprogrammable device designed to both manipulate and transport parts, tools or specialised manufacturing implements through variable programmed motions for the performance of specific manufacturing tasks. At is simplest it might be considered to be a device to pick up objects from perhaps one machine and place them in correct orientation in some other machine. 1960 saw the first industrial use of a robot, it being used to unload a die-casting machine at the General Motors car factory in New Jersey. Robots are now used in industry for a range of tasks, e.g. materials handling robots, assembly robots, welding robots, paint-spraying robots, etc.

This use of a computer numerical control began to lead, in the 1970s, to the integration of automatic machines, automatic handling of material and computer control and the term *computer-aided manufacturing* (CAM) used. We can now have *computer-aided design* (CAD), computer-aided engineering, computer-aided operations management, computer-aided manufacturing, computer-aided assembly, inspection and testing, and computer controlled warehousing. Computer-aided engineering involves CAD, programming of numerical control tools, tool design and process planning. Operations management involves such items as production planning and control, cost accounting, and purchasing. It has thus become possible to make parts of a factory, or indeed a complete factory, automated.

Problems

1 Produce a report describing how some aspect of engineering technology, e.g. the microprocessor, is applied in (a) the home/leisure activities, (b) industry/commerce, (c) medicine. Illustrate your answer with examples.

2 Compare the types of materials that are currently widely used with those used fifty years ago and describe how the changes have affected the products around us.

3 Describe how the advent of the microprocessor has affected (a) home/leisure activities, (b) industry/commerce, (c) medicine.

12 Environmental effects

12.1 Industrial processes

Industrial processes consume resources and as well as producing the required products can give rise to pollution. The products themselves may also end up as pollutants. The term *resources* is used to include both raw materials and energy. *Pollution* can be defined as the introduction into the environment of substances or energy which are liable to affect people's enjoyment or use of the environment, harm human health, livestock, wild life, crops, or ecological systems, and damage buildings or other man made structures. Pollution is thus defined as something that damages things of importance to humans.

This chapter deals with these issues of consumption of resources and pollution.

12.2 Resource consumption

Consider the resources involved in the production of steel (Figure 12.1). To start with there is the mining of the iron ore. This is then used in a blast furnace with coke and limestone, also mined, and air to produce pig iron and slag, a waste product. The molten pig iron is then mixed with steel scrap, oxygen and limestone in a converter to produce steel ingots. These may then be further processed by rolling, forging, etc. in order to produce products which might still require some machining and finishing.

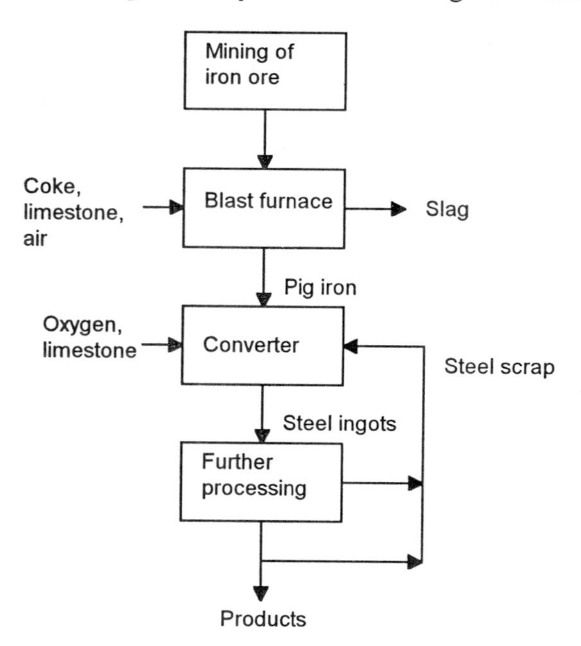

Figure 12.1 *Steel making*

But Figure 12.1 gives only part of the picture. There will be other outcomes at each stage, such as mine waste, pollution, waste heat, etc., and other inputs, in particular energy. It has been reckoned that energy currently accounts for about 11% of the costs in steel making. In 1980, the energy needed to produce 1 kg of liquid steel amounted to about 29 million joules. By the early 1990s, improved efficiency had reduced this figure to about 23 million joules. To put this in perspective, a 1 kW electric fire requires 1000 joules every second. Thus to produce 1 kg of liquid steel is equivalent to running the electric fire for 23 000 seconds, or about 6.4 hours. When you think of all the steel articles being produced per year, a lot of energy is being used in their production.

Consider another example, the production of PVC. The basic materials from which PVC is made are brine and oil. The brine is used to produce salt which in turn is then used to produce chlorine. The oil is used to produce naphtha which in turn is used to produce ethylene. The ethylene and chlorine are then reacted together to produce dichlorethylene. This is then turned into vinyl chloride which is then used to form the PVC. Figure 11.2 summarises the key stages in the production. What has not been included at each stage is all the energy involved and the other products formed.

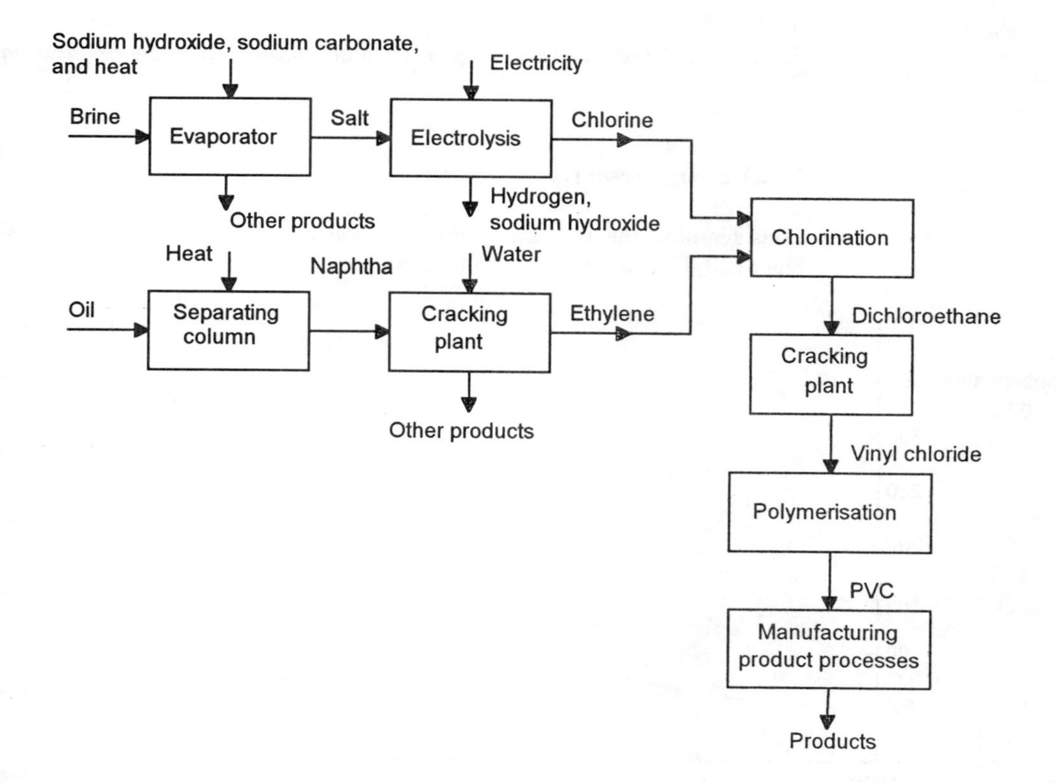

Figure 12.2 *The production of PVC*

Figures 12.1 and 12.2 give only a brief indication of all the raw materials and energy involved in the production of steel and PVC. When we take into account all the materials that are processed to make products, and the requirements we have for energy for domestic use in heating our homes, cooking, etc. and for transportation, then a great deal of raw materials and energy is consumed each year. Of all the energy consumed per year in Britain, some 39% is estimated as being used by industry, 26% for domestic use and 22% for transport (Figure 12.3).

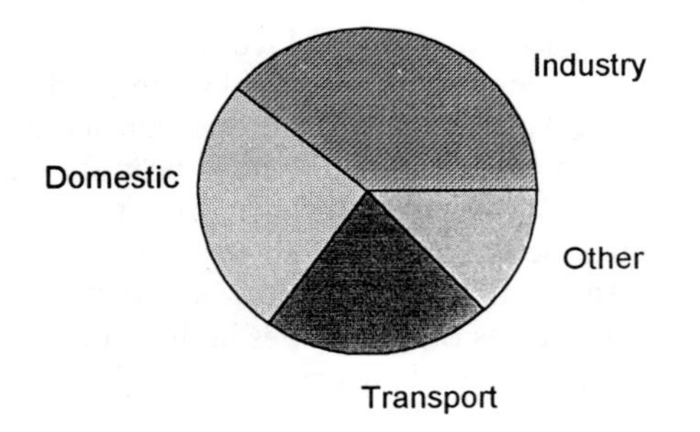

Figure 12.3 *Pie chart showing percentage breakdown of energy used in Britain*

12.2.1 Energy resources

Consider how the world fuel consumption has changed over the years. Figure 12.4 illustrates the trend that has occurred over recent years.

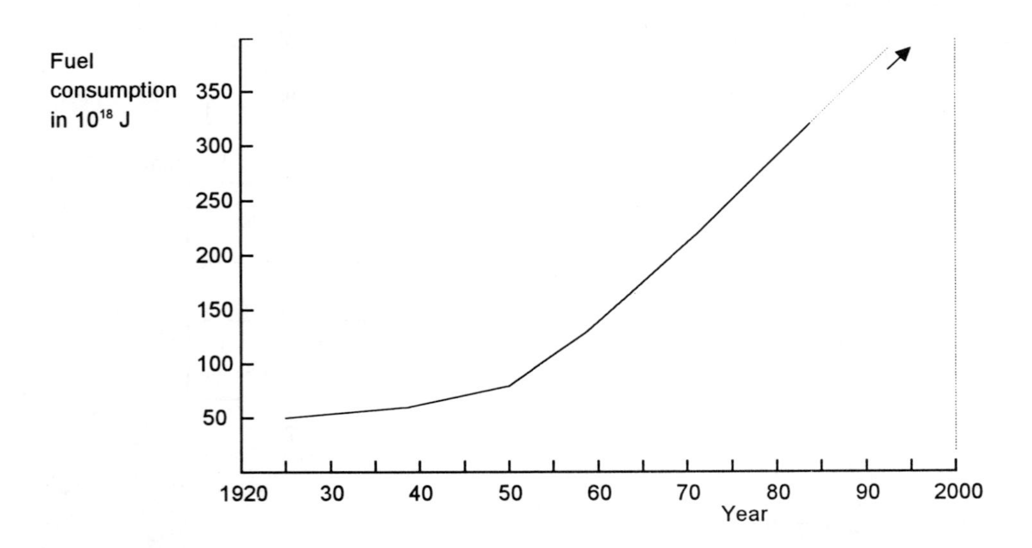

Figure 12.4 *World fuel consumption*

Back in 1925 the demand was for fuels to supply about 47×10^{18} J, i.e. 47 followed by 18 noughts. By 1930 this had risen to about 56×10^{18} J, by 1935 to 58×10^{18} J, by 1940 to 60×10^{18} J, by 1945 to 66×10^{18} J, by 1950 to 81×10^{18} J. The rate at which energy was being used was increasing ever more rapidly. By 1955 it had risen to 105×10^{18} J, by 1960 to131×10^{18} J, by 1965 to 165×10^{18} J, by 1970 to 217×10^{18} J, by 1975 to 250×10^{18} J, by 1980 to 290×10^{18} J, by 1985 to over 300×10^{18} J and still it keeps on rising. It has been estimated that by the year 2000, if it continues at this rate, it could be more than 450×10^{18} J.

Figure 12.5 shows the current sources of this energy. Oil provides the greatest percentage, about 40%. Coal supplies about 30%, natural gas about 20%, hydroelectric power about 7% and nuclear about 3%.

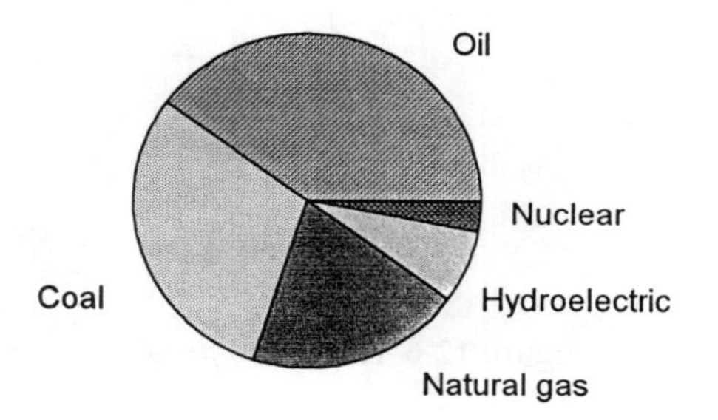

Figure 12.5 *The sources of world energy*

In 1980 in Britain we used about 10×10^{18} J. This is an average of about 1.8×10^{11} J for each person in the country. To put this in perspective, each day each person effectively used about 500 million joules or about 6000 joules each second. One bar of an electric fire tends to use 1000 joules every second (a 1 kW element). Thus every person was effectively using about six electric fires every second of every day. This figure was even higher in 1990 and is expected to be even higher in the year 2000. By then it is expected that each person in Britain would be effectively using seven or more electric fires every second of every day. It is not that we are directly using this energy, much of it is used to produce the products we use (see Figure 12.3).

12.2.1 Energy conservation

In Britain, industry is the largest consumer of energy. The industries consuming the greatest amounts are those concerned with the processing of raw materials. The chemical industry and the iron and steel industry are thus very big users. Both industries have become more energy efficient in recent years. The steel industry has been able to use less energy by technological improvements in furnace design, the use of better quality ores and the recycling of scrap iron. The chemical industry has been able to use less energy through more efficient use of the heat developed in some chemical reactions being used to heat other parts of the system where heat is required and thus a better design of the chemical plant to avoid wasting heat.

12.2.2 Recycling

Figure 12.6 shows an estimate of the relative amounts of energy needed to produce 1 kg of metals from their ores. Though iron requires the least, the great amount of steel products used means that the total amount of energy devoted to steel making is large.

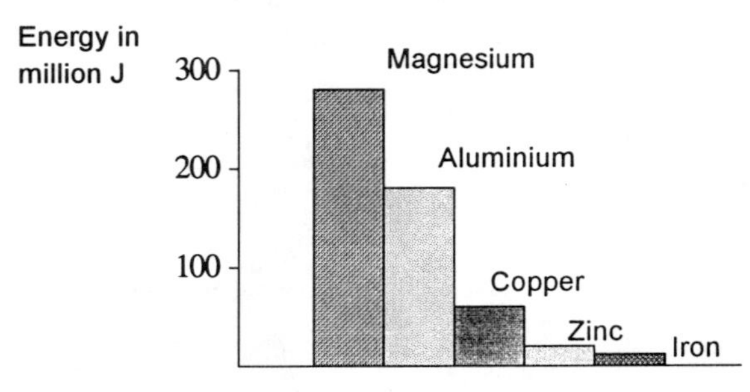

Figure 12.6 *Energy to produce 1 kg of metals from ores*

The above figures can often, however, be very markedly reduced if waste materials are recycled. Thus the recycling of aluminium can reduce the energy requirements for 1 kg of aluminium by about 80%, the recycling of copper the energy requirements for it by also about 80%. Steel produced from scrap requires only about 25% of the energy required with ore (this recycling using scrap is illustrated in Figure 12.1). Thus considerable energy savings can be made by recycling metals.

The term *recycling* is used for the feeding of waste material, i.e. scrap, back into the processing cycle so that it can be reused again in the production of fresh metal for fabrication into products. In the case of steel making an obvious example is the crushing of old car bodies to form steel cubes which can be fed into the steel furnaces, along with the iron ore, to make fresh steel. Another example, as indicated above, is the recycling of aluminium cans in the production of aluminium. It has been estimated that of the order of 40% of aluminium in Britain is recycled. Because aluminium requires a large amount of energy to produce it from its ore (see Figure 12.6), the energy savings, as well as the saving in aluminium, can be very significant. Figure 12.7 illustrates the stages involved in the production of aluminium cans when recycled material is used. Another example is the recycling of glass bottles. Typically of the order of 30% of the glass in a glass bottle comes from recycled glass. There is, however, no great energy saving in using scrap glass as using the raw materials. There is, however, a big saving if bottles are reused.

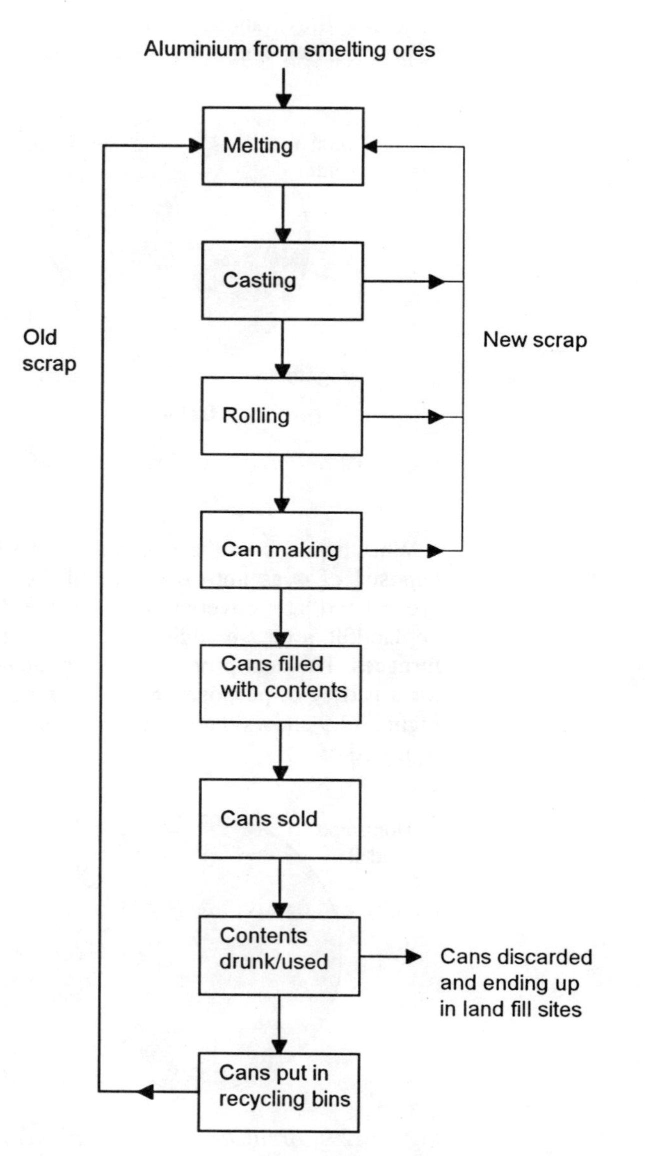

Figure 12.7 *Using recycled aluminium in the production of aluminium cans*

12.2.3 Disposal of domestic waste

What do you throw away in your dustbin? Domestic refuse typically contains about a third paper, of the order of 18% vegetable matter, a similar amount as dust and cinders (though this figure is declining as less houses have coal fires), about 10% metal, about 10% glass, and small percentages of plastics, rags and other items. Figure 12.8 illustrates this breakdown.

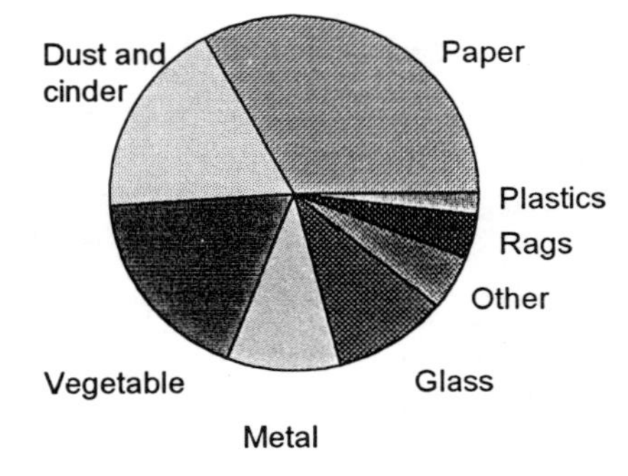

Figure 12.8 *Percentage breakdown of domestic waste*

What happens to the waste from the dustbins? The main way that it is deposed of is as untreated landfill, i.e. it is just dumped in a hole in the ground and later covered with soil. In a few areas of the country there may be landfill after shredding or compaction or burning of the waste in furnaces. The heat produced during incineration can be recovered and used for a variety of purposes, e.g.. electrical power generation, district heating. Figure 12.9 shows the percentage breakdown of the way domestic waste is disposed of.

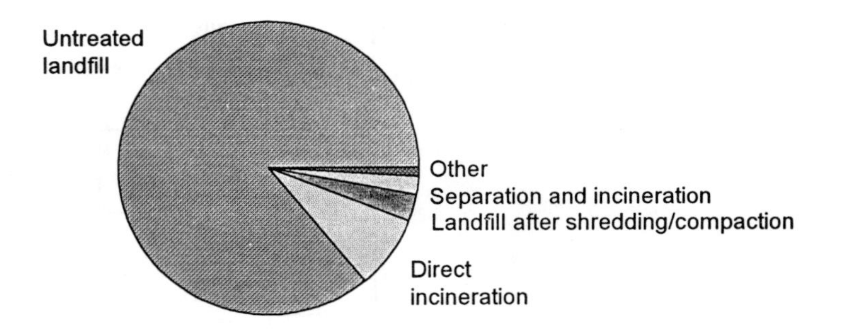

Figure 12.9 *Breakdown of disposal methods used with domestic waste*

In recent years, disposal as a method of dealing with domestic waste has been increasingly questioned. There is an enormous waste of materials which could be useful resources. The question is – how can the useful materials be economically separated from the waste and put to use? In some areas the method is being tried of each householder having a number of separate waste containers, e.g. one for paper, one for glass, one for tins, etc. The containers can then be directed towards appropriate recycling centres.

12.3 Pollution

Industrial processes are a source of many pollutants. The air, rivers, lakes, oceans and land are all, in one place or other, polluted by industry. Table 12.1 shows some of the pollutants commonly encountered, their sources and effects.

Table 12.1 Common pollutants

Pollutant	Common source	Effects of the pollution
Air pollutants		
Sulphur dioxide	Combustion of coal and oil (see section 12.3.1)	Involved in smog, lung ailments, acid rain (see 12.3.1), corrosion of metal and stone
Dust particles of various materials, including metals	Combustion ,e.g. in an industrial process such as steel making (see section 12.3.1)	Smog. Also the metals particles may be toxic. When deposited on the land they may get into the food chain
Carbon dioxide	Combustion of wood, coal, oil and gas. Forest clearing by burning	This may modify the climate as a result of the greenhouse effect (see section 12.3.1)
Nitrous oxides	Combusion of coal, oil and gas (see section 12.3.1)	May contribute to acid rain. Involved in smog problem
Hydrocarbons	Car exhausts	Involved in smog problem
Chlorofluorocarbons	The gas has been used as propellants in many aerosols, e.g. hair sprays, tough-up paint sprays for cars	May decompose ozone in the stratosphere and so lead to skin health hazards from increased exposure to unfiltered sun's rays
Land pollutants		
Metals such as lead and mercury	Industrial processes, e.g. mining and smelting. Lead from car exhausts	Health hazard to humans and animals
Radioactive materials	Accidential leaks from industrial processes using radioactive materials	Health hazard. Some cancers, leukaemia. Mutations in cells
Pesticide residues, e.g. DDT	Crop spraying against pests or spraying against disease carring insects	Harms wild life. Leaves residues which can get into the food chain
Water pollutants		
Oil and hydrocarbons	Spillage from shipwrecks and on land	Damage to wild life
PCBs (polychlorinated biphenyls)	Industry waste	Damage to wild life
Phosphates and nitrates	Sewage effluents. Excessive use of artifical fertilizers	Ecological changes, e.g. excessive water plant growth
Metals such as lead and mercury	Industrial processes	Can get into the food chain

12.3.1 Fossil fuel burning and airborne pollutants

Industry runs largely on energy from coal, oil and natural gas, the so-called fossil fuels. When such fuels are burnt they release carbon dioxide, sulphur dioxide, nitrous oxides, smoke and heat. The amounts produced by the different fuels differ with coal producing, by far, the greater amounts of pollutants. Table 12.2 shows the levels of such emissions from the burning of fossil fuels. Carbon dioxide contributes to the greenhouse effect (see later this section). Sulphur dioxide in the air promotes lung ailments and contributes to acid rain. Nitrous oxides also contributes to acid rain. Acid rain leads to corrosion of metals and stone and damage to trees.

Table 12.2 Pollutants released from fossil fuels
(Based on Table 1 in Energy and the environment by J.Skea, *Physics Education*, page 193, vol.27, 1992)

Fuel	Grams per millions of joules of energy released		
	Carbon dioxide	Sulphur dioxide	Nitrous oxides
Coal	24.5	1.24	0.37
Oil	19.0	0.89	0.16
Natural gas	13.8	0	0.05

In the 1950s when most countries relied on coal as the main source of energy, smoke and sulphur dioxide were the main problems. Smoke in combination with fog produces smog and adding to this sulphur dioxide and nitrous oxides results in a mixture which affects breathing and can result in lung ailments, in particular bronchitis. In the winter of 1952 the weather conditions resulted in the accumulation of the smoke and the other pollutants from fossil fuels close to the ground. The resulting smog is considered to have resulted in the loss of some 4000 lives. As a consequence the *1956 Clean Air Act* was introduced. This created smokeless zones and required smoke abatement equipment to be fitted to industrial plant. The greater use since that time of sulphur-free natural gas for energy supplies and the consequential reduction in the amount of coal burnt has also resulted in an improvement.

As a result of the 1956 Clean Air Act a number of processes in the steel industry were recognised as requiring cleaning processes to be developed and used. One of the main problems is the emission of smoke containing small particles of iron oxide and waste gases. The installation of air cleaning plant involving filters, wash towers and precipitators is used to reduce the emission of such smoke by extracting the dust and reducing the volume of waste gases. Power stations face similar problems, needing to reduce the amounts of sulphur dioxide and nitrous oxides emitted. Thus sulphur dioxide is removed from flue gases by spraying wet limestone through the gases. The sulphur dioxide combines with the limestone to produce gypsum which can be used as a building material. Nitrous oxides can be removed by injecting ammonia into the flue gas as is passes through a honeycomb grid of a titanium oxide catalyst. Another way that the pollutant gases can be removed is by improving the technology of combustion of the fuels so that less is emitted. A change to low sulphur coal and natural gas also reduces the amount of pollutant gases.

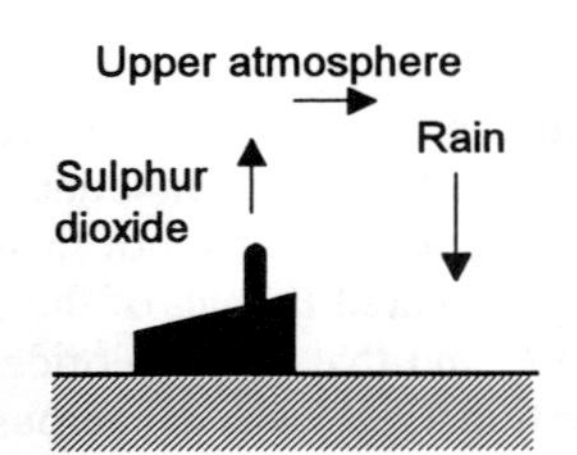

Figure 12.10 *The production of acid rain*

Sulphur dioxide emissions from the combustion of fuels can rise through the atmosphere into the upper atmosphere where it can then be transported by winds many hundreds of kilometres (Figure 12.10). It then can become deposited in the oceans, lakes, rivers and on the ground in rain as sulphuric acid. This is what is termed *acid rain*. On the land, such acid rain can corrode metal and brickwork, make soil acidic and affect the chemical balance within the soil with, for example, damage to forests.

It is not only industry that burns fossil fuels and emits pollutant gases, a major contributor is the exhaust fumes from cars. The use of catalytic converters (see section 11.2.2) can, however, significantly reduce the amounts of such gases emitted.

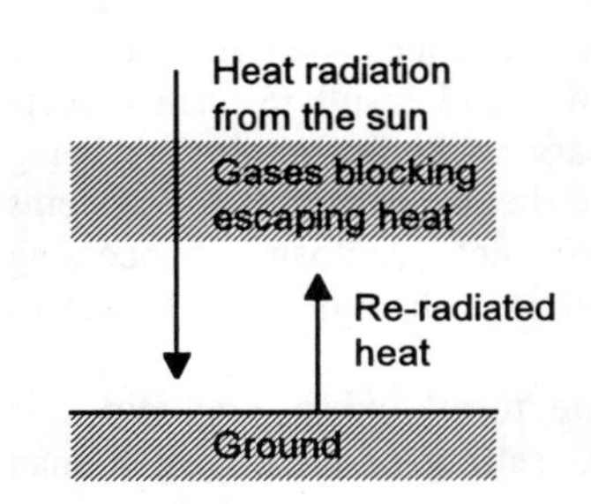

Figure 12.11 *The greenhouse effect*

The temperature of the earth is maintained at its current value by receiving radiation from the sun to warm the earth's surface and the existence in the upper atmosphere of gases which block heat being radiated back out into space. The situation is rather like a greenhouse. With a greenhouse the radiation from the sun passes through the glass and warms up the soil. The warm soil then radiates heat but this is blocked from escaping by the glass. Thus heat becomes trapped within the green house. The gases in the upper atmosphere of the earth which act like the greenhouse glass and block heat escaping are carbon dioxide, water vapour, methane, ozone and nitrous oxides. By far the biggest contributor is carbon dioxide. The emission of carbon dioxide from the burning of fossil fuels is increasing the amount of carbon dioxide in the atmosphere and so it is envisaged that more heat will be trapped on the earth and so the temperature should increase with consequential climatic changes. This is the effect termed the *greenhouse effect*.

12.3.2 Water-borne pollution

Industry uses vast quantities of water. Water is used for cooling, washing, mixing, diluting and carrying away waste products. It has been estimated that to produce 1 kg of steel requires 200 kg of water, such water mainly being used for cooling purposes. Since much of the water is now recirculated, the actual amount of water taken in for each 1 kg of steel is much less, probably about 13 kg. The water is generally taken in directly from a river, used for cooling and then passed directly back into the river. Such water has generally little contamination, although the fact that the water enters the river at a higher temperature than when it was extracted can change the local ecology. Some of the water used in the steel making process is, however, highly contaminated and has to be purified before it can be discharged back into a river.

Another example of where water is used as part of an industrial process is the manufacture of paper. Water is used in the pulping process when the raw materials are broken down into a fibrous pulp. The entire paper making process then involves a constant flow of water through the various parts of the process. The resulting contaminated water can end up back in the river from where it originated.

12.3.3 Health at work

In 1974 employers at a plastics factory in Kentucky in America revealed that three employees had died of a rare cancer of the liver. The announcement led to other plastic companies letting it be known that some of their employees had died of the same illness. Research indicated that all of them in some way or other had been involved with PVC and that vinyl chloride, the substance used in the making of PVC (see section 12.2), was dangerous and that very short exposures to it could produce cancer and a variety of other illnesses.

Over recent years a considerable amount of effort has become applied to the problems of *health at work* to ensure that the workplace does not pose risks to health (see section 4.4). Contact with dust and other materials can produce irritation of the skin (dermatitis); some chemicals can, when present in sufficient quantities, be irritating to the eyes, nose and throat and cause headaches and possibly dizziness; some chemicals might affect the reproductive systems of both men and women and result in birth defects and infertility. Thus precautions are necessary to avoid employees being exposed to such chemicals. These can involve the use of ventilation systems to remove fumes from a working environment, enclosing processing equipment, employees wearing protective clothing, changing the process to a less hazardous one, etc.

Radioactive materials emit radiations, the term ionising radiations is used, that can pose a hazard to people by upsetting or destroying human cells. The effects can take many years to become apparent and then show itself as skin problems, leukaemia or other forms of cancer. If radioactive materials enter the body there can be further hazards. If deposited in the lungs they can cause diseases of the respiratory system, if in the bones damage to the bone marrow and lead to leukaemia and bone cancers. Röntgen, the inventor of X-rays, died from bone cancer. Many of the early workers with X-rays and radioactivity suffered as a consequence of their work. If radiation penetrates the reproductive system it can damage subsequent children. Safety in the handling of radioactive materials is thus of high concern. Precautions are taken to minimise exposure to radiations. The practice is to specify a permissible dose and monitor workers to ensure that they do not receive radiations in excess of this dose.

In Britain, the Ionising Radiations Regulations 1985 apply to all work activities involving ionising radiations. These lay down very detailed requirements to keep exposure to a minimum. Before work starts with any ionising radiations or any radioactive source is taken onto the work premises, the radiochemicals inspector of the Department of the Environment has to be notified and authorisation might be needed. An employer might need to arrange for medical examinations and routine dose assessments for employees working with such substances; obtain authorisation for the use, storage and safe disposal of radioactive materials; appoint a radiation protection adviser; have contingency arrangements for when problems occur such as spillage of radioactive materials, etc.

Problems

1 Describe, with examples, some of the environmental impact of the engineering activities involved in (a) extracting natural resources, (b) manufacturing products.

2 Describe the impact of engineering activities on resource (raw materials and energy) consumption.

3 Describe the impact of engineering activities on pollution (air, water and land).

4 Give examples of the requirements for managing the environmental impact of engineering activities and how they might be achieved.

5 Figure 12.12 gives the sequence of events involved in (a) the production of plastic bottles which end up as waste, (b) the production of glass bottles when they end up as waste, are recirculated or the glass recycled, and (c) the production of cans when they end up as waste or the aluminium is recirculated.

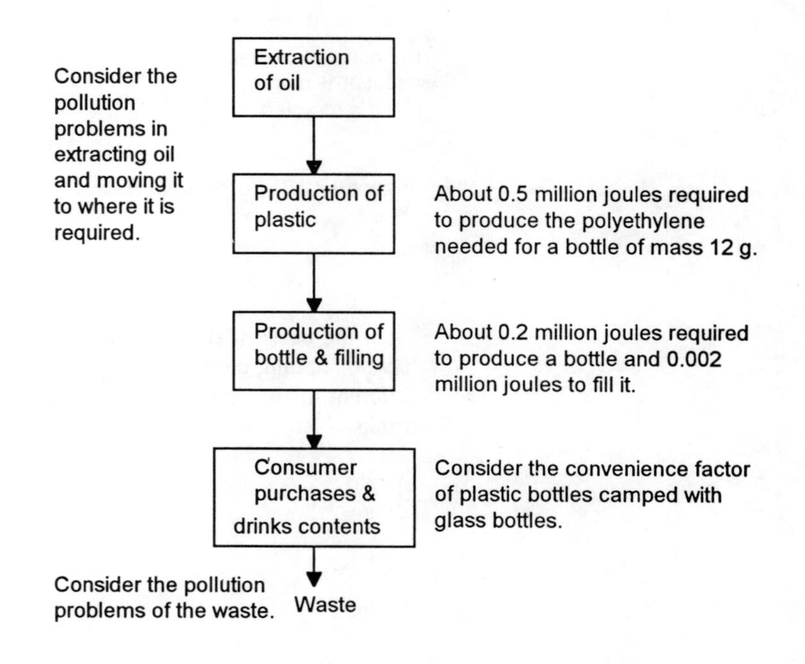

Figure 12.12 *(a) Plastic bottles*

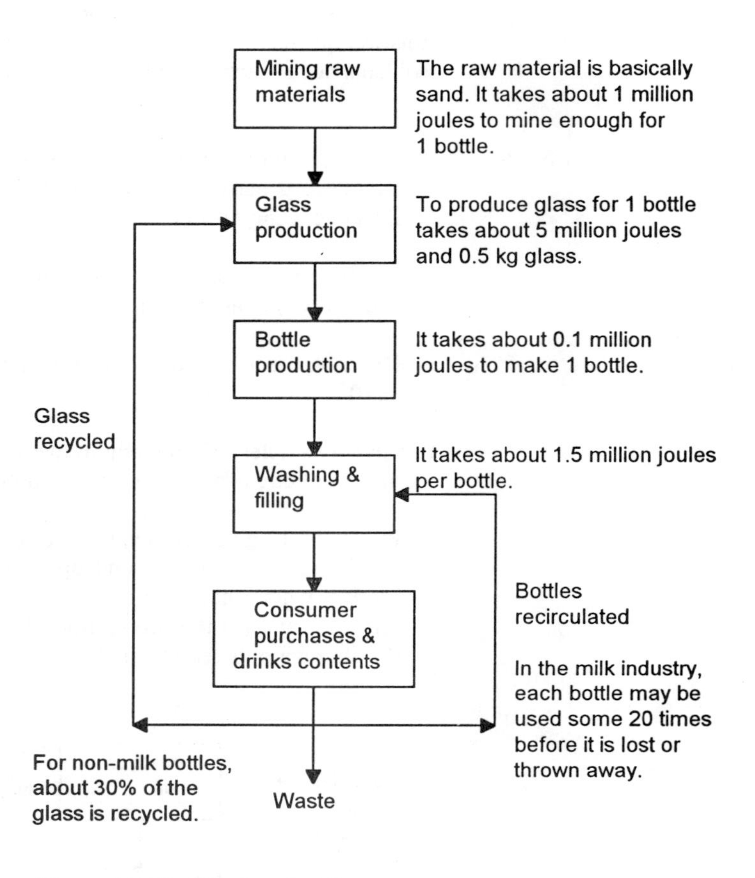

Figure 12.12 *(b) glass bottles*

On the basis of that information, and any other information you may have or obtain, compare the 'cost' in materials and energy for when the different materials are used as containers for liquids such as milk or soft drinks. Also discuss the possible pollution problems that might arise.

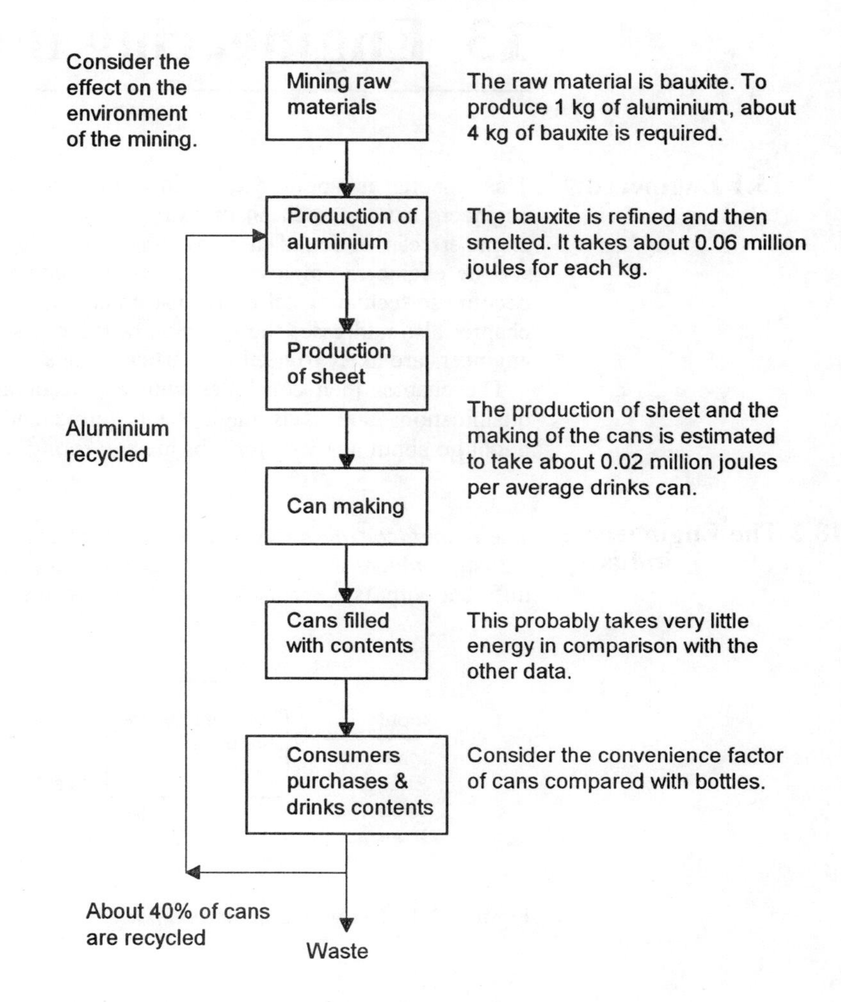

Figure 12.12 *(c) aluminium cans*

13 Engineering industry

13.1 Engineering

This chapter is about the engineering industry and the types of jobs engineers have in such an industry. It must, however, be recognised that engineers can be found in many other industries. Thus, for example, there will be engineers employed in the health industry, particularly now it has become so technological in the instrumentation and equipment it uses. This chapter also addresses the question of the types of qualifications and skills engineers are likely to need for particular jobs.

The chapter then concludes with a consideration of how the required qualifications and skills might be acquired and how potential engineers might go about applying for jobs in engineering.

13.2 The Engineering industry

The *manufacturing industry* can be considered to be defined as being that industry which converts or transforms physical inputs into physically different outputs, i.e. makes things from raw materials (Figure 13.1).

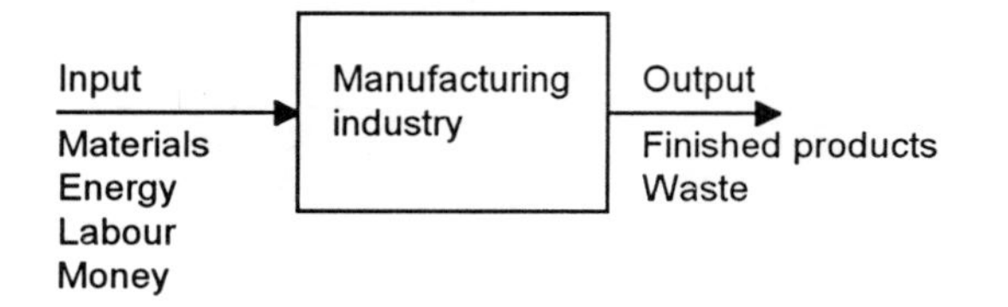

Figure 13.1 *A manufacturing company*

A common way that is used to classify manufacturing industry is in terms of the types of products they produce. Thus we have:

1 *Heavy engineering*
Such companies manufacture products that are not sold in large quantities on a regular basis but rather infrequently, each being a major work of engineering and expensive. They sell to such industries as the large processing companies, e.g. steel making, the transportation industry, e.g. ships, airlines, railway trains and carriages, the power generation industry, e.g. generators.

2 *Medium engineering*
The products produced by these companies are less costly but sold on a more regular basis. Examples are machine tools and cars.

3 *Light engineering*
The products produced by these companies are widely used by both industrial and domestic consumers and produced in large quantities. These include such products as TV sets, washing machines, and small components used by other engineering companies.

4 *Processing*
These companies produce products which involve materials in a continuous process. An example of such a form of production is the manufacture of chemicals or materials, there being a continuous mixing and processing of the raw ingredients and consequently a continuous production of the product. The food and drink industry, e.g. the production of beer, the tobacco industry, the textile industry, might be considered to also be a form of processing industry.

Another way of considering such industry is in terms of the type of engineering they tend to specialise in, these being:

1 *Mechanical engineering*
Examples of such companies are those engaged in the production of machinery, e.g. machine tools, construction and earth-moving equipment, office machinery, pumps, valves, textile machinery, agricultural machinery.

2 *Electrical and electronic engineering*
Examples of such companies are those concerned with the production of radio and electronic components, radio equipment, telephone equipment, electrical appliances for domestic used, e.g. TV sets, computers, electrical machinery, etc.

3 *Motor vehicle engineering*
Examples of such companies are those concerned with the production of motor cars or commercial vehicles.

4 *Aerospace engineering*
Examples of such companies are those concerned with the production of planes, satellites, launch vehicles, etc.

5 *Metals products engineering*
Examples of such companies are those concerned with the production of such items as engineers' small tools, hand tools in general, cutlery, bolts, nuts, screws, rivets, cans and metal boxes, etc.

6 *Chemical engineering*
Examples of such companies are those concerned with the refining of oil and the production of plastics.

13.3 Employment of engineers

The people working in the engineering industry can be classified according to the type of job they have. Table 13.1 shows the type of breakdown into various general job categories that might occur. Note that the data should only be taken as given a general indication, the percentages will vary from company to company, from one type of engineering company to another, and with time. The data is taken from the EITB Research Report 9, The technician in Engineering: Part 1 Patterns of technician employment.

Table 13.1 Employment in the engineering industry

Type of job	% employed in that job
Managers	5.4%
Scientists and technologists	3.0%
Technicians	8.3%
Administrative and professional staff	6.0%
Supervisors	5.1%
Craftsmen	19.1%
Operators	33.8%
Clerical staff	11.1%
Others	8.2%

To illustrate how the data might change from one type of industry to another, Figure 13.2 shows the breakdown by percentage in the above categories for two industries. Figure 13.2(a) gives the breakdown given in table 13.1 for the average across all engineering. Figure 13.2(b) gives the data for just the electronics industry sector and Figure 13.2(c) that for the aerospace industry sector. Note that the electronics industry has fewer craftsmen and significantly more scientists and technologists than engineering in general. This reflects its greater dependence on the technical and professional skills of its workforce. The aerospace industry has, however, more craftsmen than the average, fewer operators but more technicians, scientists and technologists. Though the industry is a high technology industry, it does involve a lot of craftsmen work in the assembling of aircraft, etc.

It is worth noting that while the greater percentage of engineering production craftsmen are considered to work in the engineering industry, a figure of 80% has been suggested, engineering maintenance craftsmen are much more widely employed. It is thought that perhaps only about 25% might actually be employed in the engineering industry, the rest employed in such occupations as television repair, computer maintenance, motor mechanics, etc.

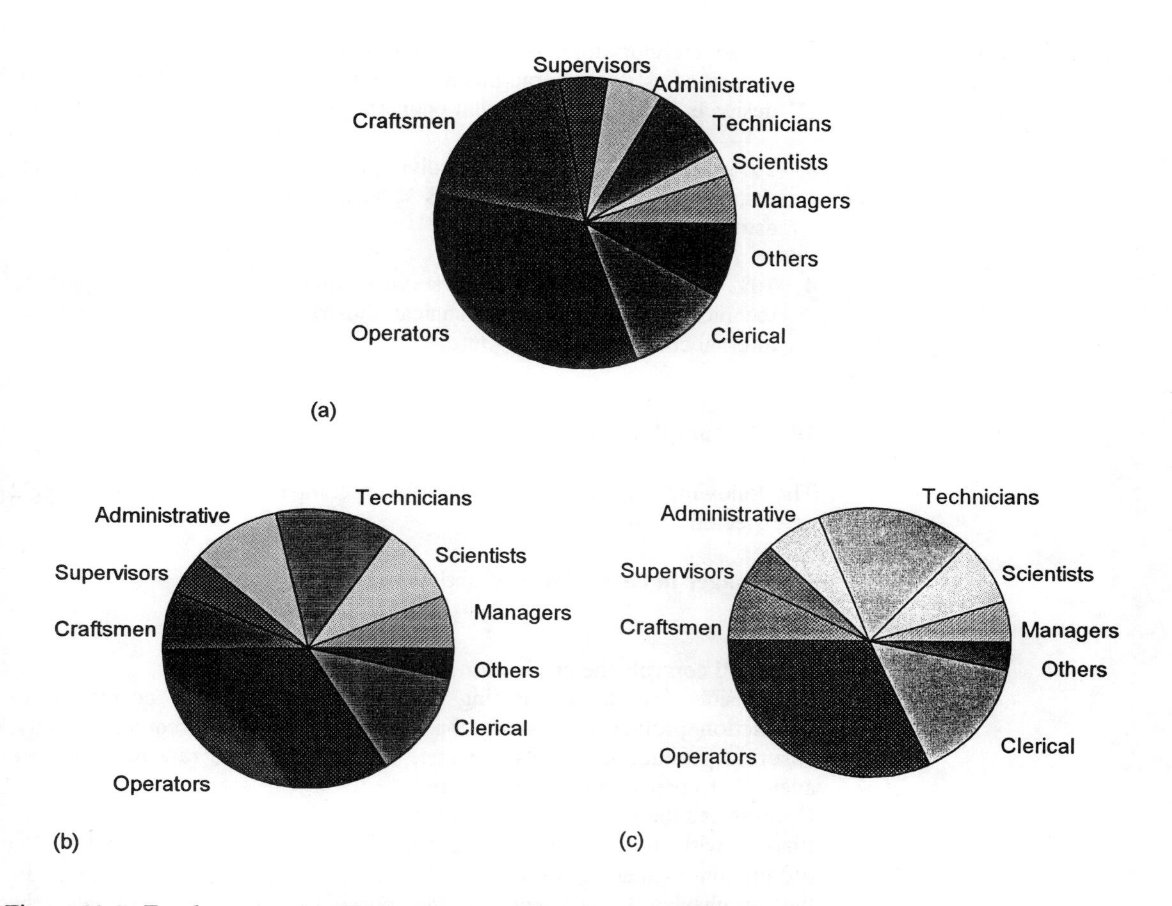

Figure 13.2 *Employees in (a) all engineering, (b) electronics, (c) aerospace*

13.3.1 Engineering technicians

As an illustration of the types of jobs undertaken by one group of employees, consider the types of jobs undertaken by technicians. *Technicians* are defined as being those persons carrying out work intermediates between that of scientists and technologists on the one hand and skilled craftsmen and operators on the other. In the survey of the industry, in general it was found that among the engineering technicians about:

1 50% were working in production as production or planning engineers/ technicians, installation or commissioning engineers, software engineers, maintenance engineers, test engineers/technicians, project engineers, quality engineers, inspectors, etc.

2 20% were working in the research, design and development area in such capacities as design draughtsmen/women, detail draughtsmen/ women, design engineers, development engineers, etc.

3 20% were working in areas involving customers or suppliers as service engineers for customer products servicing, contract engineers, buyers, estimators, technical sales, etc.

4 10% were working in central services such as work study engineers, technical writers/authors, technical illustrators, service engineers for internal company servicing, etc.

13.3.2 Examples of job specifications

The following are examples of the jobs in companies that engineers might be expected to do.

1. Manager in charge of production control

Tasks
Plans and controls the production operations.
Responsible for the monitoring of production progress, correction of production problems, ensuring the products are of the required quality, ensuring production targets are met, ensuring that the raw materials are available to production when required.
Coordinates the work of subordinates.
Liaises with the marketing department on production schedules, the production engineering department on the devising of production methods, the purchasing department on the purchasing of raw materials, the personnel department on staffing and the training of staff, sub-contractors when placing outside production orders, union representatives on industrial relations matters.
Plans for future production schedules, material requirements, capital expenditure on new plant, equipment and staff.

Knowledge
Production control methods.
Engineering knowledge of the production plant and the processes used.
Interpretation of engineering specifications and engineering drawings.
Preparation and interpretation of financial plans.
Management skills.
Up-to-date knowledge of production techniques.

Supervision received
Obtains overall direction from more senior management on policy and finance. Work is only assigned in terms of broad production objectives, the detailed planning is left to the manager, though the manager is responsible to senior management for attaining those objectives.

Supervision exercised
Responsible for all the staff engaged in production. This is exercised through section heads/supervisors.
Responsible for the maintenance of good labour relations.

2. Scientist/technologist engaged in research

Tasks
Makes recommendations for research projects.
Assists, as part of a team, in the planning of research projects.
Devises research methods for projects.
Sets up test equipment and instrumentation.
Conducts investigations and tests.
Analyses the results.
Prepares test and project reports.
Participates in conferences with other scientists and technologists.
Keeps up to data by reading scientific and technological papers in journals and books.

Knowledge
Science and technology in the relevant field.
Research methods.
Test equipment and instrumentation.
Use of a computer to record, analyse and present results.
Company report writing procedures and standards.
Current developments in the relevant scientific/technological field.

Supervision received
Receives direction from the Research and Development Manager. The research work is not supervised in detail but the planned approach and results are subject to scrutiny.

Supervision exercised
Responsible for technicians employed as project assistants. Delegates setting-up, testing, recording, results analysis of projects to assigned technicians. Gives them technical guidance and checks the results.

3. Technician engaged in mechanical testing

Tasks
Carries out mechanical testing of materials.
Checks whether they conform to specification.
Interprets engineering drawings.
Test parts using test rigs.
Calibrates instruments used.
Interprets the results of tests.
Uses a calculator.
Plots graphs.
Uses a computer to enter and process data.

Uses a computer to plot graphs.
Completes standard report forms of the results of tests.
Passes the results of tests to engineers.

Knowledge
Terminology associated with material specifications.
Company drawing office standards.
Use of testing machines.
Use of instrumentation involved in testing.
Use of a calculator.
Plotting and interpretation of graphs.
Use of computer and the appropriate software.
Company report writing procedures and standards.

Supervision received
Receives direction from and reports to the head of the testing section.

Supervision exercised
Not responsible for any other employees.

4. Technician engaged in development

Tasks
Receives circuit diagrams and instructions from the Project Engineer.
Plans the component layout and submits it for approval to the Project Engineer.
Constructs a prototype unit.
Tests the prototype.
Informs the Project Engineer of the test results.
Develops printed circuit layout and submits it for approval to the Project Engineer.
Passes the master layout drawing to the printed circuit production department.
Receives the resulting printed circuit and constructs a prototype.
Tests the prototype.
Informs the Project Engineer of the test results.

Knowledge
Interpretation of electric circuit diagrams.
Can draw layout diagrams.
Wiring of circuits using soldering.
Use of test meters, oscilloscope and systematic fault diagnosis techniques.
Drawing of printed circuit layout diagrams.
Company standards for drawing.
Company procedures for testing.

Supervision received
Receives direction from and reports to the Project Engineer.

Supervision exercised
Not responsible for any other employees.

In general (EITB Booklet No 9: The Training of Technician Engineers), technicians require the ability to:

1 Use and communicate information.
2 Measure and make use of measurements involving a variety of tools and/or instruments.
3 Choose materials and components and understand the processing of materials.
4 Understand manufacturing activities and the general organisation and practices of their company.
5 Carry out diagnosis of situations.
6 Organise and give direction to the work of others..

13.4 National qualifications

National engineering qualifications in England, Wales and Northern Ireland fall into three main groups:

1 Occupational competencies
2 Vocational education
3 Academic education

The national occupational qualifications are the *National Vocational Qualifications (NVQs)*. These are standards set for functions performed in employment and developed by industry lead bodies. They are concerned with the assessment of competence. The competencies are graded from level 1 through to level 5.

The national vocational education qualifications are the *General National Vocational Qualifications (GNVQs)*. These are national standards for broad based vocational education and set for skills, knowledge and understanding within broad vocational areas. They are designed to provide progression to further and higher education, access to employment and developmental opportunities for those in employment.

The academic qualifications follow the sequence National curriculum, GCSEs, A-levels, degrees and post-graduate studies.

In order to align qualifications within a coherent national framework, the three types of qualification are designed to be comparable in their demand at similar levels. Figure 13.3 gives a general indication of the progression routes through these qualifications and their alignment at the various levels.

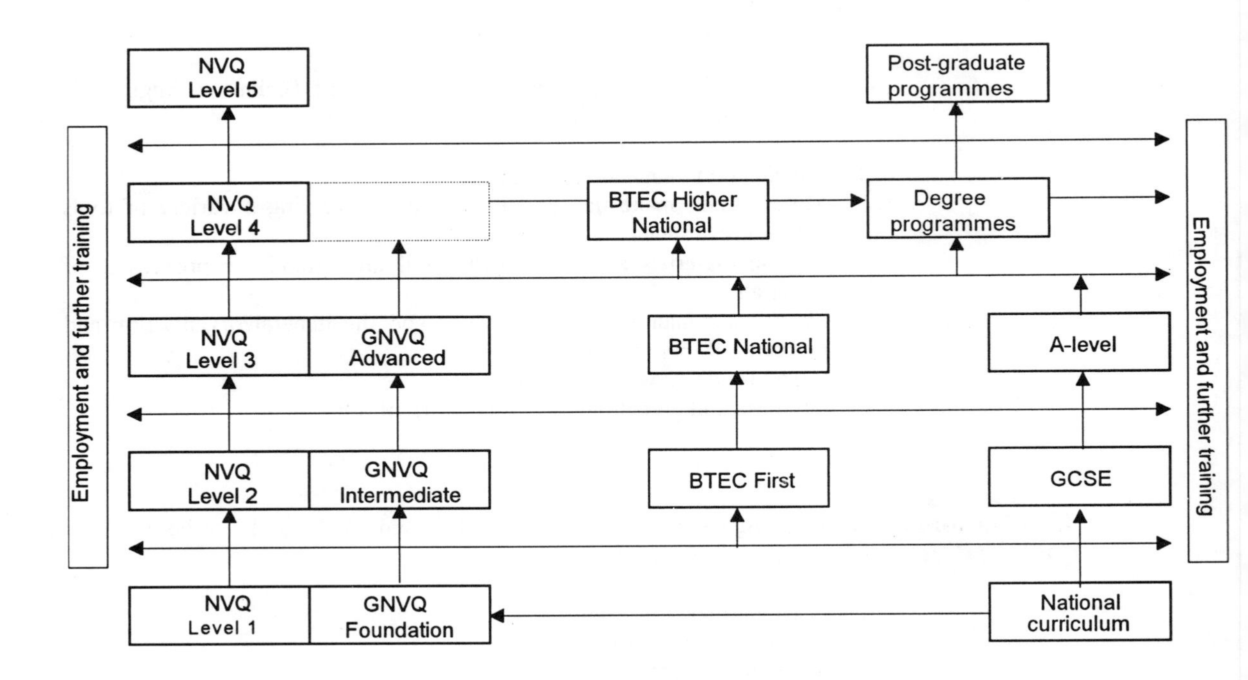

Figure 13.3 *Progression routes through the BTEC portfolio of qualifications (based on Figure 1.6 in BTEC GNVQ: Implementing BTEC GNVQs, June 1993)*

13.4.1 Entry qualifications for jobs

Scientists and technologists have typically been recruited direct from universities after completing a degree. Technicians seem to follow a number of possible routes. Some will enter employment direct from school and be trained as craftsmen, later progressing to technician. Others, with good GCSE results, might enter employment direct from school and become technicians as a result of part time study at a college combined with on the job training. Yet another route is full time study of a general vocational qualification and then employment. The qualifications expected for a particular technician job do, however, depend to some extent on the type of technician job concerned. Technicians working in research, design and development tend to have higher qualifications than those in other types of technician jobs. In all cases, however, there is generally a need for continuing education and training while working. For those wanting to ascend a career ladder it can be vital.

13.5 Applying for jobs

Figure 13.4 illustrates the type of sequence involved in a company determining the need for a post to be filled, finding a suitable applicant and then filling the post. It also shows the consequential sequence adopted an applicant applying for the post.

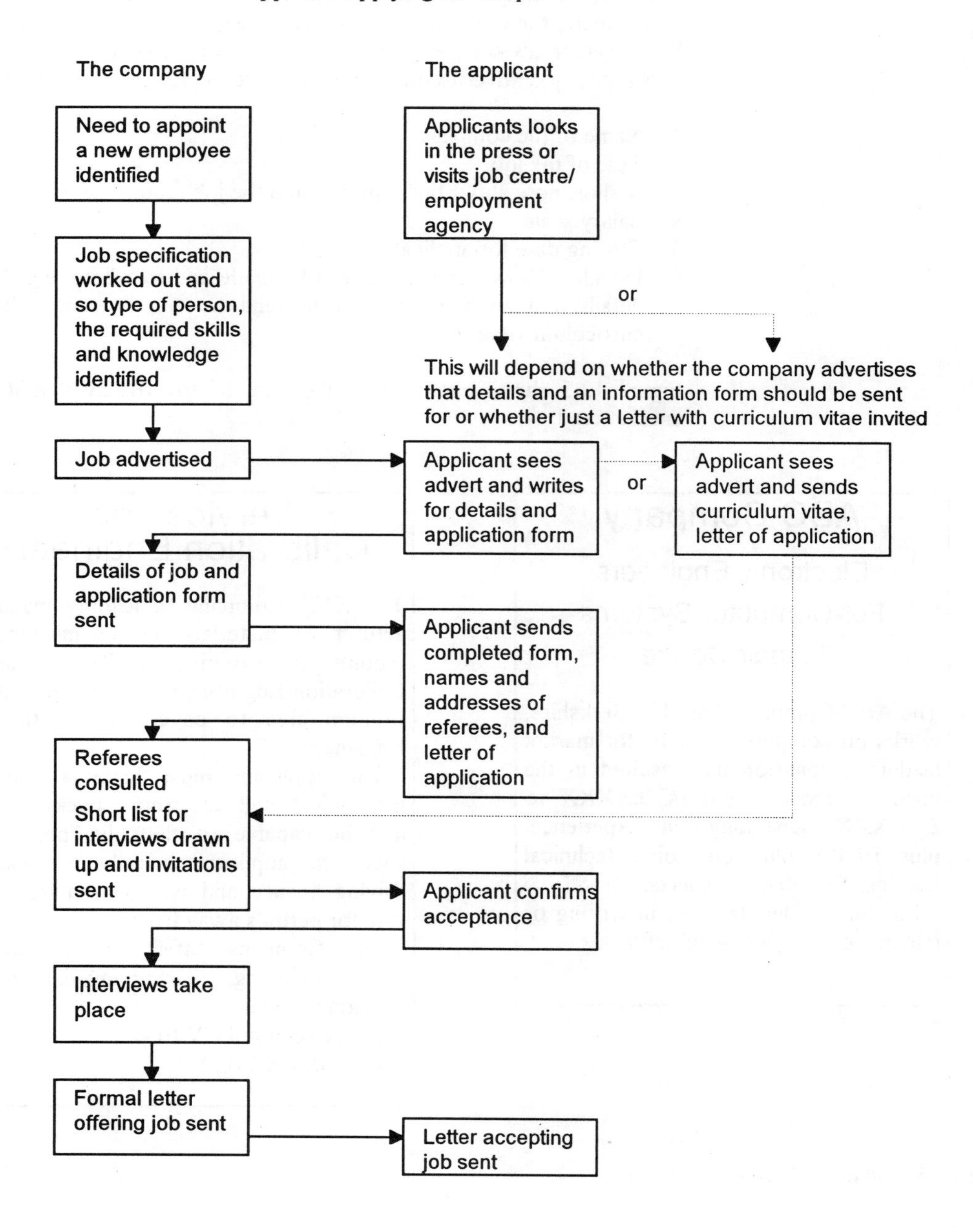

Figure 13.4 *Stages in getting a job*

13.5.1 Sources of information about jobs

Details of jobs might be publicised by companies in a number of ways. For example, it might be though adverts in local newspapers, national newspapers, the trade press, specialist magazines, or on the radio. It might be via career advisors, employment agencies or job centres.

A typical job advertisement might include the following information:

1 Name of the company
2 Title of the job
3 A short note about the company and the job requirements
4 Salary scale
5 Closing date for applications
6 Details of how applications can be made and to where, e.g. for further details and an application form send to ....., or send full CV (i.e. curriculum vitae) to ....

Figure 13.5 shows examples of the type of advertisement that might be encountered.

ABC Company

Electronic Engineers
For Computer Systems
Repair Centre

The ABC Company, based in Berkshire, works on computer systems for market leaders. Expansion has resulted in the need for more engineers. £XXXX to £XXXXX depending on experience, plus BUPA plus on going technical training. Excellent prospects.

For further details apply in writing or telephone ......, Personnel Officer, ...

Service and
Calibration Engineer

The XYZ Company, a leading manufacturer of materials testing machines, require an experienced Service and Calibration Engineer, ideally based in the London area to cover the South of England.

The applicant must have a sound mechanical and electronic background and be capable of fault finding. The successful applicant must have a valid driving licence and will be required to stay for periods away from home.

The Company offers an attractive salary including a company car and pension scheme.

Please send full CV to:-
Mr. ...., XYX Co., ...

Figure 13.5 *Job advertisements*

13.5.2 The application form

With regard to the application form, the following information is generally required:

Title of the job applied for
Name
Address
Telephone number
Age
Date of birth
Status: married, single
Education
Qualifications gained
Current job: title and brief details, current salary, name and address of employer
Previous jobs: details with dates of employment
Outline of interests and hobbies
Names, occupations and addresses of referees
Date of availability for the post
Signature of applicant indicating that the information given is an accurate record.

13.5.3 The curriculum vitae

A curriculum vitae is a biographical sketch of your life. It is thus likely to include such information as:

1 *Personal details*
Full name and current address
Telephone number
Age
Date of birth
Nationality
Status: married or single
Dependants: wife, husband, children.

2 *Education*
With dates and in chronological order
Secondary school(s)
College(s): with courses taken
University: with course taken.

3 *Qualifications*
Examinations passed, with dates, grades and names of examining boards/organisations.

4 *Work experience*
This is in chronological order, usually starting from the current post and working backwards.
Dates, name of company, location, job designation, range of duties, extent of responsibilities, reasons for leaving.

5 *Interests*
Leisure activities.
Any posts of responsibility, e.g. honorary secretary of cricket club.

6 *Personal circumstances*
Any other information about your circumstances which would assist with the job in question, e.g. car ownership if mobility is required for the job.
The period of notice that has to be given from your current job.
Any other factors limiting your potential employment with the company.

13.5.4 Interviews

The purpose of interviews are:

1 To enable the employer to obtain enough information about the applicant to determine his/her suitability for the job.

2 To give the applicant all the relevant information about the job and the company for him/her to make a decision about whether they want the job.

Many interviewers adopt a systematic approach to interviewing called the *Seven-point plan* (N.I.I.P. Paper No.1 by A.Rodger, published by the National Institute of Industrial Psychology). This plan calls for the interviewer to obtain and answer the following questions in order to obtain a comprehensive view of the applicant's occupational assets and liabilities. However, before the interview, the interviewer will have had to carry out a detailed analysis of the job in order to be able to list its requirements under the following headings and then compare what the applicant has to offer against those requirements.

1 *Physical make-up*
What does the job demand in the way of general health, strength, appearance, manner, etc. and how does the applicant meet these criteria? Thus an issue might be whether the applicant has any defects of health or physique that could hinder his/her carrying out the job.

2 *Attainments*
What does the job demand by way of education, training and previous experience and to what extent does the applicant meet these criteria?

This involves the interviewer determining the type of education the applicant has had and how well they did, what occupational training that have had, the work experience they have had and how well they did.

3 *General intelligence*
What level of general intelligence is required for the job and to what extent does the applicant meet the requirement? A general intelligence test might be used.

4 *Special aptitudes*
Does the job require any special aptitudes such as manual dexterity, verbal dexterity, artistic ability, etc. and to what extent does the candidate meet these criteria? A special test might be used.

5 *Interests*
To what extent does the job require special interests, e.g. outdoor life, being with other people, and to what extent does the applicant show such interests, e.g. via hobbies?

6 *Disposition*
Does the job require and qualities such as leadership, acceptability to others, reliability, sense of responsibility, self-reliance, etc. and to what extent does that applicant show such qualities?

7 *Circumstances*
How will such factors as the pay, prestige, status of job, etc. affect the applicant's personal life? Could, for example, domestic circumstances allow the applicant to travel widely?

It is not only the interviewer that needs to prepare for the interview, the applicant also needs to prepare. This can mean that he/she has:

1 Prepared and made copies of a curriculum vitae. Even if one is not requested, it provides a basis for filling in forms or answering questions at an interview.
2 Kept a copy of the completed application form for reference.
3 Considered what questions might be asked and considered the answers that you could give.
4 Done some research on the company with regard to its performance and prospects.
5 Compiled a list of questions to ask with regard to the performance of the company, job prospects, conditions of service, duties, place of work, salary, etc.

In addition, the applicant should turn up for the interview with smart appearance and in good time.

Problems

1. Identify two employers in each of the main sectors of the engineering industry.

2. Describe the qualifications, skills and experience necessary for some job.

3. Identify a career path for yourself within the engineering industry, describing the qualifications, skills and experience that will be needed at each point in the path and how you might acquire them.

4. Prepare a curriculum vitae for yourself.

5. Select a job advert from the local press and draft an application letter which could be used to apply for that job. Take into account the information presented in the advert.

Appendix: Units

The *International System (SI)* of units has seven base units, these being:

Length	metre	m
Mass	kilogram	kg
Time	second	s
Electric current	ampere	A
Temperature	kelvin	K
Luminous intensity	candela	cd
Amount of substance	mole	mol

In addition there are two supplementary units, the radian and the steradian.

The SI units for other physical quantities are formed from the base units via the equation defining the quantity concerned. Thus, for example, volume is defined by the equation

volume = length cubed

The unit of volume is therefore that of length unit cubed

unit of volume = unit of length cubed

Thus with the unit of length as the metre, the unit of volume is metre cubed, i.e. m^3.

Density is defined by the equation

$$\text{density} = \frac{\text{mass}}{\text{volume}}$$

Thus with the unit of density is the unit of mass divided by the unit of volume

$$\text{unit of density} = \frac{\text{unit of mass}}{\text{unit of volume}}$$

Thus, since the unit of mass is the kg and the unit of volume m^3, the unit of density is kg/m^3.

Velocity is defined, for motion in a straight line, by the equation

$$\text{velocity} = \frac{\text{change in distance covered}}{\text{time taken}}$$

Thus the unit of velocity is

$$\text{unit of velocity} = \frac{\text{unit of distance}}{\text{unit of time}}$$

Since the unit of distance, i.e. length, is the metre and the unit of time the second, then the unit of velocity is metres/second, i.e. m/s.

Acceleration is defined by the equation

$$\text{acceleration} = \frac{\text{change in velocity}}{\text{time taken}}$$

Thus the unit of acceleration is

$$\text{unit of acceleration} = \frac{\text{unit of velocity}}{\text{unit of time}}$$

Since the unit of velocity is metres per second and that of time the second, the unit of acceleration is

$$\text{unit of acceleration} = \frac{\text{m/s}}{\text{s}} = \frac{\text{m}}{\text{s} \times \text{s}} = \text{m/s}^2$$

Some of the derived units are given special names. Thus, for example, the unit of force is defined by the equation

$$\text{force} = \text{mass} \times \text{acceleration}$$

and is thus

$$\text{unit of force} = \text{unit of mass} \times \text{unit of acceleration}$$

and is thus kg m/s^2 or kg m s^{-2} . This unit is given the name newton (N). Thus 1 N is 1 kg m/s^2. The unit of pressure is given by the defining equation

$$\text{pressure} = \frac{\text{force}}{\text{area}}$$

and is thus

$$\text{unit of pressure} = \frac{\text{unit of force}}{\text{unit of area}}$$

Hence the derived unit of pressure is N/m^2. This unit is given the name pascal (Pa). Thus 1 Pa = 1 N/m^2.

Certain quantities are defined as the ratio of two comparable quantities. Thus, for example, strain is defined as change in length/length. It thus is expressed as a pure number with no units because the derived unit would be m/m.

Standard prefixes are used for multiples and submultiples of units, the SI preferred ones being multiples of 1000, i.e. 10^3, or divided by multiples of 1000. The following table shows the standard prefixes:

Table Standard unit prefixes

Multiplication factor	Prefix	
1 000 000 000 000 000 000 000 000 = 10^{24}	yotta	Y
1 000 000 000 000 000 000 000 = 10^{21}	zetta	Z
1 000 000 000 000 000 000 = 10^{18}	exa	E
1 000 000 000 000 000 = 10^{15}	peta	P
1 000 000 000 000 = 10^{12}	tera	T
1 000 000 000 = 10^9	giga	G
1 000 000 = 10^6	mega	M
1 000 = 10^3	kilo	k
100 = 10^2	hecto	h
10 = 10	deca	da
0.1 = 10^{-1}	deci	d
0.01 = 10^{-2}	centi	c
0.001 = 10^{-3}	milli	m
0.000 001 = 10^{-6}	micro	μ
0.000 000 001 = 10^{-9}	nano	n
0.000 000 000 001 = 10^{-12}	pico	p
0.000 000 000 000 001 = 10^{-15}	femto	f
0.000 000 000 000 000 001 = 10^{-18}	atto	a
0.000 000 000 000 000 000 001 = 10^{-21}	zepto	z
0.000 000 000 000 000 000 000 001 = 10^{-24}	yocto	y

Thus, for example, 1000 N can be written as 1 kN, 1 000 000 Pa as 1 MPa, 1 000 000 000 Pa as 1 GPa, 0.001 m as 1 mm, and 0.000 001 A as 1 μA.

Other units which the reader may come across are fps (foot-pound-second) units which still are often used in the USA. On that system the unit of length is the foot (ft), with 1 ft = 0.3048 m. The unit of mass is the pound (lb), with 1 lb = 0.4536 kg. The unit of time is the second, the same as the SI system. With this system the derived unit of force, which is given a special name, is the poundal (pdl), with 1 pdl = 0.1383 N. However, a more common unit of force is the pound force (lbf). This is the gravitational force acting on a mass of 1 lb and consequently, since the standard value of the acceleration due to gravity is 32.174 ft/s^2, then

1 lbf = 32.174 pdl = 4.448 N

The similar unit the kilogram force (kgf) is sometimes used. This is the gravitational force acting on a mass of 1 kg and consequently, since the standard value of the acceleration due to gravity is 9.806 65 m/s^2, then

1 kgf = 9.806 65 N

A unit often used for pressure in the USA is the psi, or pound force per square inch. 1 psi = 6.895×10^3 Pa.

Answers

The following are the answers to multiple-choice problems and numerical problems with brief clues to the form of answers for other problems.

Chapter 1

1 C	2 B	3 A	4 A	5 D	6 C
7 D	8 B	9 C	10 A	11 C	12 B
13 C					

14 (a) Thermoset, insulator, cheap to form; (b) medium carbon steel or stainless steel, reasonable strength and hardness; (c) copper, ease of bending, corrosion resistance; (d) polymer, tough, reasonably stiff, coloured, cheap to form; (e) wood, stiff, strong in bending, cheap.

Chapter 2

1 A	2 C	3 B	4 A	5 D	6 C
7 B	8 C	9 D	10 A	11 A	12 A
13 C	14 A	15 A	16 D	17 D	18 C
19 B	20 C				

21 The following illustrate the types of points that might be considered: (a) the construction of the bulb holder with its clamps for cables to the metal contacts, the way the lamp shade is fixed with a dismountable clamp, the material used for the lamp stand, the form of cable used; (b) the materials used for the various parts, the clamping system used for the mains cable, the way individual conductors are connected to the plug prongs, the fuse mounting; (c) the casing material, the way it is held together, the dismountable mounting for the drill, the way the circuitry inside the drill is mounted; (d) the material used for the bonnet, the form of the hinges, the form of the bonnet lock; (e) the material used for the casing (prevent build up of static), the way the casing can be opened up, the way the components are mounted inside the case, the mounting of the circuit components.

Chapter 3

1 C	2 B	3 A	4 C	5 C	6 B
7 A	8 C	9 D	10 D	11 B	12 A
13 B	14 C	15 A	16 C	17 B	18 B
19 A	20 A	21 D	22 B	23 B	24 A
25 C	26 B	27 A	28 C	29 B	30 C

31 The following illustrate some of the key aspects that might be involved: (a) thermoforming sheet plastic, (b) rolling, (c) injection moulding, (d)

circuit board, soldering, (e) die casting, (f) turning by lathe, (g) bending of sheets and welding or soldering/brazing.

Chapter 4

1 A	2 B	3 D	4 C	5 D	6 C
7 C	8 A	9 C	10 A	11 D	12 C
13 A	14 D	15 B	16 B	17 D	18 A
19 A	20 A				

21 See section 4.1
22 (a) Improvements are required, (b) the process or equipment is prohibited from being used.
23 (a) Contravenes HSW Act duties imposed on employees, (b) employers are failing to ensure the welfare of employees in an accident or emergency, (c) contravenes HSW Act duties imposed on employees, (d) the inspector has the right, (e) worker cannot refuse, (f) should report it.
24 Could be: (a) spectacles, goggles, face screens; (b) earplugs or muffs, (c) safety boots, gloves.

Chapter 5

1 B	2 A	3 D	4 C	5 B	6 C
7 B	8 A	9 B	10 C	11 A	12 A

13 (a) Sketch, (b) flow diagram, (c) technical drawing, (d) circuit diagram, (e) flow diagram, (f) circuit layout diagram, (g) assembly drawing, (h) flow diagram.

Chapter 6

1 C	2 B	3 B	4 B	5 C	6 A
7 D	8 A	9 A	10 A	11 C	12 D
13 B	14 A	15 D	16 B	17 A	

18 See Figure A.1
19 See Figure A.2
20 See Figure A.3
21 See Figure A.4
22 See Figure A.5
23 See Figure A.6
24/25/26/27 The answer depends on the item chosen.
28 See section 6.5.1
29 The answer depends on the software used. See section 6.5.2.

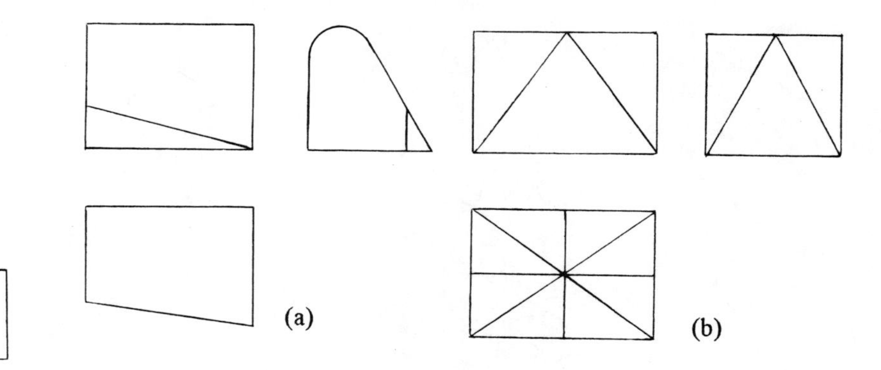

Figure A.1 *Chapter 6 Problem 18*

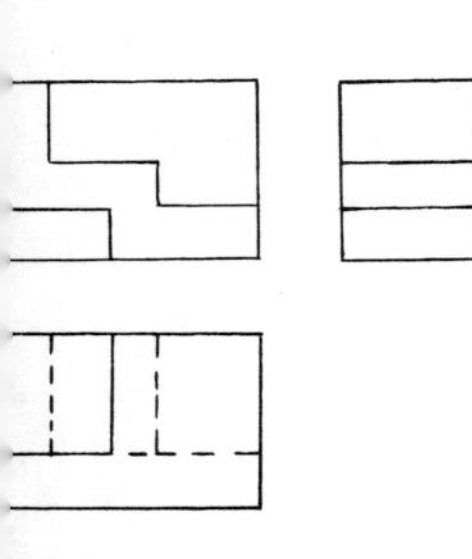

Figure A.2 *Chapter 6 Problem 19*

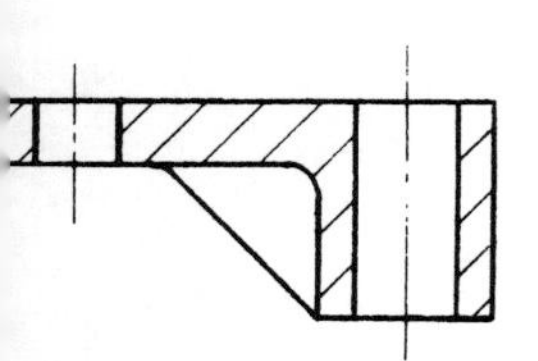

Figure A.4 *Chapter 6 Problem 21*

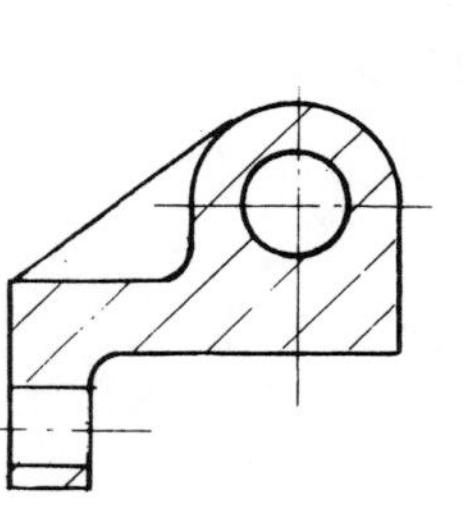

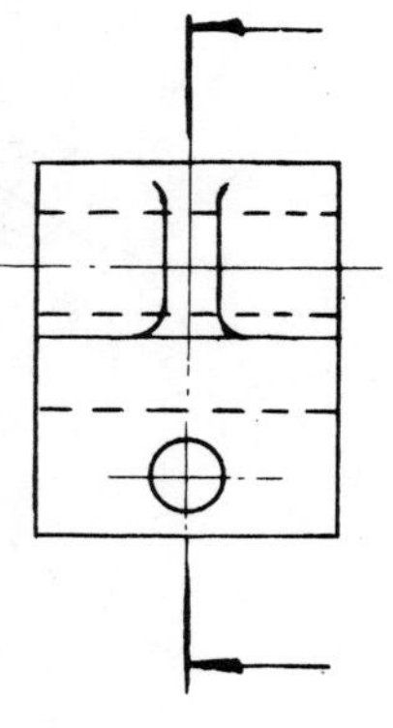

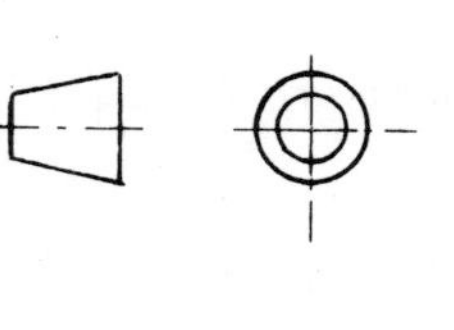

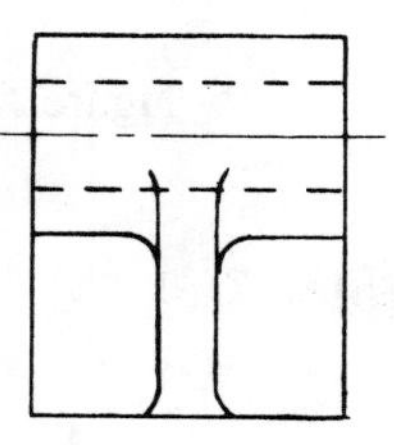

Figure A.3 *Chapter 6 Problem 20*

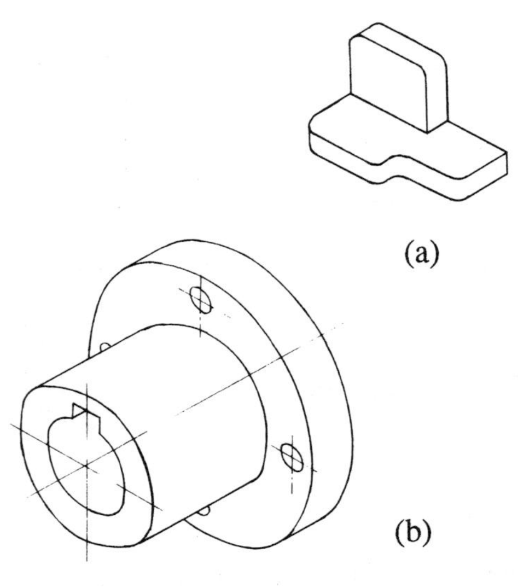

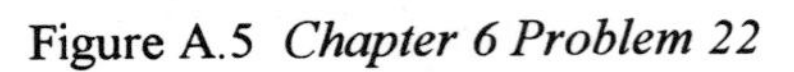
Figure A.5 *Chapter 6 Problem 22*

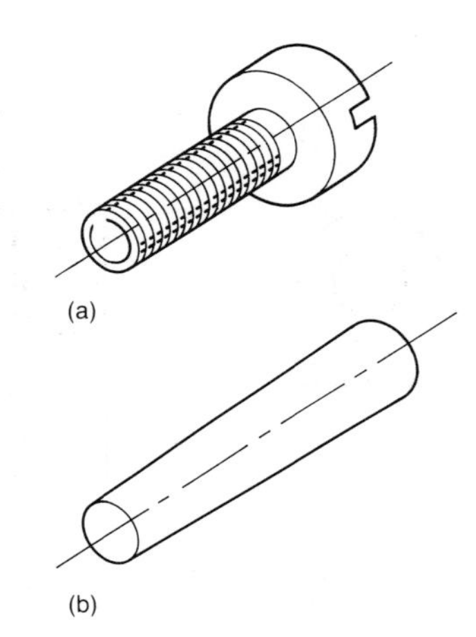

Figure A.6 *Chapter 6 Problem 23*

Chapter 7

1 A	2 C	3 D	4 C	5 B	6 B
7 C	8 B	9 A	10 A	11 D	12 B
13 A	14 A	15 A	16 C	17 A	18 B
19 A	20 D	21 C	22 B	23 C	24 D
25 C	26 C	27 A	28 A	29 A	30 B

31 See (a) (b) Figure 7.3, (c) (d) (e) Figure 7.4, (f) Figure 7.5, (g) Figure 7.6, (h) Figure 7.7, (i) Figure 7.9, (j) (k) Figure 7.11

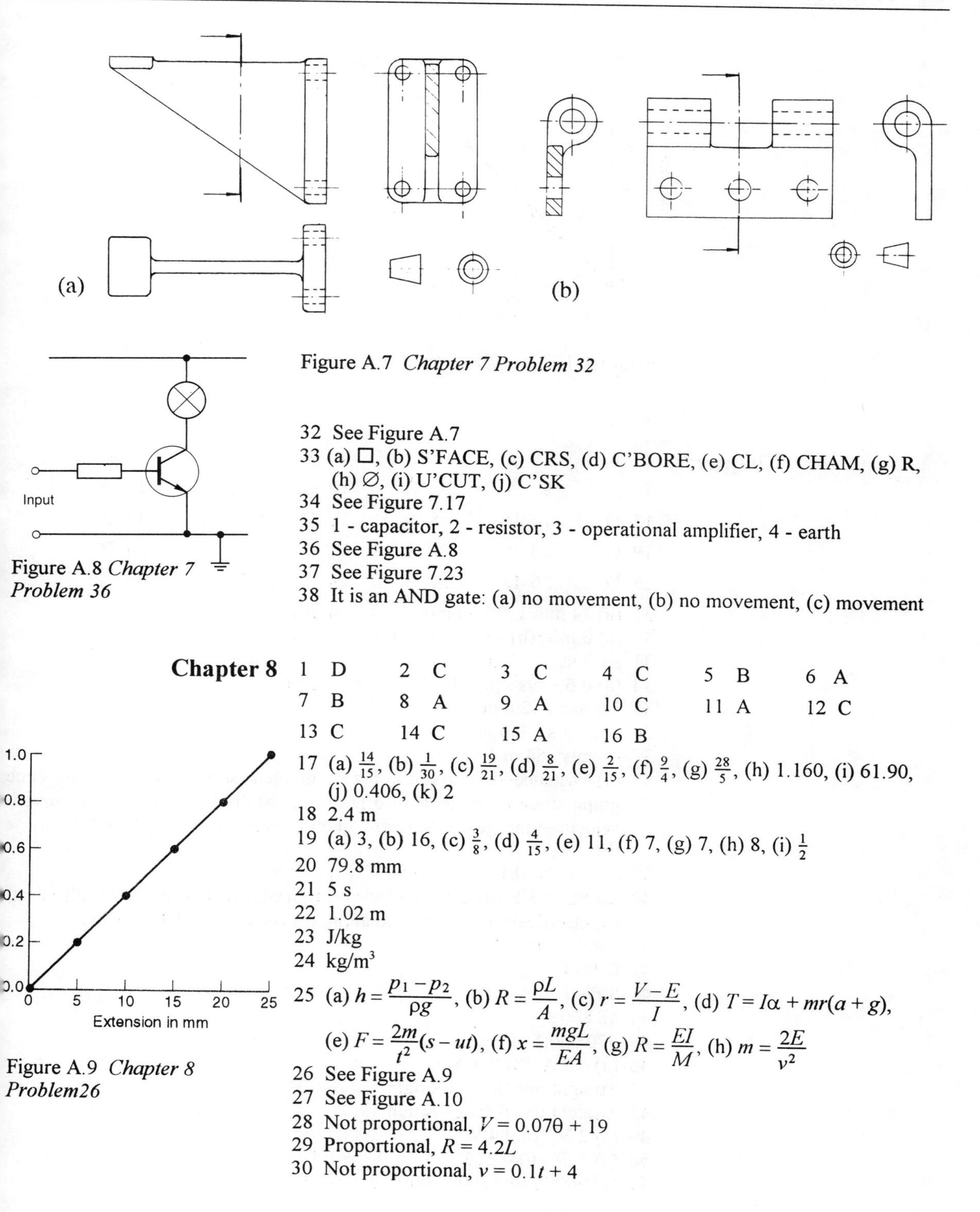

Figure A.7 *Chapter 7 Problem 32*

Figure A.8 *Chapter 7 Problem 36*

32 See Figure A.7
33 (a) □, (b) S'FACE, (c) CRS, (d) C'BORE, (e) CL, (f) CHAM, (g) R, (h) ∅, (i) U'CUT, (j) C'SK
34 See Figure 7.17
35 1 - capacitor, 2 - resistor, 3 - operational amplifier, 4 - earth
36 See Figure A.8
37 See Figure 7.23
38 It is an AND gate: (a) no movement, (b) no movement, (c) movement

Chapter 8

1 D	2 C	3 C	4 C	5 B	6 A
7 B	8 A	9 A	10 C	11 A	12 C
13 C	14 C	15 A	16 B		

17 (a) $\frac{14}{15}$, (b) $\frac{1}{30}$, (c) $\frac{19}{21}$, (d) $\frac{8}{21}$, (e) $\frac{2}{15}$, (f) $\frac{9}{4}$, (g) $\frac{28}{5}$, (h) 1.160, (i) 61.90, (j) 0.406, (k) 2
18 2.4 m
19 (a) 3, (b) 16, (c) $\frac{3}{8}$, (d) $\frac{4}{15}$, (e) 11, (f) 7, (g) 7, (h) 8, (i) $\frac{1}{2}$
20 79.8 mm
21 5 s
22 1.02 m
23 J/kg
24 kg/m^3
25 (a) $h = \frac{p_1 - p_2}{\rho g}$, (b) $R = \frac{\rho L}{A}$, (c) $r = \frac{V - E}{I}$, (d) $T = I\alpha + mr(a + g)$, (e) $F = \frac{2m}{t^2}(s - ut)$, (f) $x = \frac{mgL}{EA}$, (g) $R = \frac{EI}{M}$, (h) $m = \frac{2E}{v^2}$
26 See Figure A.9
27 See Figure A.10
28 Not proportional, $V = 0.07\theta + 19$
29 Proportional, $R = 4.2L$
30 Not proportional, $v = 0.1t + 4$

Figure A.9 *Chapter 8 Problem26*

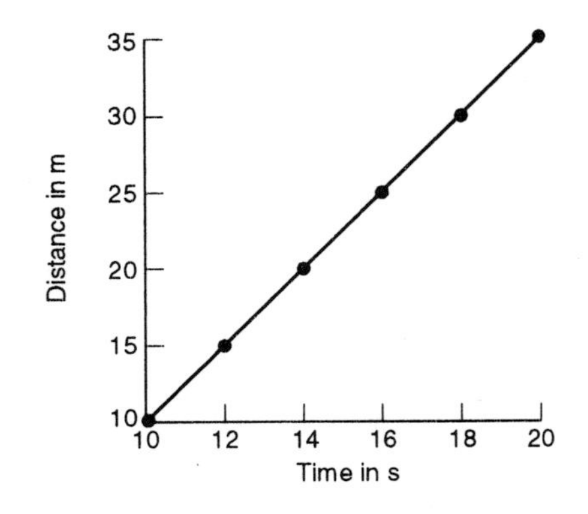

Figure A.10 *Chapter 8 Problem 27*

Chapter 9

1 B	2 A	3 C	4 A	5 B	6 B
7 B	8 A	9 D	10 C	11 B	12 A
13 D	14 D	15 A	16 C	17 A	18 B
19 C	20 A	21 A	22 D	23 D	24 B
25 D	26 D	27 B	28 A	29 C	30 A

31 (a) 34 mm/s, (b) 50 mm/s, (c) 66 mm/s
32 (a) 8 m/s, (b) 11 m/s, (c) 16 m/s
33 -1.0 m/s^2, 11 m
34 (a) 0.69 m/s^2, (b) -1.39 m/s^2, (c) 903 m
35 1.2 m/s^2, 12.6 m
36 1.25 m/s^2, 256 m
37 -4 m/s^2, 54 m
38 (a) Distance–time graph non-zero gradient straight line, velocity–time graph straight line with zero gradient, (b) distance–time graph curved, velocity–time graph non-zero gradient straight line, (c) as (a), (d) as (b).
39 (a) 200 N, (b) 4.4 N, (c) 58 N, (d) 100 N
40 (a) PE to KE to heat, (b) electrical to heat, (c) chemical to PE plus heat, (d) electrical to heat plus radiation, (e) electrical to KE plus heat.
41 196 000 J
42 6000 J
43 80%
44 46 800 J
45 670 000 J
46 (a) 0.2 A, (b) 0.3 A
47 Straight line through origin.
48 Straight line through origin.
49 (a) 2 V, (b) 0.4 W
50 (a) 5 Ω, (b) 4.2 Ω
51 (a) 6000 J, (b) 3000 000 J/kg

52 (a) B, (c) A 4 Ω, B 2 Ω
53 0.022 m

Chapter 10

1 B	2 B	3 C	4 A	5 B	6 B
7 C	8 A	9 A	10 C	11 D	12 D
13 D	14 C	15 C			

16 Your answers should be in the form of a report and judged according to the quality of your experiment as indicated by the report.

Chapter 11 1, 2, 3, 4 The answers to these questions can draw from the information presented in the chapter.

Chapter 12 1, 2, 3, 4 The answers to these questions can draw from the information presented in the chapter.

Chapter 13 1, 2, 3, 4, 5 The answers to these questions can draw from the information presented in the chapter.

Index